T0231808

PARTIAL DIFFERENTIAL EQUATIONS
Methods And Applications

THIS BOOK IS DEDICATED TO

My wife, our son and our three daughters
for supporting me in all my endeavors

Partial Differential Equations
Equations
Methods and Applications

ABDUL-MAJID WAZWAZ
Saint Xavier University, Chicago, Illinois, USA

A.A. BALKEMA PUBLISHERS / LISSE / ABINGDON / EXTON (PA) / TOKYO

Library of Congress Cataloging-in-Publication Data

Applied for

Cover design: Studio Jan de Boer, Amsterdam, The Netherlands.
Typesetting: ...
Printed by: Gorter, Steenwijk, The Netherlands.

ISBN 90 5809 3697

Contents

Preface

Partial Differential Equations: Methods and Applications is designed to serve as a text and a reference. The book can be used for advanced undergraduate and beginning graduate students in Mathematics, Science, and Engineering. The available Partial Differential Equations texts only use classical methods without incorporating the newly developed methods that have appeared in research journals. This text is different from other texts in that it explains classical methods in a non abstract way and it introduces and explains how the newly developed methods provide more concise methods to provide efficient results.

Partial Differential Equations: Methods and Applications has been created to meet the growing need for change and to bring together the remarkable developments previously scattered among a variety of journals. This book is designed to focus readers' attentions on these recently developed valuable techniques that have proven their effectiveness and reliability over existing classical methods.

This text also explains the necessary classical methods of the characteristics, the separation of variables, and D'Alembert method, because the aim is that new methods would complement the traditional methods in order to improve the understanding of the material. *Partial Differential Equations for Mathematics, Science, and Engineering* is a text that transports and uses the valuable research work in addition to the classic methods in order to have the advanced and classic approaches reinforce and complement each other.

The book avoids approaching the subject through the compact and classical methods that make the material impossible to be grasped, especially by students who do not have the background in these abstract concepts. Concepts of existence, uniqueness, and well-posedness are not presented in this text.

The book was developed as a result of many years of experience in teaching partial differential equations and conducting research work in this field. I have translated my means of introducing and teaching this subject into this book so that the reader can grasp a broader picture of the subject and the easily presented material. Numerous examples and exercises, ranging in level from easy to difficult, but consistent with the material, are given in each section to give the reader the knowledge, practice and skill in partial differential equations. There is plenty of material in this text to be covered in two semesters for senior undergraduates and beginning graduates of Mathematics, Science, and Engineering.

The book consists of 11 chapters. Each chapter is divided into sections and each section contains well-explained examples and exercises along with their answers. The book thoroughly covers recent developments and investigates the pertinent and most commonly used classical methods such that each approach can be investigated independently.

Chapter 1 provides the basic definitions and introductory concepts. Classifications of second order partial differential equations are also addressed.

In Chapter 2, the first order partial differential equations are handled independently by using the newly developed decomposition method and the classic method of characteristics. The noise terms phenomena and the modified decomposition method are thoroughly presented to accelerate the convergence of the solution obtained by using the decomposition method.

Chapter 3 deals with the one-dimensional heat flow where homogeneous and inhomogeneous problems are handled by using the newly developed decomposition methods and the traditional method of separation of variables.

Chapter 4 is entirely devoted to higher dimensional heat flow are new and classic methods were employed independently.

Chapter 5 provides the reader with a comprehensive discussion of the one dimensional wave equation. The powerful decomposition method and the classic method of separation of variables and D'Alembert method are presented.

Chapter 6 presents the higher dimensional wave equation. A comprehensive study is introduced supported by many illustrative examples using the newly developed and classic methods.

Chapter 7 is devoted to Laplace equation in two and three dimensional rectangular coordinates and in polar coordinates. The chapter stresses the reliable decomposition method and the traditional methods as well.

Chapter 8 introduces a comprehensive study on nonlinear differential equations, ordinary and partial. The decomposition method is thoroughly used with promising results.

Chapter 9 provides the reader with a variety of linear and nonlinear applications, selected from physical phenomena, population growth models, and evolution concepts. A detailed and clear explanation of every application is introduced and supported by fully explained examples and exercises of every application.

Chapter 10 is concerned with numerical applications. The Padé approximants are introduced to handle boundary value problems where the radius of convergence may not contain the boundaries. Emphasis in this chapter will be on combining the decomposition series solution with the Padé approximants that can be used for further applications in physics and engineering.

Chapter 11 is concerned with solitons and compactons: solitons with finite wave length. Emphasis in this chapter will be on determining the solitary wave solutions for some of the well known nonlinear equations. Moreover, the recently discovered concept of compactons: solitons free of exponential tails, is investigated.

The book has five useful appendices. Moreover, the book introduces the traditional methods in the same amount of concern to provide the reader with the knowledge needed to make a comparison.

I am deeply indebted to my wife, my son and my daughters who provided me with their encouragement, patience and support while this book was being written.

The author would highly appreciate any note concerning any constructive suggestion.

Saint Xavier University   Abdul-Majid Wazwaz
Chicago, IL 60655   e-mail: wazwaz@sxu.edu
U.S.A
2002

PARTIAL DIFFERENTIAL EQUATIONS
Methods and Applications

ABDUL-MAJID WAZWAZ

Saint Xavier University

PARTIAL DIFFERENTIAL EQUATIONS
Methods and Applications

ABDUL-MAJID WAZWAZ

Saint Xavier University

Chapter 1

Basic Concepts

1.1 Introduction

It is well known that most of the phenomena that arise in mathematical physics and engineering fields can be described by partial differential equations (PDE). In physics for example, the heat flow and the wave propagation phenomena are well described by partial differential equations. In ecology, most population models are governed by partial differential equations. The dispersion of a chemically reactive material is characterized by partial differential equations. In addition, most physical phenomena of fluid dynamics, quantum mechanics, electricity and many other models are controlled within its domain of validity by partial differential equations.

Partial differential equations have become a useful tool for describing these natural phenomena of science and engineering models. Therefore, it becomes increasingly important to be familiar with all traditional and recently developed methods for solving partial differential equations, and the implementation of these methods.

However, in this text, we will restrict our analysis to solve the partial differential equations along with the given conditions that characterize the initial conditions and the boundary conditions of the dependent variable. We fill focus our concern on deriving solutions to PDEs and not on the derivation of these equations. In this text, our presentation will be based on applying the recent developments in this field and on applying some of the traditional methods as well. The formulation of the partial differential equations and the scientific interpretation of the models will not be discussed.

It is to be noted that several methods are usually used in solving PDEs. The recently developed *Adomian decomposition method* and the related improvements of the *modified technique* and the *noise terms* phenomena will be effectively used. The recently developed techniques have been proved to be reliable, accurate and effective in both the analytic and the numerical purposes. The decomposition method was formally proved to provide the solution in terms of a rapid convergent infinite series that may yield the exact solution in many cases. The other related modifications were shown to be powerful in that it accelerate the rapid convergence of the solution. However, some of the traditional methods, such as the separation of variables method and the method of characteristics will be implemented as well.

Moreover, the other techniques, such as integral transforms, perturbation methods, numerical methods are among other methods that are usually used in other texts.

1.2   Definitions

1.2.1   Definition of a PDE

A partial differential equation PDE is an equation that contains the dependent variable (the unknown function), and its partial derivatives. It is known that in the ordinary differential equations ODEs, the dependent variable $u = u(x)$ depends only on one independent variable x. Unlike the ODEs, the dependent variable in the PDEs, such as $u = u(x, t)$ or $u = u(x, y, t)$, must depend on more than one independent variable. If $u = u(x, t)$, then the function u depends on the independent variable x, and on the time variable t. However, if $u = u(x, y, t)$, then the function u depends on the space variables x, y, and on the time variable t.

Examples of the PDEs are given by

$$u_t = k u_{xx}, \tag{1}$$
$$u_t = k(u_{xx} + u_{yy}), \tag{2}$$
$$u_t = k(u_{xx} + u_{yy} + u_{zz}), \tag{3}$$

that describe the heat flow in one dimensional space, two dimensional space, and three dimensional space respectively. In (1), the dependent variable $u = u(x, t)$ depends on the position x and on the time variable t. However, in (2), $u = u(x, y, t)$ depends on three independent variables, the space variables x, y and the time variable t. In (3), the dependent variable $u = u(x, y, z, t)$ depends on four independent variables, the space variables x, y, and z, and the time variable t.

Other examples of PDEs are given by

$$u_{tt} = c^2 u_{xx}, \tag{4}$$
$$u_{tt} = c^2 (u_{xx} + u_{yy}), \tag{5}$$
$$u_{tt} = c^2 (u_{xx} + u_{yy} + u_{zz}), \tag{6}$$

that describe the wave propagation in one dimensional space, two dimensional space, and three dimensional space respectively. Moreover, the unknown functions in (4), (5), and (6) are defined by $u = u(x, t)$, $u = u(x, y, t)$, and $u = u(x, y, z, t)$ respectively.

The well known Laplace equation is given by

$$u_{xx} + u_{yy} = 0, \tag{7}$$
$$u_{xx} + u_{yy} + u_{zz} = 0, \tag{8}$$

so that the function u does not depend on the time variable t. As will be seen later, the Laplace equation in polar coordinates is given by

$$u_{rr} + \frac{1}{r}u_r + \frac{1}{r^2}u_{\theta\theta} = 0, \tag{9}$$

so that $u = u(r, \theta)$.

1.2.2 Order of a PDE

The order of a PDE is the order of the highest partial derivative that appears in the equation. For example, the following equations

$$u_x - u_y = 0,$$
$$u_{xx} - u_t = 0, \tag{10}$$
$$u_y - uu_{xxx} = 0,$$

are PDEs of first order, second order, and third order respectively.

Example 1. Find the order of the following PDEs:

(a) $u_t = u_{xx} + u_{yy}$

(b) $u_x + u_y = 0$

(c) $u^4 u_{xx} + u_{xxy} = 2$

(d) $u_{xx} + u_{yyyy} = 1$

Solution.

(a) The highest partial derivative contained in this equation is u_{xx} or u_{yy}. The PDE is therefore of order two.

(b) The highest partial derivative contained in this equation is u_x or u_y. The PDE is therefore of order one.

(c) The highest partial derivative contained in this equation is u_{xxy}. The PDE is therefore of order three.

(d) The highest partial derivative contained in this equation is u_{yyyy}. The PDE is therefore of order four.

1.2.3   Linear and Nonlinear PDEs

Partial differential equations are classified as **linear** or **nonlinear**. A partial differential equation is called linear if:

(i) the power of the dependent variable and each partial derivative contained in the equation is one, and

(ii) the coefficients of the dependent variable and the coefficients of each partial derivative are constants or independent variables. However, if any of these conditions is not justified, the equation is called nonlinear.

Example 2. Classify the following PDEs as *linear* or *nonlinear*:

(a) $x u_{xx} + y u_{yy} = 0$

(b) $u u_t + x u_x = 2$

(c) $u_x + \sqrt{u} = x$

(d) $u_{rr} + \frac{1}{r} u_r + \frac{1}{r^2} u_{\theta\theta} = 0$

Solution.

(a) The power of each partial derivative u_{xx} and u_{yy} is one. In addition, the coefficients of the partial derivatives are the independent variables x and y respectively. Accordingly, the PDE is linear.

(b) Although the power of each partial derivative is one, but u_t has the dependent variable u as its coefficient. Therefore, the PDE is nonlinear.

(c) The equation is nonlinear because of the term $\sqrt{u}$.

(d) The equation is linear because it satisfies the two necessary conditions.

1.2.4   Some Linear Partial Differential Equations

As stated before, linear partial differential equations arise in many areas of scientific applications, such as diffusion equation and wave equation. In what follows, we list some of the well-known models that are of important concern:

1. The *heat equation* in one dimensional space is given by

$$u_t = k u_{xx}, \tag{11}$$

where k is a constant.

2. The *wave equation* in one dimensional space is

$$u_{tt} = c^2 u_{xx}, \tag{12}$$

where c is a constant.

3. The *Laplace equation* is

$$u_{xx} + u_{yy} = 0. \tag{13}$$

4. The *Klein-Gordon equation* is

$$\nabla^2 u - \frac{1}{c^2} u_{tt} = \mu^2 u. \tag{14}$$

5. The *Linear Schrodinger's equation* is

$$i u_t + u_{xx} = 0. \tag{15}$$

It is to be noted that most of these models will be studied in details in the forth-coming chapters.

1.2.5   Some Nonlinear Partial Differential Equations

It was mentioned earlier that partial differential equations arise in different areas of mathematical physics and engineering, including fluid dynamics, plasma physics, quantum field theory, nonlinear wave propagation and nonlinear fiber optics. In what follows we list some of the well-known nonlinear models that are of great interest:

1. The *Advection problem* is

$$u_t + u u_x = f(x, t). \tag{16}$$

2. The *Burgers' equation* is

$$u_t + u u_x = \alpha u_{xx}. \tag{17}$$

3. The *Korteweg de Vries (KdV) equation* is

$$u_t + a u u_x + b u_{xxx} = 0. \tag{18}$$

4. The *Modified KdV equation (mKdV)* is

$$u_t - 6u^2 u_x + u_{xxx} = 0. \tag{19}$$

5. The *Boussinesq equation* is

$$u_{tt} - u_{xx} + 3(u^2)_{xx} - u_{xxxx} = 0. \tag{20}$$

6. The *Sine-Gordon equation* is

$$u_{tt} = c^2 u_{xx} + \alpha \sin u. \tag{21}$$

7. The *Fisher's equation* is

$$u_t = D u_{xx} + u(1 - u). \tag{22}$$

8. The *Kadomtsev-Petviashvili equation* is

$$(u_t + a u u_x + b u_{xxx})_x + u_{yy} = 0. \tag{23}$$

9. The *K(n,n) equation* is

$$u_t + a(u^n)_x + b(u^n)_{xx} = 0, n > 1. \tag{24}$$

10. The *Nonlinear Schrodinger equation* is

$$iu_t + u_{xx} + \gamma|u|^2 u = 0. \tag{25}$$

It is to be noted that most of these nonlinear partial differential equations will be examined in the forthcoming chapters.

1.2.6   Homogeneous and Inhomogeneous PDEs

Partial differential equations can also be classified as **homogeneous** or **inhomo-geneous**. A partial differential equation of any order is called homogeneous if every term of the PDE contains the dependent variable u or one of its derivatives, other-wise, it is called an inhomogeneous PDE. This can be illustrated by the following example.

Example 3. Classify the following partial differential equations as homogeneous or inhomogeneous:

(a) $u_t = 4u_{xx}$

(b) $u_t = u_{xx} + x$

(c) $u_{xx} + u_{yy} = 0$

(d) $u_x + u_y = u + 4$

Solution.

(a) The terms of the equation contain partial derivatives of u only, therefore it is a homogeneous PDE.

(b) The equation is an inhomogeneous PDE, because one term contains the inde-pendent variable x.

(c) The equation is a homogeneous PDE.

(d) The equation is an inhomogeneous PDE.

1.2.7   Solution of a PDE

A solution of a PDE is a function u such that it satisfies the equation under discus-sion and satisfies the given conditions as well. In other words, for u to satisfy the equation, the left hand side of the PDE and the right hand side should be the same upon substituting the resulting solution. This concept will be illustrated by exam-ining the following examples. Examples of partial differential equations subject to specific conditions will be examined in the coming chapters.

Example 4. Show that $u(x,t) = \sin x\, e^{-4t}$ is a solution of the following PDE

$$u_t = 4u_{xx}. \tag{26}$$

Solution.

Left Hand Side (LHS)$= u_t = -4\sin x\, e^{-4t}$
Right Hand Side (RHS)$=4u_{xx} = -4\sin x\, e^{-4t}=$LHS

Example 5. Show that $u(x,y) = \sin x \sin y + x^2$ is a solution of the following PDE

$$u_{xx} = u_{yy} + 2. \tag{27}$$

Solution.

Left Hand Side (LHS)$= u_{xx} = -\sin x \sin y + 2$
Right Hand Side (RHS)$=u_{yy} + 2 = -\sin x \sin y + 2=$LHS

Example 6. Show that $u(x,t) = \cos x \cos t$ is a solution of the following PDE

$$u_{tt} = u_{xx}. \tag{28}$$

Solution.

Left Hand Side (LHS)$= u_{tt} = -\cos x \cos t$
Right Hand Side (RHS)$=u_{tt} = -\cos x \cos t=$LHS

Example 7. Show that

(a) $u(x,y) = xy$

(b) $u(x,y) = x^2 y^2$

(c) $u(x,y) = \sin(xy)$

are solutions of the equation

$$xu_x - yu_y = 0. \tag{29}$$

Solution.

(a) $u = xy,\ u_x = y,\ u_y = x,$
LHS $= xy - yx=0$

(b) $u = x^2 y^2,\ u_x = 2xy^2,\ u_y = 2x^2 y,$
LHS $= 2x^2 y^2 - 2x^2 y^2=0$

(c) $u = \sin(xy),\ u_x = y\cos(xy),\ u_y = x\cos(xy),$
LHS $= xy\cos(xy) - xy\cos(xy)=0$

Consequently, we conclude that the general solution is of the form

$$u = f(xy). \tag{30}$$

Remarks:

The following remarks can be drawn here in discussing the concept of a solution of a PDE:

1. For a linear homogeneous ordinary differential equation, it is well-known that if $u_1, u_2, u_3, \cdots, u_n$ are solutions of the equation, then a linear combination of $u_1, u_2, u_3, \cdots$ given by

$$u = c_1 u_1 + c_2 u_2 + c_3 u_3 + \cdots + c_n u_n, \tag{31}$$

is also a solution. The concept of combining two or more of these solutions is called the **superposition principle**.

It is interesting to note that the superposition principle works effectively for linear homogeneous PDEs in a given domain. The concept will be explained in Chapter 3 when using the method of separation of variables.

2. For a linear ordinary differential equation, the *general* solution depends mainly on *arbitrary constants*. Unlike ODEs, in linear partial differential equations, the *general* solution depends on *arbitrary functions*. This can be easily examined by noting that the PDE

$$u_x + u_y = 0, \tag{32}$$

has its solution given by

$$u = f(x - y), \tag{33}$$

where $f(x - y)$ is an arbitrary differentiable function. This means that the solution of (32) can be any of the following functions:

$$
\begin{aligned}
u &= x - y, \\
u &= e^{x-y}, \\
u &= \sinh(x - y), \\
u &= \ln(x - y),
\end{aligned}
\tag{34}
$$

and any function of the form $f(x - y)$. However, the general solution of a PDE is of little use. In fact a particular solution is always required that will satisfy prescribed conditions.

1.2.8 Boundary Conditions

As stated above, the general solution of a PDE is of little use. A particular solution is frequently required that will justify prescribed conditions. Given a PDE that controls the mathematical behavior of a physics phenomenon in a bounded domain D, the dependent variable u is usually prescribed at the boundary of the domain D. The boundary data is called *boundary conditions*. The boundary conditions are given in three types defined as follows:

1. **Dirichlet Boundary Conditions**: In this case, the function u is usually prescribed on the boundary of the bounded domain. For a rod of length L, where $0 < x < L$, the boundary conditions are defined by $u(0) = \alpha$, $u(L) = \beta$, where α and β are constants. For a rectangular plate, $0 < x < L_1, 0 < y < L_2$, the boundary conditions $u(0, y), u(L_1, y), u(x, 0)$, and $u(x, L_2)$ are usually prescribed. The boundary conditions are called homogeneous if the dependent variable u at any point on the boundary is zero, otherwise the boundary conditions are called inhomogeneous.

2. **Neumann Boundary Conditions**: In this case, the normal derivative $\frac{du}{dn}$ of u along the outward normal to the boundary is prescribed. For a rod of length L, Neumann boundary conditions are of the form $u_x(0, t) = \alpha, u_x(L, t) = \beta$.

3. **Mixed Boundary Conditions**: In this case, a linear combination of the dependent variable u and the normal form $\frac{du}{dn}$ is prescribed on the boundary.

It is important to note that it is not always necessary for the domain to be bounded, but instead one or more parts of the boundary may be at infinity. This type of problems will be discussed in the coming chapters.

1.2.9 Initial Conditions

It was indicated before that the PDEs mostly arise to govern physics phenomena such as heat distribution, wave propagation phenomena and phenomena of quantum mechanics. Most of the PDEs, such as the diffusion equation and the wave equation, depend on the time t. Accordingly, the initial values of the dependent variable u at the starting time $t = 0$ should be prescribed. It will be discussed later that for the heat case, the initial value $u(t = 0)$, that defines the temperature at the starting time, should be prescribed. For the wave equation, the initial conditions $u(t = 0)$ and $u_t(t = 0)$ should also be prescribed.

1.2.10 Well-Posed PDEs

A partial differential equation is said to be *well-posed* if a unique solution, that satisfies the equation and the prescribed conditions, exists, provided that the unique solution obtained is *stable*. The solution to a PDE is said to be stable if a small change in the conditions or the coefficients of the PDE results in a small change in the solution.

Exercises 1.2

1. Find the order of the following PDEs:

(a) $u_{xx} = u_{xxx} + u + 1$

(b) $u_{tt} = u_{xx} + u_{yy} + u_{zz}$
(c) $u_x + u_y = 0$
(d) $u_t + u_{xxyy} = u$

2. Classify the following PDEs as linear or nonlinear:

(a) $u_t = u_{xx} - u$
(b) $u_{tt} = u_{xx} + u^2$
(c) $u_x + u_y = u$
(d) $u_t + uu_{xxyy} = 0$

3. Classify the following PDEs as homogeneous or inhomogeneous:

(a) $u_t = u_{xx} + x$
(b) $u_{tt} = u_{xx} + u_{yy} + u_{zz}$
(c) $u_x + u_y = 1$
(d) $u_t + u_{xxy} = u$

4. Verify that the given function is a particular solution of the corresponding PDE:

(a) $u_x + u_y = x + y$, $u(x, y) = xy$
(b) $u_x - u_y = 0$, $u(x, y) = x + y$
(c) $u_x + u_y = u$, $u(x, y) = e^x + e^y$
(d) $xu_x + u_y = u$, $u(x, y) = x + e^y$

5. Verify that the given function is a particular solution of the corresponding PDE:

(a) $u_t = u_{xx}$, $u(x, t) = x + e^{-t} \sin x$
(b) $u_t = u_{xx} - 2u$, $u(x, t) = e^{-t} \sinh x$
(c) $u_{tt} = u_{xx}$, $u(x, t) = \sin x \sin t$
(d) $u_{tt} = 2(u_{xx} + u_{yy})$, $u(x, y, t) = \cos x \cos y \cos(2t)$

6. Show that the functions

(a) $u(x, y) = \frac{x}{y}$

(b) $u(x, y) = \sin\left(\frac{x}{y}\right)$

(c) $u(x, y) = \cosh\left(\frac{x}{y}\right)$

are solutions of the equation $xu_x + yu_y = 0$. Show that $u = f\left(\frac{x}{y}\right)$ is a general solution of the equation where f is an arbitrary differentiable function.

7. Show that the functions

(a) $u(x, y) = x + y^2$

(b) $u(x, y) = \sin(x + y^2)$

(c) $u(x, y) = e^{x + y^2}$

are solutions of the equation

$$2yu_x - u_y = 0.$$

Show that $u = f(x + y^2)$ is a general solution of the equation where f is an arbitrary differentiable function.

8. Verify that the given function is a general solution of the corresponding PDE:

(a) $u_{tt} = u_{xx}$, $u(x, t) = f(x + t) + g(x - t)$

(b) $u_x - u_y = 0$, $u(x, y) = f(x + y)$

(c) $4y^2 u_{xx} + \frac{1}{y} u_y - u_{yy} = 0$, $u(x, y) = f(x + y^2) + g(x - y^2)$

(d) $u_{xx} - \frac{1}{x} u_x - 4x^2 u_{yy} = 0$, $u(x, y) = f(x^2 + y) + g(x^2 - y)$ given that f and g are twice differentiable functions.

1.3 Classifications of a Second Order PDE

A second order linear partial differential equation in two independent variables x and y in its general form is given by

$$Au_{xx} + Bu_{xy} + Cu_{yy} + Du_x + Eu_y + Fu = G, \tag{35}$$

where A, B, C, D, E, F, and G are constants or functions of the variables x and y. A second order partial differential equation (35) is usually classified into three basic classes of equations, namely:

1. **Parabolic.** Parabolic equation is an equation that satisfies the property

$$B^2 - 4AC = 0. \tag{36}$$

Examples of parabolic equations are heat flow and diffusion processes equations. The heat transfer equation

$$u_t = ku_{xx}, \tag{37}$$

will be discussed in details in Chapters 3 and 4.

2. **Hyperbolic.** Hyperbolic equation is an equation that satisfies the property

$$B^2 - 4AC > 0. \tag{38}$$

Examples of hyperbolic equations are wave propagation equations. The wave equation

$$u_{tt} = c^2 u_{xx}, \tag{39}$$

will be discussed in details in Chapters 5 and 6.

3. **Elliptic.** Elliptic equation is an equation that satisfies the property

$$B^2 - 4AC < 0. \tag{40}$$

Examples of Elliptic equations are Laplace's equation and Schrodinger equation. The Laplace equation in a two dimensional space

$$u_{xx} + u_{yy} = 0, \tag{41}$$

will be discussed in details in Chapter 7. The Laplace's equation is often called the potential equation because $u(x, y)$ defines the potential function.

Example 1. Classify the following second order partial differential equations as *parabolic, hyperbolic* or *elliptic*:

(a) $u_t = 4u_{xx}$

(b) $u_{tt} = 4u_{xx}$

(c) $u_{xx} + u_{yy} = 0$

Solution.

(a) $A = 4, B = C = 0$
This means that

$$B^2 - 4AC = 0. \tag{42}$$

Hence, the equation in (a) is parabolic.

(b) $A = 4, B = 0, C = -1$
This means that

$$B^2 - 4AC = 16 > 0. \tag{43}$$

Hence, the equation in (b) is hyperbolic.

(c) $A = 1, B = 0, C = 1$
This means that

$$B^2 - 4AC = -4 < 0. \tag{44}$$

Hence, the equation in (a) is elliptic.

Example 2. Classify the following second order partial differential equations as *parabolic, hyperbolic* or *elliptic*:

(a) $u_{tt} = u_{xx} - u_t$

(b) $u_t = u_{xx} - u_x$

(c) $u_{xx} + xu_{yy} = 0$

Solution.

(a) $A = 1, B = 0, C = -1$
This means that

$$B^2 - 4AC = 4 > 0. \tag{45}$$

Hence, the equation in (a) is hyperbolic.

(b) $A = 1$, $B = C = 0$
This means that
$$B^2 - 4AC = 0. \tag{46}$$
Hence, the equation in (b) is parabolic.

(c) $A = 1$, $B = 0$, $C = x$
This means that
$$B^2 - 4AC = -4x. \tag{47}$$
Hence, the equation in (a) is parabolic if $x = 0$, hyperbolic if $x < 0$, and elliptic if $x > 0$.

Exercises 1.3

Classify the following second order partial differential equations as *parabolic, hyperbolic* or *elliptic*:

1. $u_{tt} = c^2 u_{xx}$

2. $u_{xx} + u_{yy} + u = 0$

3. $u_t = 4u_{xx} + xt$

4. $u_{tt} = u_{xx} + xt$

5. $u_t = u_{xx} + 2u_x + u$

6. $u_{xy} = 0$

7. $u_{xx} + u_{yy} = 4$

8. $u_{xx} + u = 0$

9. $u_{tt} = u_{xx} - u_t$

10. $yu_{xx} + u_{yy} = 0$

Chapter 2

First Order Partial Differential Equations

2.1 Introduction

In this chapter we will discuss the first order linear partial differential equations, homogeneous and inhomogeneous. First order partial differential equations appear in a variety of physical models. Partial differential equations of first order are used to model traffic flow on a crowded road, blood flow through an elastic-walled tube, shock waves and as special cases of the general theories of gas dynamics and hydraulics.

It is the concern of this text to introduce the recently developed methods to handle partial differential equations in an accessible manner. Some of the traditional techniques will be used as well. In this text we will apply the recently developed *Adomian decomposition method* and the related phenomenon of the noise terms that will accelerate the rapid convergence of the solution. The decomposition method and the improvements made by the noise terms phenomenon and the modified decomposition method are reliable and effective techniques of promising results. In addition, the classic *method of characteristics* will be implemented in this chapter. A comparative study between the method of characteristics and the Adomian decomposition method will be carried out through illustrative examples.

2.2 Adomian Decomposition Method

In this section we will discuss the newly developed Adomian decomposition method. The Adomian decomposition method has been receiving much attention in recent years in applied mathematics in general, and in the area of series solutions in particular. The method proved to be powerful, effective, and can easily handle a wide

class of linear or nonlinear, ordinary or partial differential equations, and integral
equations. The decomposition method demonstrates fast convergence of the solution and therefore provides several significant advantages. In this text, the method
will be successfully used to handle most types of partial differential equations that
appear in several physical models and scientific applications. The method attacks
the problem in a direct way and in a straightforward fashion without using linearization, perturbation or any other restrictive assumption that may change the
physical behavior of the model under discussion.

The Adomian decomposition method was introduced and developed by George
Adomian in [4–12] and is well addressed in many references. A considerable amount
of research work has been invested recently in applying this method to a wide class
of linear and nonlinear ordinary differential equations, partial differential equations
and integral equations as well. For more details, the reader is advised to see the
references [4–12,36,39,40,44,58,67–69,104–106,120–132] and the references therein.

The Adomian decomposition method consists of decomposing the unknown
function $u(x, y)$ of any equation into a sum of an infinite number of components
defined by the decomposition series

$$u(x, y) = \sum_{n=0}^{\infty} u_n(x, y), \tag{1}$$

where the components $u_n(x, y), n \geq 0$ are to be determined in a recursive manner.
The decomposition method concerns itself with finding the components $u_0, u_1, u_2, \cdots$
individually. As will be seen through the text, the determination of these components can be achieved in an easy way through a recursive relation that usually
involve simple integrals.

To give a clear overview of Adomian decomposition method, we first consider
the linear differential equation written in an operator form by

$$Lu + Ru = g, \tag{2}$$

where L is the higher order derivative which is assumed to be invertible, R is a
linear differential operator of order less than L, and g is a source term. It is to be
noted that the nonlinear differential equations will approached in Chapter 8. We
next apply the inverse operator L^{-1} to both sides of equation (2) and using the
given condition to obtain

$$u = f - L^{-1}(Ru), \tag{3}$$

where the function f represents the terms arising from integrating the source term g
and from using the given conditions, all are assumed to be prescribed. As indicated
before, Adomian method defines the solution u by an infinite series of components
given by

$$u = \sum_{n=0}^{\infty} u_n, \tag{4}$$

where the components $u_0, u_1, u_2, \cdots$ are usually recurrently determined. Substituting (4) into both sides of (3) leads to

$$\sum_{n=0}^{\infty} u_n = f - L^{-1} \left(R(\sum_{n=0}^{\infty} u_n) \right). \tag{5}$$

For simplicity, Eq. (5) can be rewritten as

$$u_0 + u_1 + u_2 + u_3 + \cdots = f - L^{-1} \left(R(u_0 + u_1 + u_2 + \cdots) \right). \tag{6}$$

To construct the recursive relation needed for the determination of the components $u_0, u_1, u_2, \cdots$, it is important to note that Adomian method suggests that the zeroth component u_0 is usually defined by the function f described above, i.e. by all terms, that are not included under the inverse operator L^{-1}, which arise from the initial data and from integrating the inhomogeneous term. Accordingly, the formal recursive relation is defined by

$$\begin{aligned} u_0 &= f, \\ u_{k+1} &= -L^{-1} \left(R(u_k) \right), \, k \geq 0, \end{aligned} \tag{7}$$

or equivalently

$$\begin{aligned} u_0 &= f, \\ u_1 &= -L^{-1} \left(R(u_0) \right), \\ u_2 &= -L^{-1} \left(R(u_1) \right), \\ u_3 &= -L^{-1} \left(R(u_2) \right), \\ &\vdots \end{aligned} \tag{8}$$

It is clearly seen that the relation (8) reduced the differential equation under consideration into an elegant determination of computable components. Having determined these components, we then substitute it into (4) to obtain the solution in a series form.

It was formally shown by many researchers that if an exact solution exists for the problem, then the obtained series converges very rapidly to that solution. The convergence concept of the decomposition series was thoroughly investigated by many researchers to confirm the rapid convergence of the resulting series. Cherruault examined the convergence of Adomian's method in [35,36]. Cherruault et. al [36] discussed the convergence of Adomian's method when applied to integral equations. In addition, Cherruault and Adomian presented a new proof of convergence of the method in [37]. Moreover, Abbaoui et. al [1] formally proved the convergence of Adomian's method when applied to differential equations in general. For more details about the proofs presented to discuss the rapid convergence, the reader is advised to see the references mentioned above and the references therein.

However, for concrete problems, where a closed form solution is not obtainable, a truncated number of terms is usually used for numerical purposes. It was also shown by many that the series obtained by evaluating few terms gives an approximation of high degree of accuracy if compared with other numerical techniques. For further reading, see [19,39,40,95,128].

It seems reasonable to give a brief outline about the works conducted by Adomian and other researchers in applying Adomian's method. Adomian in [4–12] and in many other works introduced his method and applied it to many deterministic and stochastic problems. He implemented his method to solve frontier problems of physics. The Adomian's achievements in this regard are remarkable and of promising results.

Adomian's method has attracted a considerable amount of research work. In [19], a comparison between the decomposition method and the perturbation technique showed the efficiency of the decomposition method compared to the tedious work required by the perturbation method. In [95], the advantage of the decomposition method over Picard's method has been emphasized. In [128], a comparative study between Adomian's method and Taylor series method has been examined to show that the decomposition method requires less computational work if compared with Taylor series. Other comparisons with traditional methods such as finite difference method have been conducted in the literature such as the works in [39,40,44,45,80].

It is to be noted that Datta [39,40] conducted a useful study to show that few terms of the decomposition series provide a numerical result of a high degree of accuracy. Rach *et. al.* [97] employed Adomian's method to solve differential equations with singular coefficients such as Legendre's equation, Bessel's equation, and Hermite's equation. Moreover, in [104], a suitable definition of the operator was used to overcome the difficulty of singular points of Lane-Emden equation. In [127], a new definition of the operator was introduced to overcome the singularity behavior for the Lane-Emden type of equations. Recently, a useful study was presented in [33] to implement the decomposition method for differential equations with discontinuities.

It is normal in differential equations that we seek a closed form solution or a series solution with a proper number of terms. Although this book is devoted to handle partial differential equations, but it seems reasonable to use the decomposition method to discuss two ordinary differential equations where an exact solution is obtained for the first equation and a series approximation is determined for the second. For the first problem we consider the equation

$$u^{'}(x) = u(x), \ u(0) = A. \tag{9}$$

In an operator form, Eq. (9) becomes

$$Lu = u, \tag{10}$$

where the differential operator L is given by

$$L = \frac{d}{dx},$$ (11)

and therefore the inverse operator L^{-1} is defined by

$$L^{-1}(.) = \int_0^x (.)dx.$$ (12)

Applying L^{-1} to both sides of (10) and using the initial condition we obtain

$$L^{-1}(Lu) = L^{-1}(u),$$ (13)

so that

$$u(x) - u(0) = L^{-1}(u),$$ (14)

or equivalently

$$u(x) = A + L^{-1}(u).$$ (15)

Substituting the series assumption (5) into both sides of (15) gives

$$\sum_{n=0}^{\infty} u_n(x) = A + L^{-1}\left(\sum_{n=0}^{\infty} u_n(x)\right).$$ (16)

In view of (16), the following recursive relation

$$\begin{aligned} u_0(x) &= A, \\ u_{k+1}(x) &= L^{-1}(u_k(x)), \ k \geq 0, \end{aligned}$$ (17)

follows immediately. Consequently, we obtain

$$\begin{aligned} u_0(x) &= A, \\ u_1(x) &= L^{-1}(u_0(x)) = Ax, \\ u_2(x) &= L^{-1}(u_1(x)) = A\frac{x^2}{2!}, \\ u_3(x) &= L^{-1}(u_2(x)) = A\frac{x^3}{3!}, \\ &\vdots \end{aligned}$$ (18)

Substituting (18) into (5) gives the solution in a series form by

$$u(x) = A(1 + x + \frac{x^2}{2!} + \frac{x^3}{3!} + \cdots),$$ (19)

and in a closed form by

$$u(x) = Ae^x. \tag{20}$$

We next consider the well-known Airy's equation

$$u''(x) = xu(x), \ u(0) = A, u'(0) = B. \tag{21}$$

In an operator form, Eq. (21) becomes

$$Lu = xu, \tag{22}$$

where the differential operator L is given by

$$L = \frac{d^2}{dx^2}, \tag{23}$$

and therefore the inverse operator L^{-1} is defined by

$$L^{-1}(.) = \int_0^x \int_0^x (.)dx\,dx. \tag{24}$$

Operating with L^{-1} on both sides of (21) and using the initial conditions we obtain

$$L^{-1}(Lu) = L^{-1}(xu), \tag{25}$$

so that

$$u(x) - xu'(0) - u(0) = L^{-1}(xu), \tag{26}$$

or equivalently

$$u(x) = A + Bx + L^{-1}(xu). \tag{27}$$

Substituting the series assumption (5) into both sides of (27) yields

$$\sum_{n=0}^{\infty} u_n(x) = A + Bx + L^{-1}(x \sum_{n=0}^{\infty} (u_n(x))). \tag{28}$$

Following the decomposition method we obtain the following recursive relation

$$\begin{aligned} u_0(x) &= A + Bx, \\ u_{k+1}(x) &= L^{-1}(xu_k(x)), \ k \geq 0. \end{aligned} \tag{29}$$

Consequently, we obtain

$$\begin{aligned} u_0(x) &= A + Bx, \\ u_1(x) &= L^{-1}(xu_0(x)) = A\frac{x^3}{6} + B\frac{x^4}{12}. \\ u_2(x) &= L^{-1}(xu_1(x)) = A\frac{x^6}{180} + B\frac{x^7}{504}. \end{aligned} \tag{30}$$

$$\vdots$$

Substituting (30) into (5) gives the solution in a series form by

$$u(x) = A(1 + \frac{x^3}{6} + A\frac{x^6}{180} + \cdots) + B(x + \frac{x^4}{12} + \frac{x^7}{504} + \cdots). \tag{31}$$

Other components can be easily computed to enhance the accuracy of the approximation.

It seems now reasonable to apply Adomian decomposition method to first order partial differential equations. For the convenience of the reader, and without loss of generality, we consider the inhomogeneous partial differential equation:

$$u_x + u_y = f(x, y),\ u(0, y) = g(y),\ u(x, 0) = h(x). \tag{32}$$

In an operator form, equation (32) can be written as

$$L_x u + L_y u = f(x, y), \tag{33}$$

where

$$L_x = \frac{\partial}{\partial x},\ L_y = \frac{\partial}{\partial y}, \tag{34}$$

where each operator is assumed easily invertible, and thus the inverse operators L_x^{-1} and L_y^{-1} exist and given by

$$\begin{aligned} L_x^{-1}(.) &= \int_0^x (.)dx, \\ L_y^{-1}(.) &= \int_0^y (.)dy. \end{aligned} \tag{35}$$

This means that

$$L_x^{-1} L_x u(x, y) = u(x, y) - u(0, y). \tag{36}$$

Applying L_x^{-1} to both sides of (33) gives

$$L_x^{-1} L_x u = L_x^{-1} f(x, y) - L_x^{-1}(L_y u), \tag{37}$$

or equivalently

$$u(x, y) = g(y) + L_x^{-1}(f(x, y)) - L_x^{-1}(L_y u), \tag{38}$$

obtained by using (36) and by using the condition $u(0, y) = g(y)$. As stated above, the decomposition method sets

$$u(x, y) = \sum_{n=0}^{\infty} u_n(x, y). \tag{39}$$

Substituting (39) into both sides of (38) we find

$$\sum_{n=0}^{\infty} u_n(x, y) = g(y) + L_x^{-1}(f(x, y)) - L_x^{-1}\left(L_y\left(\sum_{n=0}^{\infty} u_n(x, y)\right)\right). \tag{40}$$

This can be rewritten as

$$u_0 + u_1 + u_2 + \cdots = g(y) + L_x^{-1}(f(x,y)) - L_x^{-1}L_y(u_0 + u_1 + u_2 + \cdots). \qquad (41)$$

The zeroth component u_0, as suggested by Adomian method is always identified by the given proper condition and the terms arising from $L_x^{-1}(f(x,y))$, both of which are assumed to be known. Accordingly, we set

$$u_0(x,y) = g(y) + L_x^{-1}(f(x,y)). \qquad (42)$$

Consequently, the other components $u_{k+1}, k \geq 0$ are defined by using the relation

$$u_{k+1}(x,y) = -L_x^{-1}L_y(u_k), \, k \geq 0. \qquad (43)$$

Combining equations (42) and (43), we obtain the recursive scheme

$$
\begin{aligned}
u_0(x,y) &= g(y) + L_x^{-1}(f(x,y)), \\
u_{k+1}(x,y) &= -L_x^{-1}L_y(u_k), \, k \geq 0,
\end{aligned} \qquad (44)
$$

that forms the basis for a complete determination of the components $u_0, u_1, u_2, \cdots$. Therefore, the components can be easily obtained by

$$
\begin{aligned}
u_0(x,y) &= g(y) + L_x^{-1}(f(x,y)), \\
u_1(x,y) &= -L_x^{-1}\left(L_y u_0(x,y)\right), \\
u_2(x,y) &= -L_x^{-1}(L_y u_1(x,y)), \\
u_3(x,y) &= -L_x^{-1}(L_y u_2(x,y)),
\end{aligned} \qquad (45)
$$

and so on. Thus the components u_n can be determined recursively as far as we like. It is clear that the accuracy of the approximation can be significantly improved by simply determining more components. Having established the components of $u(x,y)$, the solution in a series form follows immediately. However, the expression

$$\phi_n = \sum_{r=0}^{n-1} u_r(x,y), \qquad (46)$$

is considered the n-term approximation to u. For concrete problems, where exact solution is not easily obtainable, we usually use the truncated series (46) for numerical purposes. As indicated earlier, the convergence of Adomian decomposition method has been established by many researchers, but will not be discussed in this text.

It is important to note that the solution can also be obtained by finding the y-solution by applying the inverse operator L_y^{-1} to both sides of the equation

$$L_y = f(x, y) - L_x u. \tag{47}$$

The equality of the x-solution and the y-solution is formally justified and will be examined through the coming examples.

It should be noted here that the series solution (39) has been proved by many researchers to converge rapidly, and a closed form solution is obtainable in many cases if a closed form solution exists.

It was found, as will be seen later, that very few terms of the series obtained in (39) provide a high degree of accuracy level which makes the method powerful when compared with other existing numerical techniques. In many cases the series representation of $u(x, y)$ can be summed to yield the closed form solution. Several works in this direction have demonstrated the power of the method for analytical and numerical applications.

The essential features of the decomposition method for linear and nonlinear equations, homogeneous and inhomogeneous, can be outlined as follows:

1. Express the partial differential equation, linear or nonlinear, in an operator form.
2. Apply the inverse operator to both sides of the equation written in an operator form.
3. Set the unknown function $u(x, y)$ into a decomposition series

$$u(x, y) = \sum_{n=0}^{\infty} u_n(x, y), \tag{48}$$

whose components are elegantly determined. We next substitute the series (48) into both sides of the resulting equation.

4. Identify the zeroth component $u_0(x, y)$ as the terms arising from the given conditions and from integrating the source term $f(x, y)$, both are assumed to be known.

5. Determine the successive components of the series solution $u_k, k \geq 1$ by applying the recursive scheme (44), where each component u_k can be completely determined by using the previous component u_{k-1}.

6. Substitute the determined components into (48) to obtain the solution in a series form. An exact solution can be easily obtained in many equations if such a closed form solution exists.

It is to be noted that Adomian decomposition method approaches any equation, homogeneous or inhomogeneous, and linear or nonlinear in a straightforward manner without any need to restrictive assumptions such as linearization, discretization or perturbation. There is no need in using this method to convert inhomogeneous conditions to homogeneous conditions as required by other techniques.

The essential steps of the Adomian decomposition method will be illustrated by discussing the following examples.

Example 1. Use Adomian decomposition method to solve the following inhomogeneous PDE

$$u_x + u_y = x + y, \ u(0, y) = 0, \ u(x, 0) = 0. \tag{49}$$

Solution.

In an operator form, equation (49) can be written as

$$L_x u = x + y - L_y u, \tag{50}$$

where

$$L_x = \frac{\partial}{\partial x}, \ L_y = \frac{\partial}{\partial y}. \tag{51}$$

It is clear that L_x is invertible, hence L_x^{-1} exists and given by

$$L_x^{-1}(.) = \int_0^x (.)dx. \tag{52}$$

The x-solution:

This solution can be obtained by applying L_x^{-1} to both sides of (50), hence we find

$$L_x^{-1} L_x u = L_x^{-1}(x + y) - L_x^{-1}(L_y u), \tag{53}$$

or equivalently

$$u(x, y) = u(0, y) + \frac{1}{2}x^2 + xy - L_x^{-1}(L_y u) = \frac{1}{2}x^2 + xy - L_x^{-1}(L_y u), \tag{54}$$

obtained upon using the given condition $u(0, y) = 0$, the equation (36) and by integrating $f(x, y) = x + y$ with respect to x. As stated above, the decomposition method identifies the unknown function $u(x, y)$ as an infinite number of components $u_n(x, y), n \geq 0$ given by

$$u(x, y) = \sum_{n=0}^{\infty} u_n(x, y). \tag{55}$$

Substituting (55) into both sides of (54) we find

$$\sum_{n=0}^{\infty} u_n(x, y) = \frac{1}{2}x^2 + xy - L_x^{-1}\left(L_y\left(\sum_{n=0}^{\infty} u_n(x, y)\right)\right). \tag{56}$$

Using few terms of the decomposition (55) we obtain

$$u_0 + u_1 + u_2 + \cdots = \frac{1}{2}x^2 + xy - L_x^{-1}\left(L_y(u_0 + u_1 + u_2 + \cdots)\right). \tag{57}$$

As presented before, the decomposition method identifies the zeroth component u_0 by all terms arising from the given condition and from integrating $f(x, y) = x + y$, therefore we set

$$u_0(x, y) = \frac{1}{2}x^2 + xy. \tag{58}$$

Consequently, the recursive scheme that will enable us to completely determine the successive components is thus constructed by

$$u_0(x, y) = \frac{1}{2}x^2 + xy,$$
$$u_{k+1}(x, y) = L_x^{-1}L_y(u_k), \ k \geq 0. \tag{59}$$

This in turn gives

$$u_1(x, y) = -L_x^{-1}(L_y u_0) = -L_x^{-1}\left(L_y\left(\frac{1}{2}x^2 + xy\right)\right) = -\frac{1}{2}x^2,$$
$$u_2(x, y) = -L_x^{-1}(L_y u_1) = -L_x^{-1}\left(L_y\left(-\frac{1}{2}x^2\right)\right) = 0. \tag{60}$$

Accordingly, $u_k = 0, k \geq 2$. Having determined the components of $u(x, y)$, we find

$$u = u_0 + u_1 + u_2 + \cdots = \frac{1}{2}x^2 + xy - \frac{1}{2}x^2 = xy, \tag{61}$$

the exact solution of the equation under discussion.

The y-solution:

It is important to note that the exact solution can also be obtained by finding the y-solution. In an operator form we can write the equation by

$$L_y = x + y - L_x u. \tag{62}$$

Assume that L_y^{-1} exists and defined by

$$L_y^{-1}(.) = \int_0^y (.)dy. \tag{63}$$

Applying L_y^{-1} to both sides of the equation (62) gives

$$u(x, y) = xy + \frac{1}{2}y^2 - L_y^{-1}(L_x u). \tag{64}$$

As mentioned above, the decomposition method sets the solution $u(x, y)$ in a series form by

$$u(x, y) = \sum_{n=0}^{\infty} u_n(x, y). \tag{65}$$

Inserting (65) into both sides of (64) we obtain

$$\sum_{n=0}^{\infty} u_n(x, y) = xy + \frac{1}{2}y^2 - L_y^{-1}\left(L_x\left(\sum_{n=0}^{\infty} u_n(x, y)\right)\right). \tag{66}$$

Using few terms only for simplicity reasons, we obtain

$$u_0 + u_1 + u_2 + \cdots = xy + \frac{1}{2}y^2 - L_y^{-1}(L_x(u_0 + u_1 + u_2 + \cdots)). \tag{67}$$

The decomposition method identifies the zeroth component u_0 by all terms arising from the given condition and from integrating $f(x, y) = x + y$, therefore we set

$$u_0(x, y) = xy + \frac{1}{2}y^2. \tag{68}$$

Consequently, to completely determine the successive components of $u(x, y)$, the recursive scheme is thus defined by

$$\begin{aligned} u_0(x, y) &= xy + \tfrac{1}{2}y^2, \\ u_{k+1}(x, y) &= L_y^{-1} L_x(u_k), \ k \geq 0. \end{aligned} \tag{69}$$

This gives

$$\begin{aligned} u_1(x, y) &= -L_y^{-1}(L_x u_0) = -L_y^{-1}\left(L_x\left(xy + \tfrac{1}{2}y^2\right)\right) = -\tfrac{1}{2}y^2, \\ u_2(x, y) &= -L_y^{-1}(L_x u_1) = -L_y^{-1}\left(L_x\left(-\tfrac{1}{2}y^2\right)\right) = 0. \end{aligned} \tag{70}$$

Consequently, $u_k = 0, k \geq 2$. Having determined the components of $u(x, y)$, we find

$$u(x, y) = u_0 + u_1 + u_2 + \cdots = xy + \frac{1}{2}y^2 - \frac{1}{2}y^2 = xy, \tag{71}$$

the exact solution of the equation under discussion.

Example 2. Solve the following homogeneous partial differential equation

$$u_x - u_y = 0, \ u(0, y) = y, \ u(x, 0) = x. \tag{72}$$

Solution.

In an operator form, equation (72) becomes

$$L_x u(x, y) = L_y u(x, y), \tag{73}$$

where the operators L_x and L_y are defined by

$$L_x = \frac{\partial}{\partial x}, \ L_y = \frac{\partial}{\partial y}. \tag{74}$$

Applying the inverse operator L_x^{-1} to both sides of (73) and using the given condition $u(0, y) = y$ yields

$$u(x, y) = y + L_x^{-1}(L_y u). \tag{75}$$

We next define the unknown function $u(x, y)$ by the decomposition series

$$u(x, y) = \sum_{n=0}^{\infty} u_n(x, y). \tag{76}$$

Inserting (76) into both sides of (75) gives

$$\sum_{n=0}^{\infty} u_n(x,y) = y + L_x^{-1}\left(L_y\left(\sum_{n=0}^{\infty} u_n(x,y)\right)\right). \tag{77}$$

By considering few terms of the decomposition of $u(x,y)$, equation (77) becomes

$$u_0 + u_1 + u_2 + \cdots = y + L_x^{-1}\left(L_y(u_0 + u_1 + u_2 + \cdots)\right). \tag{78}$$

Proceeding as before, we identify the zeroth component u_0 by

$$u_0(x,y) = y. \tag{79}$$

Having identified the zeroth component $u_0(x,y)$, we obtain the recursive scheme

$$\begin{aligned}
u_0(x,y) &= y, \\
u_{k+1}(x,y) &= L_x^{-1}L_y(u_k), \ k \geq 0.
\end{aligned} \tag{80}$$

The components $u_0, u_1, u_2, \cdots$ are thus determined as follows:

$$\begin{aligned}
u_0(x,y) &= y, \\
u_1(x,y) &= L_x^{-1}L_y\, u_0 = L_x^{-1}L_y(y) = x, \\
u_2(x,y) &= L_x^{-1}L_y\, u_1 = L_x^{-1}L_y(x) = 0.
\end{aligned} \tag{81}$$

It is obvious that all components $u_k(x,y) = 0, k \geq 2$. Consequently, the solution is given by

$$u(x,y) = u_0(x,y) + u_1(x,y) + \cdots = u_0(x,y) + u_1(x,y) = y + x, \tag{82}$$

the exact solution obtained by using the decomposition series (76).

It is important to note here that the exact solution given by (82) can also be obtained by determining the y-solution as discussed above. This is left as an exercise to the reader.

Example 3. Solve the following homogeneous partial differential equation

$$xu_x + u_y = 3u, \ u(x,0) = x^2, \ u(0,y) = 0. \tag{83}$$

Solution.

In an operator form, equation (83) becomes

$$L_y u(x,y) = 3u(x,y) - xL_x u(x,y). \tag{84}$$

Applying the inverse operator L_y^{-1} to both sides of (84) and using the given condition $u(x,0) = x^2$ yields

$$u(x,y) = x^2 + L_y^{-1}(3u - xL_x u). \tag{85}$$

Substituting $u(x,y) = \sum_{n=0}^{\infty} u_n(x,y)$ into both sides of (85) gives

$$\sum_{n=0}^{\infty} u_n(x,y) = x^2 + L_y^{-1}\left(3(\sum_{n=0}^{\infty} u_n(x,y)) - xL_x\left(\sum_{n=0}^{\infty} u_n(x,y)\right)\right). \tag{86}$$

By considering few terms of the decomposition of $u(x,y)$, equation (86) becomes

$$u_0 + u_1 + u_2 + \cdots = x^2 + L_y^{-1}\left(3(u_0 + u_1 + \cdots) - xL_x(u_0 + u_1 + u_2 + \cdots)\right). \tag{87}$$

Proceeding as before, we identify the recursive scheme

$$\begin{aligned} u_0(x,y) &= x^2, \\ u_{k+1}(x,y) &= L_y^{-1}(3u_k - xL_x(u_k), \ k \geq 0. \end{aligned} \tag{88}$$

The components $u_0, u_1, u_2, \cdots$ are thus determined as follows:

$$\begin{aligned} u_0(x,y) &= x^2, \\ u_1(x,y) &= L_y^{-1}(3u_0 - xL_x u_0) = x^2 y, \\ u_2(x,y) &= L_y^{-1}(3u_1 - xL_x u_1)L_y = \frac{x^2 y^2}{2!}, \\ &\vdots \end{aligned} \tag{89}$$

It is obvious that all components $u_k(x,y) = 0, k \geq 2$. Consequently, the solution is given by

$$u(x,y) = u_0 + u_1 + u_2 + \cdots = x^2(1 + y + \frac{y^2}{2!} + \cdots) = x^2 e^y. \tag{90}$$

Example 4. Solve the following homogeneous partial differential equation

$$u_x - yu = 0, \ u(0,y) = 1. \tag{91}$$

Solution.

In an operator form, equation (91) becomes

$$L_x u(x,y) = yu(x,y), \tag{92}$$

where the operator L_x is defined as

$$L_x = \frac{\partial}{\partial x}. \tag{93}$$

Applying the integral operator L_x^{-1} to both sides of (92) and using the given condition that $u(0, y) = 1$ gives

$$u(x, y) = 1 + L_x^{-1}(yu(x, y)). \tag{94}$$

Following the discussion presented above, we define the unknown function $u(x, y)$ by the decomposition series

$$u(x, y) = \sum_{n=0}^{\infty} u_n(x, y). \tag{95}$$

Inserting (95) into both sides of (94) gives

$$\sum_{n=0}^{\infty} u_n(x, y) = 1 + L_x^{-1} \left(y \sum_{n=0}^{\infty} u_n(x, y) \right), \tag{96}$$

or equivalently

$$u_0 + u_1 + u_2 + \cdots = 1 + L_x^{-1}(y(u_0 + u_1 + u_2 + \cdots)), \tag{97}$$

by considering few terms of the decomposition of $u(x, y)$. The components $u_0, u_1, u_2, \cdots$ are thus determined by using the recursive relationship as follows:

$$
\begin{aligned}
u_0(x, y) &= 1, \\
u_1(x, y) &= L_x^{-1}(yu_0) = xy, \\
u_2(x, y) &= L_x^{-1}(yu_1) = \tfrac{1}{2!}x^2y^2, \\
u_3(x, y) &= L_x^{-1}(yu_2) = \tfrac{1}{3!}x^3y^3,
\end{aligned} \tag{98}
$$

and so on for other components. Consequently, the solution in a series form is given by

$$
\begin{aligned}
u(x, y) &= u_0(x, y) + u_1(x, y) + u_2(x, y) + \cdots, \\
&= 1 + xy + \frac{1}{2!}x^2y^2 + \frac{1}{3!}x^3y^3 + \cdots,
\end{aligned} \tag{99}
$$

and in a closed form

$$u(x, y) = e^{xy}. \tag{100}$$

Example 5. Solve the following homogeneous PDE

$$u_t + cu_x = 0, \ u(x, 0) = x, \tag{101}$$

where c is a constant.

Solution.

In an operator form, equation (101) can be rewritten as

$$L_t\, u(x,t) = -cL_x u,\qquad\qquad (102)$$

where the operator L_t is defined as

$$L_t = \frac{\partial}{\partial t}.\qquad\qquad (103)$$

It is clear that the operator L_t is invertible, and the inverse operator L_t^{-1} is an indefinite integral from 0 to t. Applying the integral operator L_t^{-1} to both sides of (102) and using the given condition that $u(x,0) = x$ yields

$$u(x,t) = x - cL_t^{-1}(L_x\, u(x,t)).\qquad\qquad (104)$$

Proceeding as before, we substitute the decomposition series for $u(x,t)$ into both sides of (104) to obtain

$$\sum_{n=0}^{\infty} u_n(x,t) = x - cL_t^{-1}\left(L_x\left(\sum_{n=0}^{\infty} u_n(x,t)\right)\right).\qquad\qquad (105)$$

Using few terms of the decomposition of $u(x,y)$, equation (105) becomes

$$u_0 + u_1 + u_2 + \cdots = x - cL_t^{-1}\left(L_x(u_0 + u_1 + u_2 + \cdots)\right).\qquad\qquad (106)$$

The components $u_0, u_1, u_2, \cdots$ can be determined by using the recursive relationship as follows:

$$
\begin{aligned}
u_0(x,t) &= x,\\
u_1(x,t) &= -cL_t^{-1}(L_x\, u_0) = -ct,\\
u_2(x,t) &= -cL_t^{-1}(L_x\, u_1) = 0.
\end{aligned}\qquad (107)
$$

We can easily observe that $u_k = 0$, $k \geq 2$. It follows that the solution in a closed form is given by

$$u(x,t) = x - ct.\qquad\qquad (108)$$

Example 6. Solve the following partial differential equation

$$
\begin{aligned}
u_x + u_y + u_z &= u,\ u(0,y,z) = 1 + e^y + e^z,\\
u(x,0,z) &= 1 + e^x + e^z,\\
u(x,y,0) &= 1 + e^x + e^y,
\end{aligned}\qquad (109)
$$

where $u = u(x,y,z)$.

Solution.

In an operator form, equation (109) can be rewritten as

$$L_x u(x, y, z) = u - L_y u - L_z u, \tag{110}$$

where the operators L_x, L_y and L_z are defined by

$$L_x = \frac{\partial}{\partial x}, L_y = \frac{\partial}{\partial y}, L_z = \frac{\partial}{\partial z}. \tag{111}$$

Assume that the operator L_x is invertible, and the inverse operator L_x^{-1} is an indefinite integral from 0 to x. Applying the integral operator L_x^{-1} to both sides of (110) and using the given condition that $u(0, y, z) = 1 + e^y + e^z$ yields

$$u(x, y, z) = 1 + e^y + e^z - L_x^{-1}(u - L_y u - L_z u). \tag{112}$$

Proceeding as before, we substitute the decomposition

$$u(x, y, z) = \sum_{n=0}^{\infty} u_n(x, y, z), \tag{113}$$

into both sides of (112) to find

$$\sum_{n=0}^{\infty} u_n(x, y, z) = 1 + e^y + e^z - L_x^{-1}\left(\sum_{n=0}^{\infty} u_n - L_y(\sum_{n=0}^{\infty} u_n) - L_z(\sum_{n=0}^{\infty} u_n)\right). \tag{114}$$

Using few terms of the decomposition of $u(x, y, z)$, equation (114) becomes

$$\begin{aligned}
u_0 + u_1 + u_2 + \cdots &= 1 + e^y + e^z - L_x^{-1}(u_0 + u_1 + u_2 + \cdots) \\
&\quad - L_x^{-1}(L_y(u_0 + u_1 + u_2 + \cdots)) \\
&\quad - L_x^{-1}(L_z(u_0 + u_1 + u_2 + \cdots)).
\end{aligned} \tag{115}$$

The components $u_0, u_1, u_2, \cdots$ can be determined recurrently as follows

$$\begin{aligned}
u_0(x, y, z) &= 1 + e^y + e^z, \\
u_1(x, y, z) &= L_x^{-1}(u_0 - L_y u_0 - L_z u_0) = x, \\
u_2(x, y, z) &= L_x^{-1}(u_1 - L_y u_1 - L_z u_1) = \frac{1}{2!}x^2, \\
u_3(x, y, z) &= L_x^{-1}(u_2 - L_y u_2 - L_z u_2) = \frac{1}{3!}x^3,
\end{aligned} \tag{116}$$

and so on. Consequently, the solution in a series form is given by

$$u(x, y, z) = (1 + x + \frac{1}{2!}x^2 + \frac{1}{3!}x^3 + \cdots) + e^y + e^z, \tag{117}$$

and in a closed form
$$u(x, y, z) = e^x + e^y + e^z. \tag{118}$$

It is interesting to note that we can easily show that the y-solution and the z-solution will provide the exact solution obtained in (118).

In closing this section, we point out that Adomian decomposition method works effectively for nonlinear differential equations. However, an algorithm is needed to express the nonlinear terms contained in the nonlinear equation. The implementation of the decomposition method to handle the nonlinear differential equations will be explained in details in Chapter 8.

Exercises 2.2

In exercises 1 – 4, use the decomposition method to show that the exact solution can be obtained by determining the x-solution or the y-solution:

1. $u_x + u_y = 2xy^2 + 2x^2y$, $u(x,0) = 0$, $u(0,y) = 0$

2. $u_x + u_y = 2x + 2y$, $u(x,0) = x^2$, $u(0,y) = y^2$

3. $u_x + yu = 0$, $u(x,0) = 1$, $u(0,y) = 1$

4. $u_x + u_y = u$, $u(x,0) = 1 + e^x$, $u(0,y) = 1 + e^y$

In exercises 5 – 12, use the decomposition method to solve the following partial differential equations:

5. $u_x + u_y = 2u$, $u(x,0) = e^x$, $u(0,y) = e^y$

6. $u_x - u_y = 2$, $u(x,0) = x$, $u(0,y) = -y$

7. $u_x + u_y = x^2 + y^2$, $u(x,0) = \frac{1}{3}x^3$, $u(0,y) = \frac{1}{3}y^3$

8. $xu_x + u_y = u$, $u(x,0) = 1 + x$, $u(0,y) = e^y$

9. $u_x + yu_y = u$, $u(x,0) = e^x$, $u(0,y) = 1 + y$

10. $xu_x + u_y = 2u$, $u(x,0) = x$, $u(0,y) = 0$

11. $u_x + yu_y = 2u$, $u(x,0) = 0$, $u(0,y) = y$

12. $u_x + u_y = 0$, $u(x,0) = e^x$, $u(0,y) = e^{-y}$

In exercises 13 – 16, use the decomposition method to solve the following partial differential equations:

13. $u_x + u_y + u_z = 3$, $u(0,y,z) = y + z$, $u(x,0,z) = x + z$, $u(x,y,0) = x + y$

14. $u_x + u_y + u_z = 3u$, $u(0,y,z) = e^{y+z}$, $u(x,0,z) = e^{x+z}$, $u(x,y,0) = e^{x+y}$

15. $u_x + yu_y + zu_z = 3u$, $u(0,y,z) = yz$, $u(x,0,z) = 0$, $u(x,y,0) = 0$

16. $u_x + yu_y + zu_z = u$, $u(0, y, z) = 1 + y + z$, $u(x, 0, z) = z + e^x$, $u(x, y, 0) = y + e^x$

In exercises 17 – 20, use the decomposition method to solve the following partial differential equations:

17. $u_x - u_y = 1 + 2x + 2y$, $u(0, y) = y + y^2$, $u(x, 0) = 2x + 3x^2$

18. $u_x - u_y = 0$, $u(0, y) = y + y^2$, $u(x, 0) = x + x^2$

19. $u_x - u_y = 0$, $u(0, y) = \sin y$, $u(x, 0) = \sin x$

20. $u_x - u_y = 0$, $u(0, y) = \cosh y$, $u(x, 0) = \cosh x$

2.3 The Noise Terms Phenomenon

In this section, we will present a useful tool that will accelerate the convergence of the Adomian decomposition method. The noise terms phenomenon provides a major advantage in that it demonstrates a fast convergence of the solution. It is important to note here that the **noise terms** phenomenon, that will be introduced in this section, may appear only for inhomogeneous PDEs. In addition, this phenomenon is applicable to all inhomogeneous PDEs of any order and will be used where appropriate in the coming chapters. The noise terms, if existed in the components u_0 and u_1, will provide the solution in a closed form with only two successive iterations.

In view of these remarks, we now outline the ideas of the noise terms :

1. The **noise terms** are defined as the identical terms with opposite signs that arise in the components u_0 and u_1. As stated above, these identical terms with opposite signs may exist only for inhomogeneous differential equations.

2. By canceling the noise terms between u_0 and u_1, even though u_1 contains further terms, the remaining non-canceled terms of u_0 may give the exact solution of the PDE. Therefore, it is necessary to verify that the non-canceled terms of u_0 satisfy the PDE under discussion.

On the other hand, if the non-canceled terms of u_0 did not justify the given PDE, or the noise terms did not appear between u_0 and u_1, then it is necessary to determine more components of u to determine the solution in a series form.

3. It was formally shown that the noise terms appear for specific cases of inhomogeneous equations, whereas homogeneous equations do not show noise terms. The conclusion about the self-canceling noise terms was based on observations drawn from solving specific models where no proof was presented. For further readings about the noise terms phenomenon, see [8,124].

4. It was formally proved by researchers that a necessary condition for the appearance of the noise terms is required. The conclusion made in this regard is that the zeroth component u_0 must contain the exact solution u among other terms. Moreover, it was shown that the nonhomogenity condition does not always guarantee

the appearance of the noise terms as examined in [124].

A useful summary about the noise terms phenomenon can be drawn as follows:
1. The noise terms are defined as the identical terms with opposite signs that may appear in the components u_0 and u_1.
2. The noise terms appear only for specific types of inhomogeneous equations whereas noise terms do not appear for homogeneous equations.
3. Noise terms may appear if the exact solution is part of the zeroth component u_0.
4. Verification that the remaining non-canceled terms justify the equation is necessary and essential.

The phenomenon of the useful noise terms will be explained by the following illustrative examples.

Example 1. Use the decomposition method and the noise terms phenomenon to solve the following inhomogeneous PDE

$$u_x + u_y = (1 + x)e^y, \quad u(0, y) = 0, \quad u(x, 0) = x. \tag{119}$$

Solution.

The inhomogeneous PDE can be rewritten in an operator form by

$$L_x = (1 + x)e^y - L_y u. \tag{120}$$

Clearly L_x is invertible and therefore the inverse operator L_x^{-1} exists. Applying L_x^{-1} to both sides of (120) and using the given condition leads to

$$u(x, y) = (x + \frac{x^2}{2!})e^y - L_x^{-1}(L_y u). \tag{121}$$

Using the decomposition series $u(x, y) = \sum_{n=0}^{\infty} u_n(x, y)$ into (121) gives

$$\sum_{n=0}^{\infty} u_n(x, y) = (x + \frac{x^2}{2!})e^y - L_x^{-1}\left(L_y\left(\sum_{n=0}^{\infty} u_n(x, y)\right)\right), \tag{122}$$

or equivalently

$$u_0 + u_1 + u_2 + \cdots = (x + \frac{x^2}{2!})e^y - L_x^{-1}(L_y(u_0 + u_1 + u_2 + \cdots)). \tag{123}$$

Proceeding as before, the components $u_0, u_1, u_2, \cdots$ are determined in a recursive manner by

$$
\begin{aligned}
u_0(x, y) &= (x + \frac{x^2}{2!})e^y, \\
u_1(x, y) &= -L_x^{-1}(L_y u_0) = -(\frac{x^2}{2!} + \frac{x^3}{3!})e^y, \\
u_2(x, y) &= -L_x^{-1}(L_y u_2) = (\frac{x^4}{4!} + \frac{x^5}{5!})e^y.
\end{aligned}
\tag{124}
$$

Considering the first two components u_0 and u_1 in (124), it is easily observed that the noise terms $\frac{x^2}{2!}e^y$ and $-\frac{x^2}{2!}e^y$ appear in u_0 and u_1 respectively. By canceling the noise term $\frac{x^2}{2!}e^y$ in u_0, and by verifying that the remaining non-canceled terms of u_0 justify the equation (119), we find that the exact solution is given by

$$u(x,y) = xe^y. \tag{125}$$

Notice that the exact solution is verified through substitution in the equation (119) and not only upon the appearance of the noise terms. In addition, the other noise terms that appear between other components will vanish in the limit.

Example 2. Use the decomposition method and the noise terms phenomenon to solve the following PDE:

$$u_x + yu_y = y(\cosh x + \sinh x), \ u(0,y) = 1, \ u(x,0) = 1. \tag{126}$$

Solution.

The inhomogeneous PDE can be rewritten in an operator form by

$$L_x u(x,y) = y(\cosh x + \sinh x) - yL_y u. \tag{127}$$

Applying L_x^{-1} to both sides of (127) and using the given condition gives

$$u(x,y) = 1 + y\sinh x + y\cosh x - y - L_x^{-1}(yL_yu). \tag{128}$$

Substituting the decomposition series of $u(x,y)$ into (128) gives

$$\sum_{n=0}^{\infty} u_n(x,y) = 1 + y\sinh x + y\cosh x - y - L_x^{-1}\left(yL_y\left(\sum_{n=0}^{\infty} u_n(x,y)\right)\right), \tag{129}$$

or equivalently

$$\begin{aligned}
u_0 + u_1 + u_2 + \cdots &= 1 + y\sinh x + y\cosh x - y \\
&\quad - L_x^{-1}(yL_y(u_0 + u_1 + u_2 + \cdots)).
\end{aligned} \tag{130}$$

Identifying the zeroth component u_0 as discussed before, the components $u_0, u_1, u_2, \cdots$ can be determined in a recursive manner by

$$\begin{aligned}
u_0(x,y) &= 1 + y\sinh x + y\cosh x - y, \\
u_1(x,y) &= -L_x^{-1}(yL_y u_0), = -y\cosh x + y - \sinh x + xy.
\end{aligned} \tag{131}$$

It is easily observed that there two noise terms $\pm y\cosh x$ and $\pm y$ appear in u_0 and u_1. By canceling the noise terms in u_0, and by verifying that the remaining non-canceled terms of u_0 justify the equation, we obtain the exact solution given by

$$u(x, y) = 1 + y \sinh x. \tag{132}$$

This can be verified through substitution in equation (126).

Example 3. Use the decomposition method and the noise terms phenomenon to solve the following PDE:

$$u_x + u_y = x^2 + 4xy + y^2, \ u(0, y) = 0, \ u(x, 0) = 0. \tag{133}$$

Solution.

We first rewrite the inhomogeneous PDE (133) in an operator form

$$L_x u = x^2 + 4xy + y^2 - L_y u. \tag{134}$$

Proceeding as before and applying the inverse operator L_x^{-1} to both sides of (134) and using the given condition we obtain

$$u(x, y) = \frac{1}{3}x^3 + 2x^2 y + xy^2 - L_x^{-1} \left(L_y u(x, y) \right). \tag{135}$$

Proceeding as before, the first two components u_0 and u_1 are given by

$$
\begin{aligned}
u_0(x, y) &= \tfrac{1}{3}x^3 + 2x^2 y + xy^2, \\
u_1(x, y) &= -L_x^{-1}(L_y u_0) = -x^2 y - \tfrac{2}{3}x^3.
\end{aligned}
\tag{136}
$$

We can easily observe that the two components u_0 and u_1 do not contain noise terms. This confirms our belief that although the PDE is an inhomogeneous equation, but the noise terms between the first two components did not exist in this problem. Unlike the previous examples, we should determine more components to obtain an insight through the solution. Therefore, other components should be determined. Hence we find

$$
\begin{aligned}
u_2(x, y) &= -L_x^{-1}(-x^2) = \tfrac{1}{3}x^3, \\
u_k(x, y) &= 0, k \geq 3.
\end{aligned}
\tag{137}
$$

Based on the result we obtained for u_2, other components of $u(x, y)$ will vanish. Consequently, we find that

$$
\begin{aligned}
u(x, y) &= u_0 + u_1 + u_2 + \cdots, \\
&= \tfrac{1}{3}x^3 + 2x^2 y + xy^2 - x^2 y - \tfrac{2}{3}x^3 + \tfrac{1}{3}x^3 = xy^2 + x^2 y,
\end{aligned}
\tag{138}
$$

the exact solution of the equation.

Exercises 2.3

In Exercises 1 – 12, use the decomposition method and the noise terms phenomenon to solve the following partial differential equations:

1. $u_x + u_y = 3x^2 + 3y^2$, $u(x,0) = x^3$, $u(0,y) = y^3$

2. $u_x + u_y = \sinh x + \sinh y$, $u(x,0) = 1 + \cosh x$, $u(0,y) = 1 + \cosh y$

3. $u_x + u_y = x + y$, $u(x,0) = u(0,y) = 0$

4. $u_x - u_y = \cos x + \sin y$, $u(x,0) = 1 + \sin x$, $u(0,y) = \cos y$

5. $u_x - u_y = \sin x + \sin y + x \cos y + y \cos x$, $u(x,0) = u(0,y) = 0$

6. $u_x - u_y = \cos x + \cos y + x \sin y + y \sin x$, $u(x,0) = x$, $u(0,y) = -y$

7. $u_x + u_y = (1+y)e^x + (1+x)e^y$, $u(x,0) = x$, $u(0,y) = y$

8. $u_x - u_y = (1+y)e^{-x} + (1+x)e^{-y}$, $u(x,0) = x$, $u(0,y) = -y$

9. $u_x + yu_y - u = 2xy^2 + x^2y^2$, $u(x,0) = u(0,y) = 0$

10. $u_x + yu_y - u = xy^2 + y^2 + 2xy$, $u(x,0) = u(0,y) = 0$

11. $u_x + u_y = \cos x + \sinh y$, $u(x,0) = 1 + \sin x$, $u(0,y) = \cosh y$

12. $u_x + u_y = u + e^y$, $u(x,0) = x$, $u(0,y) = 0$

In Exercises 13 – 18, show that the noise terms do not appear in the first two components of the solution of the inhomogeneous partial differential equations. Find the exact solution.

13. $u_x + u_y = 2xy^3 + 6x^2y^2 + 2x^3y$, $u(x,0) = u(0,y) = 0$

14. $u_x - u_y = 3x^2y^4 - 3x^4y^2$, $u(x,0) = u(0,y) = 0$

15. $u_x + u_y = 4(x + y)$, $u(x,0) = x^2$, $u(0,y) = y^2$

16. $u_x + u_y = 4x + 4y$, $u(x,0) = x^2$, $u(0,y) = y^2$

17. $u_x + u_y = x + y$, $u(x,0) = x^2$, $u(0,y) = y^2$

18. $u_x + u_y = 1 + u$, $u(x,0) = 1 + x + e^x$, $u(0,y) = 1 + e^y$

2.4 The Modified Decomposition Method

In this section we will introduce a reliable modification of the Adomian decomposition method presented in [129]. The modified decomposition method will further accelerate the convergence of the series solution. It is to be noted that the modified decomposition method will be applied, wherever it is appropriate, to all partial

differential equations of any order. The modification will be outlined in this section
and will be employed in this section and in other chapters as well.

To give a clear description of the technique, we consider the partial differential
equation in an operator form

$$Lu + Ru = g, \tag{139}$$

where L is the highest order derivative, R is a linear differential operator of less
order or equal order to L, and g is the source term. Operating with the inverse
operator L^{-1} on (139) we obtain

$$u = f - L^{-1}(Ru), \tag{140}$$

where f represents the terms arising from the given condition and from integrating
the source term g. We then proceed as discussed in Section 2.2 and define the
solution u as an infinite sum of components defined by

$$u = \sum_{n=0}^{\infty} u_n. \tag{141}$$

The aim of the decomposition method is to determine the components u_n, $n \geq 0$
recurrently and elegantly. To achieve this goal, the decomposition method admits
the use of the recursive relation

$$\begin{aligned} u_0 &= f, \\ u_{k+1} &= -L^{-1}(Ru_k),\ k \geq 0. \end{aligned} \tag{142}$$

In view of (142), the components u_n, $n \geq 0$ are readily obtained.

The modified decomposition method introduces a slight variation to the recursive relation (142) that will lead to the determination of the components of u in a
faster and easier way. For specific cases, the function f can be set as the sum of
two partial functions, namely f_1 and f_2. In other words, we can set

$$f = f_1 + f_2. \tag{143}$$

Using (143), we introduce a qualitative change in the formation of the recursive
relation (142). To reduce the size of calculations, we identify the zeroth component
u_0 by one part of f, namely f_1 or f_2. The other part of f can be added to the
component u_1 among other terms. In other words, the modified recursive relation
can be identified by

$$\begin{aligned} u_0 &= f_1, \\ u_1 &= f_2 - L^{-1}(Ru_0), \\ u_{k+1} &= -L^{-1}(Ru_k),\ k \geq 1. \end{aligned} \tag{144}$$

An important point can be made here in that we suggest a change in the formation of the first two components u_0 and u_1 only. Although this variation in the formation of u_0 and u_1 is slight, but it plays a major role in accelerating the convergence of the solution and in minimizing the size of calculations.

Two important remarks related to the modified method [129] can be made here. First, by proper selection of the functions f_1 and f_2, the exact solution u may be obtained by using very few iterations, and sometimes by evaluating only two components. The success of this modification depends only on the choice of f_1 and f_2, and this can be made through trials. Second, if f consists of one term only, the standard decomposition method should be employed in this case.

It is worth mentioning that the modified decomposition method will be used for linear and nonlinear equations of any order. In the coming chapters, it will be used wherever it is appropriate.

The modified decomposition method will be illustrated by discussing the following examples.

Example 1. Use the modified decomposition method to solve the first order partial differential equation:

$$u_x + u_y = 3x^2y^3 + 3x^3y^2, \quad u(0, y) = 0. \tag{145}$$

Solution.

In an operator form Eq. (145) becomes

$$L_x u = 3x^2y^3 + 3x^3y^2 - u_y, \tag{146}$$

where L_x is a first order partial derivative with respect to x. Applying the inverse operator L_x^{-1} to both sides of (146) gives

$$u(x, y) = x^3y^3 + \frac{3}{4}x^4y^2 - L_x^{-1}(u_y). \tag{147}$$

The function $f(x, y)$ consists of two terms, hence we set

$$\begin{aligned}
f_1(x, y) &= x^3y^3, \\
f_2(x, y) &= \tfrac{3}{4}x^4y^2.
\end{aligned} \tag{148}$$

In view of (148) we introduce the modified recursive relation

$$\begin{aligned}
u_0(x, y) &= x^3y^3, \\
u_1(x, y) &= \tfrac{3}{4}x^4y^2 - L^{-1}(u_0)_y, \\
u_{k+1}(x, y) &= -L^{-1}(u_k)_y, \ k \geq 1.
\end{aligned} \tag{149}$$

This gives

$$u_0(x,y) = x^3 y^3,$$

$$u_1(x,y) = \tfrac{3}{4}x^4 y^2 - L^{-1}(3x^3 y^2) = 0, \tag{150}$$

$$u_{k+1}(x,y) = 0, \; k \geq 1.$$

It then follows that the solution is

$$u(x,y) = x^3 y^3. \tag{151}$$

This example clearly shows that the solution can be obtained by using two iterations, and hence the volume of calculations is reduced.

Example 2. Use the modified decomposition method to solve the first order partial differential equation:

$$u_x - u_y = x^3 - y^3, \; u(0,y) = \frac{1}{4}y^4. \tag{152}$$

Solution.

In an operator form Eq. (152) becomes

$$L_x u = x^3 - y^3 + u_y, \tag{153}$$

where L_x is a first order partial derivative with respect to x. Proceeding as before we obtain

$$u(x,y) = \frac{1}{4}y^4 + \frac{1}{4}x^4 - xy^3 + L_x^{-1}(u_y). \tag{154}$$

We next split the function $f(x,y)$ as follows

$$f_1(x,y) = \tfrac{1}{4}y^4 + \tfrac{1}{4}x^4,$$

$$f_2(x,y) = -xy^3. \tag{155}$$

Consequently, we set the modified recursive relation

$$u_0(x,y) = \tfrac{1}{4}y^4 + \tfrac{1}{4}x^4,$$

$$u_1(x,y) = -xy^3 + L_x^{-1}(u_{0_y}), \tag{156}$$

$$u_{k+1}(x,y) = -L_x^{-1}(u_{k_y}), \; k \geq 1.$$

This gives

$$u_0(x,y) = \tfrac{1}{4}y^4 + \tfrac{1}{4}x^4,$$

$$u_1(x,y) = -xy^3 - L^{-1}(y^3) = 0, \tag{157}$$

$$u_{k+1}(x,y) = 0, \; k \geq 1.$$

The exact solution

$$u(x,y) = \frac{1}{4}x^4 + \frac{1}{4}y^4, \tag{158}$$

follows immediately.

Example 3. Use the modified decomposition method to solve the first order partial differential equation:

$$u_x + u_y = u, \ u(0,y) = 1 + e^y. \tag{159}$$

Solution.

Operating with the inverse operator L_x^{-1} on (159) and using the given condition gives

$$u(x,y) = 1 + e^y + L_x^{-1}(u - u_y). \tag{160}$$

We next split function $f(x,y)$ as follows

$$\begin{aligned}
f_1(x,y) &= e^y, \\
f_2(x,y) &= 1.
\end{aligned} \tag{161}$$

To determine the components of $u(x,y)$, we set the modified recursive relation

$$\begin{aligned}
u_0(x,y) &= e^y, \\
u_1(x,y) &= 1 + L_x^{-1}\left(u_0 - (u_0)_y\right), \\
u_{k+1}(x,y) &= L_x^{-1}\left(u_k - (u_k)_y\right), \ k \geq 1.
\end{aligned} \tag{162}$$

This gives

$$\begin{aligned}
u_0(x,y) &= e^y, \\
u_1(x,y) &= 1, \\
u_2(x,y) &= x, \\
u_3(x,y) &= \frac{x^2}{2!},
\end{aligned} \tag{163}$$

and so on. The solution in a series form is given by

$$u(x,y) = e^y + \left(1 + x + \frac{x^2}{2!} + \frac{x^3}{3!} + \cdots\right), \tag{164}$$

and in a closed form by

$$u(x,y) = e^x + e^y. \tag{165}$$

Example 4. Use the modified decomposition method to solve the first order partial differential equation:

$$u_x + u_y = \cosh x + \cosh y, \quad u(x,0) = \sinh x. \tag{166}$$

Solution.

To effectively use the given condition, we rewrite (166) in an operator form by

$$L_y u = \cosh x + \cosh y - u_x. \tag{167}$$

Applying the inverse operator L_y^{-1} on (167) and using the given condition gives

$$u(x,y) = \sinh x + \sinh y + y \cosh x - L_y^{-1}(u_x). \tag{168}$$

The function $f(x,y)$ can be written as $f_1 + f_2$ where

$$
\begin{aligned}
f_1(x,y) &= \sinh x + \sinh y, \\
f_2(x,y) &= y \cosh x.
\end{aligned}
\tag{169}
$$

To determine the components of $u(x,y)$, we set the modified recursive relation

$$
\begin{aligned}
u_0(x,y) &= \sinh x + \sinh y, \\
u_1(x,y) &= y \cosh x - L^{-1}((u_0)_x) = 0, \\
u_{k+1}(x,y) &= -L^{-1}((u_k)_x) = 0, \ k \geq 1.
\end{aligned}
\tag{170}
$$

The exact solution

$$u(x,y) = \sinh x + \sinh y, \tag{171}$$

follows immediately.

It is interesting to point out that two iterations only were used to determine the exact solution. However, using the following formation

$$
\begin{aligned}
f_1(x,y) &= \sinh x, \\
f_2(x,y) &= \sinh y + y \cosh x,
\end{aligned}
\tag{172}
$$

for $f(x,y)$ will give the following recursive relation

$$
\begin{aligned}
u_0(x,y) &= \sinh x, \\
u_1(x,y) &= \sinh y + y \cosh x - L_y^{-1}(\cosh x) = \sinh y, \\
u_{k+1}(x,y) &= 0, \ k \geq 1.
\end{aligned}
\tag{173}
$$

It is obvious from (173) that all components $u_j = 0, j \geq 2$.

Consequently, the exact solution is

$$u(x, y) = \sinh x + \sinh y, \tag{174}$$

obtained by using the first two components only.

Exercises 2.4

Use the modified decomposition method to solve the following first order partial differential equations:

1. $u_x + u_y = 3x^2 + 3y^2$, $u(0, y) = y^3$

2. $u_x - u_y = 2x + 2y$, $u(0, y) = -y^2$

3. $u_x + u_y = 4x + 4y$, $u(0, y) = y^2$

4. $u_x + u_y = \sinh x + \sinh y$, $u(0, y) = 1 + \cosh y$

5. $u_x - u_y = \cos x + \sin y$, $u(0, y) = \cos y$

6. $u_x + yu_y - u = 2xy^2 + x^2y^2$, $u(0, y) = 0$

7. $u_x - u_y = \cos x - \cos y$, $u(0, y) = \sin y$

8. $u_x + u_y = u$, $u(0, y) = 1 - e^y$

9. $xu_x + u_y = 2x^2 + 3y^2$, $u(0, y) = y^3$

10. $u_x + u_y = 2x + \cos y$, $u(0, y) = 1 + \sin y$

11. $u_x + xu_y = 1 + x \cosh y$, $u(0, y) = 1 + \sinh y$

12. $u_x - xu_y = \cos x - x \cosh y$, $u(0, y) = \sinh y$

2.5 Method of Characteristics

In this section, the first order partial differential equation

$$au_x + bu_y = f(x, y) + ku, \quad u(0, y) = h(y), \tag{175}$$

will be investigated by using the traditional method of characteristics. It is important to note that a, b, and f depend on x, y, and u but not on the derivatives of u. In addition, we also assume that a, b, and f are continuously differentiable of their arguments.

Assuming that $u(x, y)$ is a solution of (175), then by using the chain rule we obtain

$$du = u_x \, dx + u_y \, dy. \tag{176}$$

A close examination of (175) and (176) leads to the system of equations

$$\frac{dx}{a} = \frac{dy}{b} = \frac{du}{f(x,y) + ku}. \tag{177}$$

The pair

$$\frac{dx}{a} = \frac{dy}{b}, \tag{178}$$

gives the solution

$$bx - ay = c, \tag{179}$$

where c is a constant. We next consider the pair

$$\frac{dx}{a} = \frac{du}{f(x,y) + ku}, \tag{180}$$

or equivalently

$$\frac{dx}{a} = \frac{du}{f(x, \frac{bx-c}{a}) + ku}. \tag{181}$$

Equation (181) can be rewritten as

$$\frac{du}{dx} - \frac{k}{a}u = \frac{1}{a}f(x, \frac{bx - c}{a}), \tag{182}$$

a first order linear ordinary differential equation. The integrating factor of (182) is given by

$$\mu = e^{-\frac{k}{a}x}. \tag{183}$$

Accordingly, the solution of (182) can be expressed in the form

$$u = G(x,c) + c_1, \tag{184}$$

where

$$c_1 = g(c), \tag{185}$$

where g is an arbitrary function. Eq. (184) can be rewritten as

$$u = G(x,c) + g(c). \tag{186}$$

Using the given condition leads to the determination of $g(c)$. Based on this and using (179), the solution $u(x,y)$ is readily obtained.

It is to be noted that first order partial differential equations in higher dimensions will not be discussed in this section. The decomposition method can handle such problems elegantly and easily if compared with the method of characteristics.

To give a clear overview of the method of characteristics, we will discuss some of the examples presented before in Section 2.2 and Section 2.3. The illustration can be used as a comparative study between the method of characteristics and the decomposition method.

Example 1. Use the method of characteristics to solve the first order partial differential equation

$$u_x + u_y = x + y,$$
$$u(x,0) = 0. \tag{187}$$

Solution.

Following the discussion presented above we set the system of equations

$$\frac{dx}{1} = \frac{dy}{1} = \frac{du}{x+y}. \tag{188}$$

Notice that the system (188) is a system of first order ordinary differential equations. The left pair of (188) gives the solution

$$x - y = c, \tag{189}$$

where c is a constant. Using (189) into the right pair of (188) gives

$$\frac{dy}{1} = \frac{du}{2y+c}, \tag{190}$$

a separable differential equation that gives the solution

$$u(x,y) = y^2 + cy + c_1. \tag{191}$$

Recall that

$$c_1 = g(c), \tag{192}$$

then (191) becomes

$$u(x,y) = y^2 + y(x-y) + g(x-y), \tag{193}$$

where $g(x-y)$ is an arbitrary function. To determine $g(x-y)$ we substitute the given condition into (193) to obtain

$$g(x) = 0, \tag{194}$$

and therefore

$$g(x-y) = 0. \tag{195}$$

Consequently, the solution is given by

$$u(x,y) = xy. \tag{196}$$

Example 2. Use the method of characteristics to solve the first order partial differential equation

$$u_x - u_y = 0,$$
$$u(x,0) = x. \tag{197}$$

Solution.

Following Example 1, we set the system of equations

$$\frac{dx}{1} = \frac{dy}{-1}, \ du = 0.$$
(198)

The first equation of (198) gives the solution

$$x + y = c,$$
(199)

where c is a constant. The second equation of (198) gives the solution

$$u(x, y) = c_1.$$
(200)

This means that

$$u(x, y) = c_1 = g(c) = g(x + y),$$
(201)

where $g(x + y)$ is an arbitrary function. Substituting the given condition into (201) gives

$$g(x) = x,$$
(202)

and therefore

$$g(x + y) = x + y.$$
(203)

The solution is therefore given by

$$u(x, y) = x + y,$$
(204)

obtained upon substituting (203) into (201).

Example 3. Use the method of characteristics to solve the first order partial differential equation

$$xu_x + u_y = x \sinh y + u, \ u(0, y) = 0.$$
(205)

Solution.

Following the examples discussed above we set the system of equations

$$\frac{dx}{x} = \frac{dy}{1} = \frac{du}{x \sinh y + u}.$$
(206)

The left pair of (206) gives the solution

$$x = ce^y.$$
(207)

Substituting $x = ce^y$ into the right pair of (206) gives

$$\frac{dy}{1} = \frac{du}{ce^y \sinh y + u},$$
(208)

that can be reduced to the first order linear ordinary differential equation

$$\frac{du}{dy} - u = c e^y \sinh y, \tag{209}$$

that gives the solution

$$u(x, y) = e^y \left(c \cosh y + c_1 \right). \tag{210}$$

Using the given condition and noting that $c = x e^{-y}$ gives the solution

$$u(x, y) = x \cosh y. \tag{211}$$

Example 4. Use the method of characteristics to solve the first order partial differential equation

$$u_x - y u = 0, u(0, y) = 1. \tag{212}$$

Solution.

First we set the system of equations

$$\frac{dx}{1} = \frac{du}{yu}, \ dy = 0. \tag{213}$$

The second equation gives the solution

$$y = c. \tag{214}$$

Substituting $y = c$ into the first equation of (213) gives

$$\frac{du}{cu} = \frac{dx}{1}, \tag{215}$$

which gives the solution

$$\ln u = cx + g(c), \tag{216}$$

or equivalently

$$\ln u = cx + g(y). \tag{217}$$

To determine $g(y)$ we substitute the given condition into (217) to obtain

$$g(y) = 0, \tag{218}$$

and therefore the solution is given by

$$u(x, y) = e^{xy}. \tag{219}$$

Exercises 2.5

Use the method of characteristics to solve the following first order partial differential equations:

1. $u_x + u_y = 2x + 2y$, $u(x,0) = x^2$

2. $u_x + u_y = u$, $u(x,0) = 1 + e^x$

3. $u_x + yu = 0$, $u(0,y) = 1$

4. $3u_x - 2u_y = 3\sin x$, $u(0,y) = 3y - 1$

5. $u_x + u_y = 2u$, $u(x,0) = e^x$

6. $xu_x + u_y = 2u$, $u(x,0) = x$

7. $u_x + yu_y = 2u$, $u(0,y) = y$

8. $u_x + u_y = \sinh x + \sinh y$, $u(0,y) = 1 + \cosh y$

9. $u_x + u_y = 2x + 2y$, $u(x,0) = 0$

10. $u_x + yu_y = 2xy^2 + 2x^2y^2$, $u(0,y) = 0$

11. $u_x + ku_y = 0$, $u(0,y) = y$

· 12. $u_x - u_y = 2$, $u(0,y) = -y$

13. $u_x + u_y = 1 + \cos y$, $u(0,y) = \sin y$

14. $xu_x + u_y = 2u$, $u(0,y) = 0$

15. $xu_x + yu_y = 2u$, $u(0,y) = 0$

16. $2u_x + 3u_y = 2u + e^y$, $u(0,y) = 1 + e^y$

17. $xu_x + yu_y = 2xyu$, $u(0,y) = 1$

18. $u_x + 4u_y = 5u$, $u(0,y) = e^y$

19. $xu_x + u_y = x\cosh y + u$, $u(0,y) = 0$

20. $u_x + yu_y = y\sinh x + u$, $u(0,y) = y$

2.6 Systems of Linear PDEs

Systems of partial differential equations, linear or nonlinear, have attracted much concern in studying evolution equations that describe wave propagation, in investigating shallow water waves, and in examining the chemical reaction-diffusion model of Brusselator. The general ideas and the essential features of these systems are of wide applicability. The commonly used methods are the method of characteristics and the Riemann invariants among other methods. The existing techniques encountered some difficulties in terms of the size of computational work needed, especially when the system involves several partial differential equations.

To avoid the difficulties that usually arise from traditional strategies, the Adomian decomposition method will form a reasonable basis for studying systems of partial differential equations. The method, as we have seen before, has a useful attraction in that it provides the solution in a rapidly convergent power series with elegantly computable terms. The Adomian decomposition method transforms the system of partial differential equations into a set of recursive relations that can be easily examined. Due to simplicity reasons, we will use in this section Adomian decomposition method.

We first consider the system of partial differential equations written in an operator form

$$
\begin{aligned}
L_t u + L_x v &= g_1, \\
L_t v + L_x u &= g_2,
\end{aligned}
\tag{220}
$$

with initial data

$$
\begin{aligned}
u(x,0) &= f_1(x), \\
v(x,0) &= f_2(x),
\end{aligned}
\tag{221}
$$

where L_t and L_x are considered, without loss of generality, first order partial differential operators, and g_1 and g_2 are inhomogeneous terms. Applying the inverse operator L_t^{-1} to the system (220) and using the initial data (221) yields

$$
\begin{aligned}
u(x,t) &= f_1(x) + L_t^{-1} g_1 - L_t^{-1} L_x v, \\
v(x,t) &= f_2(x) + L_t^{-1} g_2 - L_t^{-1} L_x u.
\end{aligned}
\tag{222}
$$

The Adomian decomposition method suggests that the linear terms $u(x,t)$ and $v(x,t)$ be decomposed by an infinite series of components

$$
\begin{aligned}
u(x,t) &= \sum_{n=0}^{\infty} u_n(x,t), \\
v(x,t) &= \sum_{n=0}^{\infty} v_n(x,t),
\end{aligned}
\tag{223}
$$

where $u_n(x,t)$ and $v_n(x,t), n \geq 0$ are the components of $u(x,t)$ and $v(x,t)$ that will be elegantly determined in a recursive manner.

Substituting (223) into (222) gives

$$
\begin{aligned}
\sum_{n=0}^{\infty} u_n(x,t) &= f_1(x) + L_t^{-1} g_1 - L_t^{-1}\left(L_x \left(\sum_{n=0}^{\infty} v_n \right) \right), \\
\sum_{n=0}^{\infty} v_n(x,t) &= f_2(x) + L_t^{-1} g_2 - L_t^{-1}\left(L_x \left(\sum_{n=0}^{\infty} u_n \right) \right).
\end{aligned}
\tag{224}
$$

Following Adomian analysis, the system (220) is transformed into a set of recursive relations given by

$$
\begin{aligned}
u_0(x,t) &= f_1(x) + L_t^{-1} g_1, \\
u_{k+1}(x,t) &= -L_t^{-1}\left(L_x v_k\right),\ k \geq 0,
\end{aligned}
\tag{225}
$$

and

$$
\begin{aligned}
v_0(x,t) &= f_2(x) + L_t^{-1} g_2, \\
v_{k+1}(x,t) &= -L_t^{-1}\left(L_x u_k\right),\ k \geq 0.
\end{aligned}
\tag{226}
$$

The zeroth components $u_0(x,t)$ and $v_0(x,t)$ are defined by all terms that arise from initial data and from integrating the inhomogeneous terms. Having defined the zeroth pair (u_0, v_0), the pair (u_1, v_1) can be determined recurrently by using (225) and (226). The remaining pairs (u_k, v_k), $k \geq 2$ can be easily determined in a parallel manner. Additional pairs for the decomposition series normally account for higher accuracy. Having determined the components of $u(x,t)$ and $v(x,t)$, the solution (u, v) of the system follows immediately in the form of a power series expansion upon using (223). The series obtained can be summed up in many cases to give a closed form solution. For concrete problems, the $n-$term approximants can be used for numerical purposes.

To give a clear overview of the content of this work, several illustrative examples have been selected to demonstrate the efficiency of the method.

Example 1.   We first consider the linear system:

$$
\begin{aligned}
u_t + v_x &= 0, \\
v_t + u_x &= 0,
\end{aligned}
\tag{227}
$$

with the initial data

$$
\begin{aligned}
u(x,0) &= e^x, \\
v(x,0) &= e^{-x},
\end{aligned}
\tag{228}
$$

To derive the solution by using the decomposition method, we follow the recursive relations (225) and (226) to obtain

$$
\begin{aligned}
u_0(x,t) &= e^x, \\
u_{k+1}(x,t) &= -L_t^{-1} L_x(v_k),\ k \geq 0,
\end{aligned}
\tag{229}
$$

and

$$v_0(x,t) \;=\; e^{-x},$$
$$v_{k+1}(x,t) \;=\; -L_t^{-1}L_x(u_k), \; k \geq 0.$$

(230)

The remaining components are thus determined by

$$u_1(x,t) \;=\; te^{-x}, \qquad v_1(x,t) \;=\; -te^{x},$$
$$u_2(x,t) \;=\; \tfrac{t^2}{2!}e^{x}, \qquad v_2(x,t) \;=\; \tfrac{t^2}{2!}e^{-x},$$
$$u_3(x,t) \;=\; \tfrac{t^3}{3!}e^{-x}, \qquad v_3(x,t) \;=\; -\tfrac{t^3}{3!}e^{x},$$

(231)

and so on. Using (231) we obtain

$$u(x,t) \;=\; e^{x}\left(1 + \tfrac{t^2}{2!} + \tfrac{t^4}{4!} + \cdots\right) + e^{-x}\left(t + \tfrac{t^3}{3!} + \tfrac{t^5}{5!} + \cdots\right),$$
$$v(x,t) \;=\; e^{-x}\left(1 + \tfrac{t^2}{2!} + \tfrac{t^4}{4!} + \cdots\right) - e^{x}\left(t + \tfrac{t^3}{3!} + \tfrac{t^5}{5!} + \cdots\right),$$

(232)

which has an exact analytical solution of the form

$$(u,v) = (e^{x}\cosh t + e^{-x}\sinh t, \; e^{-x}\cosh t - e^{x}\sinh t).$$

(233)

Example 2. Consider the linear system of partial differential equations

$$u_t + u_x + 2v \;=\; 0,$$
$$v_t + v_x - 2u \;=\; 0,$$

(234)

with the initial data

$$u(x,0) \;=\; \cos x,$$
$$v(x,0) \;=\; \sin x.$$

(235)

Operating with L_t^{-1} on (234) and using (235) we obtain

$$u(x,t) \;=\; \cos x - L_t^{-1}(2v + L_x u),$$
$$v(x,t) \;=\; \sin x + L_t^{-1}(2u - L_x v).$$

(236)

Using the series representation (223) into (236) admits the use of the system of recursive relations

$$u_0(x,t) \;=\; \cos x,$$
$$u_{k+1}(x,t) \;=\; -L_t^{-1}\left(2v_k + L_x(u_k)\right), \; k \geq 0,$$

(237)

and

$$v_0(x,t) = \sin x,$$

$$v_{k+1}(x,t) = L_t^{-1}\left(2u_k - L_x(v_k)\right), \ k \geq 0. \tag{238}$$

Consequently, the pair of zeroth components is defined by

$$(u_0, v_0) = (\cos x, \sin x). \tag{239}$$

Using (239) into (237) and (238) gives

$$u_1(x,t) = -t\sin x,$$

$$v_1(x,t) = t\cos x. \tag{240}$$

In a like manner we obtain the pairs

$$(u_2, v_2) = (-\tfrac{t^2}{2!}\cos x, -\tfrac{t^2}{2!}\sin x),$$

$$(u_3, v_3) = (\tfrac{t^3}{3!}\sin x, -\tfrac{t^3}{3!}\cos x). \tag{241}$$

Combining the results obtained above we obtain

$$u(x,t) = \cos x(1 - \tfrac{t^2}{2!} + \tfrac{t^4}{4!} - \cdots) - \sin x(t - \tfrac{t^3}{3!} + \tfrac{t^5}{5!} - \cdots),$$

$$v(x,t) = \sin x(1 - \tfrac{t^2}{2!} + \tfrac{t^4}{4!} - \cdots) + \cos x(t - \tfrac{t^3}{3!} + \tfrac{t^5}{5!} - \cdots), \tag{242}$$

so that the pair (u,v) is known in a closed form by

$$(u,v) = (\cos(x+t), \sin(x+t)). \tag{243}$$

Exercises 2.6

Use Adomian decomposition method to solve the following systems of first order partial differential equations:

1. $u_t + u_x - (u+v) = 0, \ v_t + u_x - (u+v) = 0,$
   with the initial data: $u(x,0) = \sinh x, \ v(x,0) = \cosh x$

2. $u_t + u_x - 2v = 0, \ v_t + v_x + 2v = 0,$
   with the initial data: $u(x,0) = \sin x, \ v(x,0) = \cos x$

3. $u_t + u_x - 2v_x = 0, \ v_t + v_x - 2u_x = 0,$
   with the initial data: $u(x,0) = \cos x, \ v(x,0) = \cos x$

4. $u_t - v_x + (u + v) \;\; = \;\; 0, \; v_t - u_x + (u + v) = 0,$
   with the initial data: $u(x, 0) = \sinh x, \; v(x, 0) = \cosh x$

5. $u_t + v_x - w_y \;\; = \;\; w, \; v_t + w_x + u_y = u, \; w_t + v_x - v_y = v,$
   $u(x, y, 0) = -w(x, y, 0) = \sin(x + y), \; v(x, y, 0) = \cos(x + y)$

6. $u_t + u_x + 2w \;\; = \;\; 0, \; v_t + v_x + 2u = 0, \; w_t + w_x - 2u = 0,$
   $u(x, y, 0) = \sin(x + y), \; v(x, y, 0) = -w(x, y, 0) = \cos(x + y)$

Chapter 3

One-Dimensional Heat Flow

3.1 Introduction

In Chapter 1, it was indicated that many phenomena of physics and engineering are expressed by partial differential equations PDEs. The PDE is termed a *Boundary Value Problem* (BVP) if the boundary conditions of the dependent variable u and some of its partial derivatives are often prescribed. However, the PDE is called an *Initial Value Problem* (IVP) if the initial conditions of the dependent variable u are prescribed at the starting time $t = 0$. Moreover, the PDE is termed *Initial-Boundary Value Problem* (IBVP) if both initial conditions and boundary conditions are prescribed.

In this chapter, we will study the one dimensional heat flow. Our concern will be focused on solving the PDE in conjunction with the prescribed initial and boundary conditions. The strategy in this chapter will be the use of the newly developed Adomian decomposition method and the use of the well-known traditional method of the separation of variables as well.

In this section we will study the physical problem of heat conduction in a rod of length L. The temperature distribution of a rod is governed by an initial-boundary value problem that is often defined by:

1. **Partial Differential Equation** (PDE) that governs the heat flow in a rod. The PDE can be formally shown to satisfy

$$u_t = \overline{k} u_{xx}, \ 0 < x < L, t > 0, \tag{1}$$

where $u \equiv u(x, t)$ represents the temperature of the rod at the position x at time t, and $\overline{k}$ is the thermal diffusivity of the material that measures the rod ability to heat conduction.

2. **Boundary Conditions** (B.C) that describe the temperature u at both ends of the rod. One form of the B.C is given by the Dirichlet boundary conditions

$$u(0, t) \ = \ 0, \ t \geq 0,$$

$$u(L, t) \;=\; 0,\; t \geq 0. \tag{2}$$

The given boundary conditions in (2) indicate that the ends of the rod are kept at 0 temperature. As indicated in Chapter 1, boundary conditions are given in three types, namely: Dirichlet boundary conditions, Neumann boundary conditions, and mixed boundary conditions. In addition, the boundary conditions come in a homogeneous or inhomogeneous type.

3. **Initial Condition** (I.C) that describes the initial temperature u at time $t = 0$. The I.C is usually defined by

$$u(x, 0) = f(x),\; 0 \leq x \leq L. \tag{3}$$

Based on these definitions, the initial-boundary value problem that controls the heat conduction in a rod is given by

$$
\begin{array}{llll}
\text{PDE} & u_t & = & \bar{k} u_{xx},\; 0 < x < L,\; t > 0 \\[4pt]
\text{B.C} & u(0, t) & = & 0,\; t \geq 0 \\[4pt]
 & u(L, t) & = & 0,\; t \geq 0 \\[4pt]
\text{I.C} & u(x, 0) & = & f(x),\; 0 \leq x \leq L
\end{array}
\tag{4}
$$

As stated before we will focus our discussions on determining a particular solution of the heat equation (4), recalling that the general solution is of little use.

It is of interest to note that the PDE in (4) arises in two different types, namely:

1. **Homogeneous Heat Equation**: This type of equations is often given by

$$u_t = \bar{k} u_{xx},\; 0 < x < L,\; t > 0. \tag{5}$$

Further, heat equation with a lateral heat loss is formally derived as a homogeneous PDE of the form

$$u_t = \bar{k} u_{xx} - u,\; 0 < x < L,\; t > 0. \tag{6}$$

2. **Inhomogeneous Heat Equation**: This type of equations is often given by

$$u_t = \bar{k} u_{xx} + g(x),\; 0 < x < L,\; t > 0, \tag{7}$$

where $g(x)$ is called the heat source which is independent of time.

3.2   The Adomian Decomposition Method

In this chapter the Adomian decomposition method will be used in a similar way to that used in the previous chapter. As shown before, the method introduces the solution of any equation in a series form, where the components of the solution are elegantly computed. Further, the resulting series may provide a closed form solution. In the case where a closed form solution is not obtainable, a truncated n-term approximation is usually used for approximations and numerical purposes. It was formally proved by many researchers that the method provides the solution in a rapidly convergent power series.

An important point can be made here in that the method attacks the problem, homogeneous or inhomogeneous, in a straightforward manner without any need for transformation formulas. Further, there is no need to change the inhomogeneous boundary conditions to homogeneous conditions as required by the method of separation of variables that will be discussed later. The formal steps of the decomposition method have been outlined before in Chapter 2. In what follows, we introduce a framework for implementing this method to solve the one dimensional heat equation.

Without loss of generality, we study the initial-boundary value problem

$$
\begin{aligned}
\text{PDE} \qquad u_t &= u_{xx}, \; 0 < x < \pi, \; t > 0 \\[1mm]
\text{B.C} \qquad u(0,t) &= 0, \; t \geq 0 \\[1mm]
u(L,t) &= 0, \; t \geq 0 \\[1mm]
\text{I.C} \qquad u(x,0) &= f(x), \; 0 \leq x \leq L
\end{aligned}
\tag{8}
$$

to achieve our goal..

To begin our analysis, we first rewrite (8) in an operator form by

$$
L_t u(x,t) = L_x u(x,t),
\tag{9}
$$

where the differential operators L_t and L_x are defined by

$$
L_t = \frac{\partial}{\partial t}, \; L_x = \frac{\partial^2}{\partial x^2}.
\tag{10}
$$

It is obvious that the integral operators L_t^{-1} and L_x^{-1} exist and may be regarded as one and two-fold definite integrals respectively defined by

$$
L_t^{-1}(.) = \int_0^t (.)dt, \; L_x^{-1}(.) = \int_0^x \int_0^x (.)\, dx\, dx.
\tag{11}
$$

This means that

$$
L_t^{-1} L_t u(x,t) = u(x,t) - u(x,0).
\tag{12}
$$

Applying L_t^{-1} to both sides of (9) and using the initial condition we find

$$
u(x,t) = f(x) + L_t^{-1}(L_x u(x,t)).
\tag{13}
$$

The decomposition method defines the unknown function $u(x,t)$ into a sum of components defined by the series

$$
u(x,t) = \sum_{n=0}^{\infty} u_n(x,t),
\tag{14}
$$

where the components $u_0(x,t), u_1(x,t), u_2(x,t), \cdots$ are to be determined. Substituting (14) into both sides of (13) yields

$$\sum_{n=0}^{\infty} u_n(x,t) = f(x) + L_t^{-1}\left(L_x\left(\sum_{n=0}^{\infty} u_n(x,t)\right)\right), \tag{15}$$

or equivalently

$$u_0 + u_1 + u_2 + \cdots = f(x) + L_t^{-1}\left(L_x(u_0 + u_1 + u_2 + \cdots)\right). \tag{16}$$

The decomposition method suggests that the zeroth component $u_0(x,t)$ is identified by the terms arising from the initial/boundary conditions and from source terms. The remaining components of $u(x,t)$ are determined in a recursive manner such that each component is determined by using the previous component. Accordingly, we set the recurrence scheme

$$
\begin{aligned}
u_0(x,t) &= f(x), \\
u_{k+1}(x,t) &= L_t^{-1}\left(L_x\left(u_k(x,t)\right)\right), \ k \geq 0,
\end{aligned} \tag{17}
$$

for the complete determination of the components $u_n(x,t), n \geq 0$. In view of (17), the components $u_0(x,t), u_1(x,t), u_2(x,t), \cdots$ are determined individually by

$$
\begin{aligned}
u_0(x,t) &= f(x), \\
u_1(x,t) &= L_t^{-1} L_x(u_0) &=& \ f''(x)\,t, \\
u_2(x,t) &= L_t^{-1} L_x(u_1) &=& \ f^{(iv)}(x)\,\tfrac{t^2}{2!}, \\
u_3(x,t) &= L_t^{-1} L_x(u_2) &=& \ f^{(vi)}(x)\,\tfrac{t^3}{3!}, \\
&\vdots
\end{aligned} \tag{18}
$$

Other components can be determined in a like manner as far as we like. The accuracy level can be effectively improved by increasing the number of components determined. Having determined the components $u_0, u_1, \cdots$, the solution $u(x,t)$ of the PDE is thus obtained in a series form given by

$$u(x,t) = \sum_{n=0}^{\infty} f^{(2n)}(x)\frac{t^n}{n!}, \tag{19}$$

obtained by substituting (18) into (14).

An important conclusion can be made here; the solution (19) is obtained by using the initial condition only without using the boundary conditions. This solution is obtained by using the inverse operator L_t^{-1}. The obtained solution can be used to show that it justifies the given boundary conditions.

However, the solution (19) can also be obtained by using the inverse operator L_x^{-1}. In fact, the solution obtained in this way requires the use of boundary conditions and initial condition as well. This leads to an important conclusion that solving the PDE in the t direction reduces the size of computational work. This important observation will be confirmed through examples that will be discussed later.

To give a clear overview of the content of the decomposition method, we have chosen several examples, homogeneous and inhomogeneous, to illustrate the discussion given above.

3.2.1   Homogeneous Heat Equations

The Adomian decomposition method will be used to solve the following homogeneous heat equations where the boundary conditions are also homogeneous.

Example 1. Use the Adomian decomposition method to solve the initial-boundary value problem

$$
\begin{array}{lrcl}
\text{PDE} & u_t & = & u_{xx},\ 0 < x < \pi,\ t > 0 \\[2mm]
\text{B.C} & u(0,t) & = & 0,\ t \geq 0 \\[2mm]
& u(\pi,t) & = & 0,\ t \geq 0 \\[2mm]
\text{I.C} & u(x,0) & = & \sin x
\end{array}
\tag{20}
$$

Solution.
In an operator form, Eq. (20) can be written as

$$
L_t u(x,t) = L_x u(x,t). \tag{21}
$$

Applying L_t^{-1} to both sides of (21) and using the initial condition we find

$$
u(x,t) = \sin x + L_t^{-1}\left(L_x u(x,t)\right). \tag{22}
$$

We next define the unknown function $u(x,t)$ by a sum of components defined by the series

$$
u(x,t) = \sum_{n=0}^{\infty} u_n(x,t). \tag{23}
$$

Substituting the decomposition (23) into both sides of (22) yields

$$
\sum_{n=0}^{\infty} u_n(x,t) = \sin x + L_t^{-1}\left(L_x\left(\sum_{n=0}^{\infty} u_n(x,t)\right)\right), \tag{24}
$$

or equivalently

$$u_0 + u_1 + u_2 + \cdots = \sin x + L_t^{-1}\left(L_x\left(u_0 + u_1 + u_2 + \cdots\right)\right). \tag{25}$$

Identifying the zeroth component $u_0(x,t)$ as assumed before and following the recursive algorithm (17) we obtain

$$
\begin{aligned}
u_0(x,t) &= \sin x, \\
u_1(x,t) &= L_t^{-1}\left(L_x\left(u_0\right)\right) = -t\,\sin x, \\
u_2(x,t) &= L_t^{-1}\left(L_x\left(u_1\right)\right) = \tfrac{1}{2!}t^2\,\sin x, \\
&\vdots
\end{aligned}
\tag{26}
$$

Consequently, the solution $u(x,t)$ in a series form is given by

$$
\begin{aligned}
u(x,t) &= u_0(x,t) + u_1(x,t) + u_2(x,t) + \cdots \\
&= \sin x\left(1 - t + \frac{1}{2!}t^2 - \cdots\right),
\end{aligned}
\tag{27}
$$

and in a closed form by

$$u(x,t) = e^{-t}\sin x, \tag{28}$$

obtained upon using the Taylor expansion of e^{-t}. The solution (28) justifies the PDE, the boundary conditions and the initial condition.

Example 2. Use the Adomian decomposition method to solve the initial-boundary value problem

$$
\begin{array}{llll}
\text{PDE} & u_t &=& u_{xx}, \ \ 0 < x < \pi, \ t > 0 \\[4pt]
\text{B.C} & u(0,t) &=& e^{-t}, \ t \geq 0 \\[4pt]
& u(\pi,t) &=& \pi - e^{-t}, \ t \geq 0 \\[4pt]
\text{I.C} & u(x,0) &=& x + \cos x
\end{array}
\tag{29}
$$

Solution.
It is important to note that the boundary conditions in this example are inhomogeneous. The decomposition method does not require any restrictive assumption on boundary conditions when approaching the problem in the t direction or in the x direction.

Applying L_t^{-1} to both sides of the operator form

$$L_t u(x,t) = L_x u(x,t), \tag{30}$$

and using the initial condition we find

$$u(x,t) = x + \cos x + L_t^{-1}\left(L_x(u(x,t))\right). \tag{31}$$

Substituting the decomposition series

$$u(x,t) = \sum_{n=0}^{\infty} u_n(x,t), \tag{32}$$

into both sides of (31) yields

$$\sum_{n=0}^{\infty} u_n(x,t) = x + \cos x + L_t^{-1}\left(L_x\left(\sum_{n=0}^{\infty} u_n(x,t)\right)\right). \tag{33}$$

Identifying the component $u_0(x,t)$ and following the recursive algorithm (17) we obtain

$$
\begin{aligned}
u_0(x,t) &= x + \cos x, \\
u_1(x,t) &= L_t^{-1}\left(L_x\left(u_0\right)\right) = -t\,\cos x, \\
u_2(x,t) &= L_t^{-1}\left(L_x\left(u_1\right)\right) = \tfrac{1}{2!}t^2\,\cos x, \\
u_3(x,t) &= L_t^{-1}\left(L_x\left(u_2\right)\right) = -\tfrac{1}{3!}t^2\,\cos x, \\
&\vdots
\end{aligned}
\tag{34}
$$

Consequently, the solution $u(x,t)$ in a series form is given by

$$
\begin{aligned}
u(x,t) &= u_0(x,t) + u_1(x,t) + u_2(x,t) + \cdots \\
&= x + \cos x\left(1 - t + \frac{1}{2!}t^2 - \frac{1}{3!}t^3 + \cdots\right),
\end{aligned}
\tag{35}
$$

and in a closed form by

$$u(x,t) = x + e^{-t}\cos x, \tag{36}$$

obtained upon using the Taylor expansion for e^{-t}.

It is important to point out that the decomposition method has been used in the last two examples in the t-dimension by using the differential operator L_t and by operating with the inverse operator L_t^{-1}. However, the method can also be used in the x-dimension. Although the x-solution can be obtained in a similar fashion, but it requires more computational work if compared with the solution in the t-dimension. This can be attributed to the fact that we use the initial condition I.C only in using the t-dimension, whereas a boundary condition and an initial condition are used to obtain the solution in the x-direction. This can be clearly illustrated by discussing the following examples.

Example 3. Use the decomposition method in the x-direction to solve the initial-boundary value problem of Example 1 given by

$$
\begin{array}{llll}
\text{PDE} & u_t & = & u_{xx}, \ 0 < x < \pi, \ t > 0 \\[2mm]
\text{B.C} & u(0,t) & = & 0, \ t \geq 0 \\[2mm]
& u(\pi,t) & = & 0, \ t \geq 0 \\[2mm]
\text{I.C} & u(x,0) & = & \sin x
\end{array}
\tag{37}
$$

Solution.

In an operator form, Eq. (37) can be written by

$$
L_x\, u(x,t) = L_t\, u(x,t), \ \ 0 < x < \pi, \ t > 0
\tag{38}
$$

where

$$
L_x = \frac{\partial^2}{\partial x^2},
\tag{39}
$$

so that L_x^{-1} is a two-fold integral operator defined by

$$
L_x^{-1}(.) = \int_0^x \int_0^x (.)\, dx\, dx.
\tag{40}
$$

This means that

$$
L_x^{-1} L_x u = u(x,t) - u(0,t) - x u_x(0,t).
\tag{41}
$$

Applying L_x^{-1} to both sides of (38) and using the proper boundary condition we obtain

$$
\begin{aligned}
u(x,t) & = u(0,t) + x u_x(0,t) + L_x^{-1}\left(L_t\, u(x,t)\right), \\
& = x h(t) + L_x^{-1}\left(L_t\, u(x,t)\right),
\end{aligned}
\tag{42}
$$

where

$$
h(t) = u_x(0,t).
\tag{43}
$$

Substituting the decomposition (23) into both sides of (42) gives

$$
\sum_{n=0}^{\infty} u_n(x,t) = x h(t) + L_x^{-1}\left(L_t\left(\sum_{n=0}^{\infty} u_n(x,t)\right)\right).
\tag{44}
$$

Proceeding as before, the components of $u(x,t)$ are determined by

$$
\begin{aligned}
u_0(x,t) & = x h(t), \\
u_1(x,t) & = L_x^{-1}\left(L_t u_0\right) = \tfrac{1}{3!} x^3 h'(t) \\
u_2(x,t) & = L_x^{-1}\left(L_t u_1\right) = \tfrac{1}{5!} x^5 h''(t) \\
& \vdots
\end{aligned}
\tag{45}
$$

Accordingly, the solution in a series form is given by

$$u(x,t) = xh(t) + \frac{1}{3!}x^3 h^{'}(t) + \frac{1}{5!}x^5 h^{''}(t) + \cdots. \tag{46}$$

The unknown function $h(t)$ should be derived so that the solution $u(x,t)$ is completely determined. This can be achieved by using the initial condition

$$u(x,0) = \sin x. \tag{47}$$

Substituting $t = 0$ into (46), using the initial condition (47), and using the Taylor expansion of $\sin x$ we find

$$xh(0) + \frac{1}{3!}x^3 h^{'}(0) + \frac{1}{5!}x^5 h^{''}(0) + \cdots = x - \frac{1}{3!}x^3 + \frac{1}{5!}x^5 - \cdots \tag{48}$$

Equating the coefficients of like powers of x in both sides gives

$$h(0) = 1,\ h^{'}(0) = -1,\ h^{''}(0) = 1,\ \cdots \tag{49}$$

Using the Taylor expansion of $h(t)$ and the result (49) we obtain

$$
\begin{aligned}
h(t) &= h(0) + h^{'}(0)t + \frac{1}{2!}h^{''}(0)t^2 - \frac{1}{3!}h^{'''}(0)t^3 + \cdots \\
&= 1 - t + \frac{1}{2!}t^2 - \frac{1}{3!}t^3 + \cdots \\
&= e^{-t}.
\end{aligned}
\tag{50}
$$

Combining (46) and (50), the solution $u(x,t)$ in a series form is

$$u(x,t) = e^{-t}\left(x - \frac{1}{3!}x^3 - \frac{1}{5!}x^5 + \cdots\right), \tag{51}$$

and in a closed form is given by

$$u(x,t) = e^{-t}\sin x. \tag{52}$$

Example 4. Use the decomposition method in the x-direction to solve the initial-boundary value problem of Example 2.

Solution.

In an operator form we set

$$L_x u(x,t) = L_t u(x,t),\ 0 < x < \pi,\ t > 0 \tag{53}$$

where

$$L_x = \frac{\partial^2}{\partial x^2}, \tag{54}$$

so that L_x^{-1} is a two-fold integral operator defined by

$$L_x^{-1}(.) = \int_0^x \int_0^x (.)dx\, dx. \tag{55}$$

Applying L_x^{-1} to both sides of (53) and using the first boundary condition

$$\begin{aligned}
u(x,t) &= u(0,t) + xu_x(0,t) + L_x^{-1}\left(L_t u(x,t)\right), \\
&= e^{-t} + xh(t) + L_x^{-1}\left(L_t u(x,t)\right),
\end{aligned} \tag{56}$$

where

$$h(t) = u_x(0,t). \tag{57}$$

Substituting the decomposition (23) into both sides of (56) gives

$$\sum_{n=0}^{\infty} u_n(x,t) = e^{-t} + xh(t) + L_x^{-1}\left(L_t\left(\sum_{n=0}^{\infty} u_n(x,t)\right)\right). \tag{58}$$

Proceeding as before, the components of $u(x,t)$ are determined by

$$\begin{aligned}
u_0(x,t) &= e^{-t} + xh(t), \\
u_1(x,t) &= L_x^{-1}\left(L_t\, u_0\right) = -\tfrac{1}{2!}x^2 e^{-t} + \tfrac{1}{3!}x^3 h'(t) \\
u_2(x,t) &= L_x^{-1}\left(L_t\, u_1\right) = \tfrac{1}{4!}x^4 e^{-t} + \tfrac{1}{5!}x^5 h''(t) \\
&\vdots
\end{aligned} \tag{59}$$

Accordingly, the solution in a series form is given by

$$\begin{aligned}
u(x,t) &= e^{-t}\left(1 - \frac{1}{2!}x^2 + \frac{1}{4!}x^4 - \cdots\right) \\
&\quad + xh(t) + \frac{1}{3!}x^3 h'(t) + \frac{1}{5!}x^5 h''(t) + \cdots, \\
&= e^{-t}\cos x + xh(t) + \frac{1}{3!}x^3 h'(t) + \frac{1}{5!}x^5 h''(t) + \cdots.
\end{aligned} \tag{60}$$

It remains to determine the function $h(t)$ in order to completely determine $u(x,t)$. This can be done by using the initial condition

$$u(x,0) = x + \cos x. \tag{61}$$

Using initial condition (61) into (60) we find

$$x + \cos x = \cos x + xh(0) + \frac{1}{3!}x^3 h'(0) + \frac{1}{5!}x^5 h''(0) + \cdots. \tag{62}$$

Equating the coefficients of like powers of x in both sides gives

$$\begin{aligned} h(0) &= 1, \\ h^{(n)}(0) &= 0,\ n \geq 1. \end{aligned} \tag{63}$$

Using the Taylor expansion of $h(t)$ and the result (63) we obtain

$$h(t) = 1. \tag{64}$$

Combining (60) and (64), the solution $u(x, t)$ in a closed form is given by

$$u(x, t) = x + e^{-t} \cos x. \tag{65}$$

For simplicity reasons, we will apply the inverse operator L_t^{-1} to obtain the solution in the following homogeneous PDEs.

Example 5. Use the Adomian decomposition method to solve the initial-boundary value problem

$$\begin{aligned} \text{PDE} \qquad u_t &= u_{xx} - u,\ \ 0 < x < \pi,\ t > 0 \\ \text{B.C} \qquad u(0, t) &= 0,\ t \geq 0 \\ u(\pi, t) &= 0,\ t \geq 0 \\ \text{I.C} \qquad u(x, 0) &= \sin x \end{aligned} \tag{66}$$

Solution.

We point out here that the homogeneous PDE in (66) defines a heat equation with a lateral heat loss. This can be attributed to the additional term $-u(x, t)$ at the right hand side of the standard heat equation.

In an operator form, Eq. (66) can be written as

$$L_t u(x, t) = L_x u(x, t) - u(x, t). \tag{67}$$

Applying L_t^{-1} to both sides of (67) gives

$$u(x, t) = \sin x + L_t^{-1}\left(L_x u(x, t) - u(x, t)\right). \tag{68}$$

Substituting the decomposition (23) into both sides of (68) yields

$$\sum_{n=0}^{\infty} u_n(x, t) = \sin x + L_t^{-1}\left(L_x\left(\sum_{n=0}^{\infty} u_n(x, t)\right) - \sum_{n=0}^{\infty} u_n(x, t)\right). \tag{69}$$

Proceeding as before we obtain

$$
\begin{aligned}
u_0(x,t) &= \sin x, \\
u_1(x,t) &= L_t^{-1}\left(L_x\left(u_0\right)-u_0\right) = -2t\sin x, \\
u_2(x,t) &= L_t^{-1}\left(L_x\left(u_1\right)-u_1\right) = \tfrac{1}{2!}(2t)^2\sin x, \\
&\vdots
\end{aligned}
\tag{70}
$$

Consequently, the solution $u(x,t)$ in a series form is given by

$$
\begin{aligned}
u(x,t) &= u_0(x,t)+u_1(x,t)+u_2(x,t)+\cdots \\
&= \sin x\left(1-2t+\frac{1}{2!}(2t)^2-\cdots\right),
\end{aligned}
\tag{71}
$$

and in a closed form by

$$
u(x,t)=e^{-2t}\sin x,
\tag{72}
$$

obtained upon using the Taylor expansion of e^{-2t}.

3.2.2   Inhomogeneous Heat Equations

A great advantage of the decomposition method is that it can provide solutions to PDE, homogeneous or inhomogeneous, without any need to use any transformation formula as required by the method of separation of variables. The advantage lies in the fact that the method is computationally convenient and provides the solution in a rapid convergent series. The method attacks the inhomogeneous problem in a similar way to that used in the homogeneous type of problems.

Example 6. Use the Adomian decomposition method to solve the inhomogeneous PDE

$$
\begin{aligned}
\text{PDE}\qquad u_t &= u_{xx}+\sin x,\ \ 0<x<\pi,\ t>0 \\[4pt]
\text{B.C}\qquad u(0,t) &= e^{-t},\ t\ge 0 \\
u(\pi,t) &= -e^{-t},\ t\ge 0 \\[4pt]
\text{I.C}\qquad u(x,0) &= \cos x
\end{aligned}
\tag{73}
$$

Solution.

In an operator form, Equation (73) becomes

$$
L_t u(x,t)=L_x u(x,t)+\sin x.
\tag{74}
$$

Operating with L_t^{-1} on both sides of (74) gives

$$u(x,t) = t \sin x + \cos x + L_t^{-1}\left(L_x u(x,t)\right). \tag{75}$$

Using the decomposition (23) we obtain

$$\sum_{n=0}^{\infty} u_n(x,t) = t \sin x + \cos x + L_t^{-1}\left(L_x\left(\sum_{n=0}^{\infty} u_n(x,t)\right)\right). \tag{76}$$

It should be noted here that the zeroth component u_0 will be defined as the sum of all terms that are not included in the operator L_t^{-1}. In fact the zeroth component is assigned the terms that arise from integrating the source term $\sin x$ and from using the initial condition.

To determine the components of $u(x,t)$, we proceed as before, hence we set

$$
\begin{aligned}
u_0(x,t) &= t \sin x + \cos x, \\
u_1(x,t) &= L_t^{-1}\left(L_x\left(u_0\right)\right), \\
&= L_t^{-1}\left(-\cos x - t \sin x\right) = -t \cos x - \tfrac{1}{2!}t^2 \sin x, \tag{77} \\
u_2(x,t) &= L_t^{-1}\left(L_x\left(u_2\right)\right), \\
&= L_t^{-1}\left(t \cos x + \tfrac{1}{2!}t^2 \sin x\right) = \tfrac{1}{2!}t^2 \cos x + \tfrac{1}{3!}t^3 \sin x,
\end{aligned}
$$

and so on. Consequently, the solution $u(x,t)$ in a series form is given by

$$
\begin{aligned}
u(x,t) &= u_0(x,t) + u_1(x,t) + u_2(x,t) + \cdots \\
&= \sin x \left(t - \frac{1}{2!}t^2 + \frac{1}{3!}t^3 \cdots\right) + \cos x \left(1 - t + \frac{1}{2!}t^2 \cdots\right), \tag{78}
\end{aligned}
$$

and in a closed form by

$$u(x,t) = \left(1 - e^{-t}\right)\sin x + e^{-t}\cos x, \tag{79}$$

obtained upon using the Taylor series for the exponential function e^{-t}.

Example 7. Use Adomian decomposition method to solve the inhomogeneous PDE

$$
\begin{aligned}
\text{PDE} \qquad u_t &= u_{xx} + \cos x, \ 0 < x < \pi, \ t > 0 \\
\text{B.C} \qquad u(0,t) &= 1 - e^{-t}, \ t \geq 0 \\
u(\pi,t) &= e^{-t} - 1, \ t \geq 0 \tag{80} \\
\text{I.C} \qquad u(x,0) &= 0, \ 0 \leq x \leq \pi
\end{aligned}
$$

Solution.

Proceeding as before we find

$$u(x,t) = t\,\cos x + L_t^{-1}\left(L_x u(x,t)\right). \tag{81}$$

This gives

$$\sum_{n=0}^{\infty} u_n(x,t) = t\,\cos x + L_t^{-1}\left(L_x\left(\sum_{n=0}^{\infty} u_n(x,t)\right)\right). \tag{82}$$

We next use the recurrence relation

$$
\begin{aligned}
u_0(x,t) &= t\,\cos x, \\
u_1(x,t) &= L_t^{-1}\left(L_x\left(u_0\right)\right) = -\tfrac{1}{2!}t^2\,\cos x, \\
u_2(x,t) &= L_t^{-1}\left(L_x\left(u_2\right)\right) = \tfrac{1}{3!}t^3\,\cos x,
\end{aligned}
\tag{83}
$$

and so on. In view of (83), the solution $u(x,t)$ in a series form is given by

$$u(x,t) = \cos x\left(t - \frac{1}{2!}t^2 + \frac{1}{3!}t^3 - \cdots\right), \tag{84}$$

and in a closed form by

$$u(x,t) = \left(1 - e^{-t}\right)\cos x. \tag{85}$$

Exercises 3.2

In Exercises 1 – 6, use Adomian method to solve the homogeneous equations:

1. $u_t = u_{xx}$, $0 < x < \pi$, $t > 0$
   $u(0,t) = 0$, $u(\pi,t) = \pi$, $t \geq 0$
   $u(x,0) = x + \sin x$

2. $u_t = u_{xx}$, $0 < x < \pi$, $t > 0$
   $u(0,t) = 4 + e^{-t}$, $u(\pi,t) = 4 - e^{-t}$, $t \geq 0$
   $u(x,0) = 4 + \cos x$

3. $u_t = u_{xx}$, $0 < x < \pi$, $t > 0$
   $u(0,t) = 0$, $u(\pi,t) = 0$, $t \geq 0$
   $u(x,0) = \sin x$

4. $u_t = u_{xx} - 4u$, $0 < x < \pi$, $t > 0$
   $u(0,t) = 0$, $u(\pi,t) = 0$, $t \geq 0$
   $u(x,0) = \sin x$

5. $u_t = u_{xx} - 2u$, $0 < x < \pi$, $t > 0$

$$u(0,t) = 0, \ u(\pi,t) = e^{-t}\sinh\pi, \ t \geq 0$$
$$u(x,0) = \sinh x$$

6. $u_t = u_{xx} - 2u, \ 0 < x < \pi, \ t > 0$
   $u(0,t) = e^{-t}, \ u(\pi,t) = e^{-t}\cosh\pi, \ t \geq 0$
   $u(x,0) = \cosh x$

In Exercises 7 – 12, solve the inhomogeneous initial-boundary value problems:

7. $u_t = u_{xx} + \sin(2x), \ 0 < x < \pi, \ t > 0$
   $u(0,t) = 0, \ u(\pi,t) = 0, \ t \geq 0$
   $u(x,0) = \sin x + \frac{1}{4}\sin(2x)$

8. $u_t = u_{xx} - 2, \ 0 < x < \pi, \ t > 0$
   $u(0,t) = 0, \ u(\pi,t) = \pi^2, \ t \geq 0$
   $u(x,0) = x^2 + \sin x$

9. $u_t = u_{xx} - 6x, \ 0 < x < \pi, \ t > 0$
   $u(0,t) = 0, \ u(\pi,t) = \pi^3, \ t \geq 0$
   $u(x,0) = x^3 + \sin x$

10. $u_t = u_{xx} - 6, \ 0 < x < \pi, \ t > 0$
    $u(0,t) = e^{-t}, \ u(\pi,t) = 3\pi^2 - e^{-t}, \ t \geq 0$
    $u(x,0) = 3x^2 + \cos x$

11. $u_t = u_{xx} - 2, \ 0 < x < \pi, \ t > 0$
    $u(0,t) = e^{-t}, \ u(\pi,t) = \pi^2 - e^{-t}, \ t \geq 0$
    $u(x,0) = x^2 + \cos x$

12. $u_t = u_{xx} - 6x, \ 0 < x < \pi, \ t > 0$
    $u(0,t) = 0, \ u(\pi,t) = \pi^3 - e^{-t}, \ t \geq 0$
    $u(x,0) = x^3 + \cos x$

In Exercises 13 – 18, solve the initial-boundary value problems:

13. $u_t = u_{xx}, \ 0 < x < 1, \ t > 0$
    $u(0,t) = 1, \ u(1,t) = 1, \ t \geq 0$
    $u(x,0) = 1 + \sin(\pi x)$

14. $u_t = 4u_{xx}, \ 0 < x < 1, \ t > 0$
    $u(0,t) = 1, \ u(1,t) = 1, \ t \geq 0$
    $u(x,0) = 1 + \sin(\pi x)$

15. $u_t = 4u_{xx}, \ 0 < x < \frac{\pi}{2}, \ t > 0$
    $u(0,t) = e^{-4t}, \ u(\frac{\pi}{2},t) = 0, \ t \geq 0$
    $u(x,0) = \cos x$

16. $u_t = 2u_{xx}, \ 0 < x < \pi, \ t > 0$
    $u(0,t) = 0, \ u(\pi,t) = \pi, \ t \geq 0$

$$u(x,0) = x + \sin x$$

17. $u_t = u_{xx}, \ 0 < x < \pi, \ t > 0$
    $u_x(0,t) = 0, \ u_x(\pi,t) = 0, \ t \geq 0$
    $u(x,0) = \cos x$

18. $u_t = u_{xx}, \ 0 < x < \pi, \ t > 0$
    $u_x(0,t) = 0, \ u_x(\pi,t) = 0, \ t \geq 0$
    $u(x,0) = 2 + \cos x$

3.3   Method of Separation of Variables

In this section the homogeneous partial differential equation that describes the heat flow in a rod will be discussed by using a well-known method called the *method of separation of variables*. The method is commonly used to solve heat conduction problems and other types of problems such as the wave equation and the Laplace equation.

The most important feature of the method of separation of variables is that it successively replaces the partial differential equation by a system of ordinary differential equations that are usually easy to handle. Unlike the decomposition method, the method of separation of variables employs specific assumptions and transformation formulas in handling partial differential equations. In particular, the method of separation of variables requires that the boundary conditions be homogeneous. For inhomogeneous boundary conditions, a transformation formula should be employed to transform inhomogeneous boundary conditions to homogeneous boundary conditions.

3.3.1   Analysis of the Method

We begin our analysis by writing the homogeneous partial differential equation, with homogeneous boundary conditions, that describes the heat flow by the partial differential equation

$$\begin{array}{lll} \text{PDE} & u_t & = \ \bar{k}u_{xx}, \ 0 < x < L, \ t > 0 \\[2mm] \text{B.C} & u(0,t) & = \ 0, \ u(L,t) = 0, \\[2mm] \text{I.C} & u(x,0) & = \ f(x). \end{array} \tag{86}$$

The method of separation of variables consists of assuming that the temperature $u(x,t)$ is identified as the product of two distinct functions $F(x)$ and $T(t)$, where $F(x)$ depends on the space variable x and $T(t)$ depends on the time variable t. In other words, this assumption allows us to set

$$u(x,t) = F(x)T(t). \tag{87}$$

Differentiating both sides of (87) with respect to t and twice with respect to x we obtain

$$
\begin{aligned}
u_t(x,t) &= F(x)T'(t), \\
u_{xx}(x,t) &= F''(x)T(t).
\end{aligned}
\tag{88}
$$

Substituting (88) into (86) yields

$$
F(x)T'(t) = \overline{k}F''(x)T(t).
\tag{89}
$$

Dividing both sides of (89) by $\overline{k}F(x)T(t)$ gives

$$
\frac{T'(t)}{\overline{k}T(t)} = \frac{F''(x)}{F(x)}.
\tag{90}
$$

It is clear from (90) that the left hand side depends only on t and the right hand side depends only on x. This means that the equality holds only if both sides are equal to the same constant. Therefore, we set

$$
\frac{T'(t)}{\overline{k}T(t)} = \frac{F''(x)}{F(x)} = -\lambda^2.
\tag{91}
$$

The selection of $-\lambda^2$, and not λ^2, in (91) is the only selection for which nontrivial solutions exist. However, we can easily show that selecting the constant to be zero or a positive value will lead to the trivial solution $u(x,t) = 0$.

It is clear that (91) gives two distinct ordinary differential equations given by

$$
\begin{aligned}
T'(t) + \overline{k}\lambda^2 T(t) &= 0, \\
F''(x) + \lambda^2 F(x) &= 0.
\end{aligned}
\tag{92}
$$

This means that the partial differential equation (86) is reduced to the more familiar ordinary differential equations ODEs (92) where each equation relies only on one variable.

To determine $T(t)$, we solve the first order linear ODE

$$
T'(t) + \overline{k}\lambda^2 T(t) = 0,
\tag{93}
$$

to find that

$$
T(t) = Ce^{-\overline{k}\lambda^2 t},
\tag{94}
$$

where C is a constant.

On the other hand, the function $F(x)$ can be easily determined by solving the second order linear ODE

$$
F''(x) + \lambda^2 F(x) = 0,
\tag{95}
$$

to find that
$$F(x) = A\cos(\lambda x) + B\sin(\lambda x), \tag{96}$$

where A and B are constants.

To determine A, B, and λ we use the homogeneous boundary conditions

$$\begin{aligned} u(0,t) &= 0, \\ u(L,t) &= 0, \end{aligned} \tag{97}$$

as given above by (86). Substituting (97) into the assumption (87) gives

$$\begin{aligned} F(0)T(t) &= 0, \\ F(L)T(t) &= 0, \end{aligned} \tag{98}$$

which gives

$$\begin{aligned} F(0) &= 0, \\ F(L) &= 0. \end{aligned} \tag{99}$$

Using $F(0) = 0$ into (96) leads to
$$A = 0, \tag{100}$$

hence Eq. (96) becomes
$$F(x) = B\sin(\lambda x). \tag{101}$$

Substituting the condition $F(L) = 0$ of (99) into (101) yields

$$B\sin(\lambda L) = 0. \tag{102}$$

This means that
$$B = 0, \tag{103}$$

or
$$\sin(\lambda L) = 0. \tag{104}$$

We ignore $B = 0$ since it gives the trivial solution $u(x,t) = 0$. It remains that

$$\sin(\lambda L) = 0. \tag{105}$$

This gives an infinite number of values for λ_n given by

$$\lambda_n L = n\pi, \ n = 1, 2, 3, \cdots, \tag{106}$$

or equivalently

$$\lambda_n = \frac{n\pi}{L}, \ n = 1, 2, 3, \cdots. \tag{107}$$

We exclude $n = 0$ since it gives the trivial solution $u(x,t) = 0$.

In view of the infinite number of values for λ_n, we therefore write

$$
\begin{aligned}
F_n(x) &= \sin(\tfrac{n\pi}{L}x), \\
T_n(t) &= e^{-\overline{k}(\frac{n\pi}{L})^2 t}, \ n = 1, 2, 3, \cdots.
\end{aligned}
\tag{108}
$$

Ignoring the constants B and C, we conclude that the functions, called the fundamental solutions

$$
\begin{aligned}
u_n(x,t) &= F_n(x)T_n(t), \\
&= \sin(\tfrac{n\pi}{L}x)e^{-\overline{k}(\frac{n\pi}{L})^2 t}, \ n = 1, 2, \cdots,
\end{aligned}
\tag{109}
$$

that satisfy Eq. (86) and the given boundary conditions.

Recall that the superposition principle admits that a linear combination of the functions $u_n(x,t)$ also satisfy the given equation and the boundary conditions. Therefore, using this principle gives the general solution by

$$
u(x,t) = \sum_{n=1}^{\infty} B_n e^{-\overline{k}(\frac{n\pi}{L})^2 t} \sin(\frac{n\pi}{L}x),
\tag{110}
$$

where the arbitrary constants $B_n, n \geq 1$, are as yet undetermined.

To determine $B_n, n \geq 1$, we substitute $t = 0$ in (110) and by using the initial condition we find

$$
\sum_{n=1}^{\infty} B_n \sin(\frac{n\pi}{L}x) = f(x).
\tag{111}
$$

The constants B_n can be determined in this case by using Fourier coefficients given by the formula

$$
B_n = \frac{2}{L} \int_0^L f(x) \sin(\frac{n\pi}{L}x)\, dx.
\tag{112}
$$

Having determined the constants B_n, the particular solution $u(x,t)$ follows immediately.

On the other hand, if the initial condition $f(x)$ is given in terms of $\sin(\frac{m\pi}{L}x)$, $m \geq 1$, the constants B_n can be completely determined by expanding (110), using the initial condition, and by equating the coefficients of like terms on both sides. The initial condition in the first two examples will be trigonometric functions.

To give a clear overview of the method of separation of variables, we have selected several examples to illustrate the analysis presented above.

Example 1. Use the separation of variables method to solve the following initial-boundary value problem

$$
\begin{array}{lrcl}
\text{PDE} & u_t & = & u_{xx},\ 0 < x < \pi,\ t > 0 \\[2mm]
\text{B.C} & u(0,t) & = & 0,\ t \geq 0 \\[2mm]
 & u(\pi,t) & = & 0,\ t \geq 0 \\[2mm]
\text{I.C} & u(x,0) & = & \sin x + 3\sin(2x)
\end{array}
\tag{113}
$$

Solution.

We first set

$$
u(x,t) = F(x)T(t).
\tag{114}
$$

Differentiating (114) once with respect to t and twice with respect to x and proceeding as before we obtain the two distinct ODEs given by

$$
T^{'}(t) + \lambda^2 T(t) = 0,
\tag{115}
$$

and

$$
F^{''}(x) + \lambda^2 F(x) = 0,
\tag{116}
$$

so that

$$
T(t) = Ce^{-\lambda^2 t}.
\tag{117}
$$

and

$$
F(x) = A\cos(\lambda x) + B\sin(\lambda x).
\tag{118}
$$

To determine the constants A, B and λ, we first use the boundary conditions to obtain

$$
\begin{array}{rcl}
u(0,t) = F(0)T(t) = 0, & \Longrightarrow & F(0) = 0. \\[2mm]
u(\pi,t) = F(\pi)T(t) = 0, & \Longrightarrow & F(\pi) = 0.
\end{array}
\tag{119}
$$

Using (119) into (118) we find

$$
A = 0,
\tag{120}
$$

and

$$
\sin(\pi\lambda) = 0,
\tag{121}
$$

which gives λ_n by

$$
\lambda_n = n,\ n = 1,2,3,\cdots.
\tag{122}
$$

Recall that $n = 0$ gives the trivial solution $u(x,t) = 0$, and therefore it is excluded from the values of λ_n. In accordance with the infinite number of values of λ_n, we therefore write

$$
\begin{array}{rcl}
F_n(x) & = & \sin(nx), \\[2mm]
T_n(t) & = & e^{-n^2 t},\ n = 1,2,\cdots.
\end{array}
\tag{123}
$$

This gives the fundamental set of solutions

$$u_n(x,t) = F_n(x)T_n(t) = \sin(nx)e^{-n^2 t}, \ n = 1, 2, 3, \cdots, \tag{124}$$

where these solutions satisfy Eq. (113) and the given boundary conditions. Using the superposition principle we obtain

$$u(x,t) = \sum_{n=1}^{\infty} B_n e^{-n^2 t} \sin(nx), \tag{125}$$

or in a series form by

$$u(x,t) = B_1 e^{-t}\sin(x) + B_2 e^{-4t}\sin(2x) + B_3 e^{-9t}\sin(3x) + \cdots, \tag{126}$$

where the arbitrary constants $B_n, n \geq 1$ are as yet undetermined. To determine the constants B_n we use the initial condition and substitute $t = 0$ in (126) to find

$$B_1 \sin(x) + B_2 \sin(2x) + B_3 \sin(3x) + \cdots = \sin(x) + 3\sin(2x). \tag{127}$$

Equating the coefficients of like terms of both sides we obtain

$$B_1 = 1, \ B_2 = 3, \ B_k = 0, \ k \geq 3. \tag{128}$$

In view of (128), it is clear that the particular solution consists of two terms only and can be obtained by substituting $n = 1, 2$ into (125) to find

$$u(x,t) = e^{-t}\sin(x) + 3e^{-4t}\sin(2x). \tag{129}$$

Example 2. Use the method of separation of variables to solve the following initial-boundary value problem

$$
\begin{array}{llll}
\text{PDE} & u_t & = & u_{xx}, \ 0 < x < \pi, \ t > 0 \\[2mm]
\text{B.C} & u_x(0,t) & = & 0, \ t \geq 0 \\[2mm]
& u_x(\pi,t) & = & 0, \ t \geq 0 \\[2mm]
\text{I.C} & u(x,0) & = & 2 + 3\cos x
\end{array} \tag{130}
$$

Solution.

It is interesting to note that the problem uses the Neumann boundary conditions, i.e the rates of flow $u_x(0,t) = 0$ and $u_x(\pi,t) = 0$ at the boundaries instead of the temperatures at both ends of the rod. This case arises when both ends of the rod are insulated. This means that no heat flows in or out at the ends of the rod.

We first set

$$u(x,t) = F(x)T(t). \tag{131}$$

Proceeding as before we find

$$T'(t) + \lambda^2 T(t) = 0, \tag{132}$$

and

$$F''(x) + \lambda^2 F(x) = 0, \tag{133}$$

so that

$$T(t) = Ce^{-\lambda^2 t}, \tag{134}$$

and

$$F(x) = A\cos(\lambda x) + B\sin(\lambda x). \tag{135}$$

To determine the constants A, B and λ, we apply the boundary conditions in (130) that can be expressed as

$$
\begin{aligned}
u_x(0,t) = F'(0)T(t) = 0, &\implies F'(0) = 0, \\
u_x(\pi,t) = F'(\pi)T(t) = 0, &\implies F'(\pi) = 0.
\end{aligned}
\tag{136}
$$

Using (136) into (135) we find

$$B = 0, \tag{137}$$

and

$$\lambda \sin(\pi\lambda) = 0, \tag{138}$$

which gives λ_n by

$$\lambda = 0, \text{ or } \lambda_n = n, \, n = 1, 2, 3, \cdots, \tag{139}$$

and therefore

$$\lambda_n = n, \, n = 0, 1, 2, 3, \cdots. \tag{140}$$

where unlike the case of Example 1, $\lambda = 0$ is included because it will not give the trivial solution $u(x,t) = 0$.

In accordance with the infinite number of values of λ_n, we therefore write

$$
\begin{aligned}
F_n(x) &= \cos(nx), \\
T_n(t) &= e^{-n^2 t}, \, n = 0, 1, 2, \cdots.
\end{aligned}
\tag{141}
$$

Using the superposition principle, the general solution is given by

$$u(x,t) = \sum_{n=0}^{\infty} A_n e^{-n^2 t} \cos(nx). \tag{142}$$

or in a series form by

$$u(x,t) = A_0 + A_1 e^{-t} \cos(x) + A_2 e^{-4t} \cos(2x) + A_3 e^{-9t} \cos(3x) + \cdots. \tag{143}$$

To determine the constants A_n we use the initial condition and replace t by zero in (143) to find

$$A_0 + A_1 \cos(x) + A_2 \cos(2x) + \cdots = 2 + 3 \cos(x). \tag{144}$$

Equating the coefficients of like terms on both sides we obtain

$$A_0 = 2, \ A_1 = 3, \ A_k = 0, \ k \geq 2. \tag{145}$$

In view of (145), the particular solution is given by

$$u(x, t) = 2 + 3e^{-t} \cos(x). \tag{146}$$

Example 3. Use the method of separation of variables to solve the following initial-boundary value problem

$$
\begin{array}{llcl}
\text{PDE} & u_t & = & u_{xx}, \ 0 < x < \pi, \ t > 0 \\
\text{B.C} & u(0, t) & = & 0, \ t \geq 0 \\
& u(\pi, t) & = & 0, \ t \geq 0 \\
\text{I.C} & u(x, 0) & = & 1
\end{array}
\tag{147}
$$

Solution.

We first set

$$u(x, t) = F(x)T(t). \tag{148}$$

Proceeding as before, we find

$$T(t) = Ce^{-\lambda^2 t}, \tag{149}$$

and

$$F(x) = A \cos(\lambda x) + B \sin(\lambda x). \tag{150}$$

Using the boundary conditions gives

$$A = 0, \tag{151}$$

and

$$\lambda_n = n, \ n = 1, 2, 3, \cdots. \tag{152}$$

Using the superposition principle, the general solution is given by

$$u(x, t) = \sum_{n=0}^{\infty} B_n e^{-n^2 t} \sin(nx). \tag{153}$$

To determine the constants B_n we use the initial condition to find

$$\sum_{n=1}^{\infty} B_n \sin(nx) = 1. \tag{154}$$

The arbitrary constants are determined by using the Fourier method, therefore we find

$$
\begin{aligned}
B_n &= \frac{2}{\pi} \int_0^\pi \sin(nx)dx, \\
&= \frac{2}{n\pi}(1 - \cos(n\pi)),
\end{aligned}
\tag{155}
$$

so that

$$
B_n =
\begin{cases}
0 & \text{if } n \text{ is even,} \\[2mm]
\frac{4}{n\pi} & \text{if } n \text{ is odd.}
\end{cases}
\tag{156}
$$

This means that we can express B_n by

$$
\begin{aligned}
B_{2m} &= 0, \\
B_{2m+1} &= \frac{4}{(2m+1)\pi}, \ m = 0, 1, 2, \cdots.
\end{aligned}
\tag{157}
$$

Combining (153) and (157), the particular solution is given by

$$u(x,t) = \frac{4}{\pi} \sum_{m=0}^{\infty} \frac{1}{(2m+1)} e^{-(2m+1)^2 t} \sin(2m+1)x. \tag{158}$$

Example 4. Use the method of separation of variables to solve the following initial-boundary value problem

$$
\begin{array}{lll}
\text{PDE} & u_t = u_{xx}, \ 0 < x < \pi, \ t > 0 \\[2mm]
\text{B.C} & u_x(0,t) = 0, \ t \geq 0 \\[2mm]
& u_x(\pi,t) = 0, \ t \geq 0 \\[2mm]
\text{I.C} & u(x,0) = x
\end{array}
\tag{159}
$$

Solution.

We first set

$$u(x,t) = F(x)T(t). \tag{160}$$

Following the previous discussions we find

$$T(t) = Ce^{-\lambda^2 t}, \tag{161}$$

and

$$F(x) = A\cos(\lambda x) + B\sin(\lambda x). \tag{162}$$

The Neumann boundary conditions give

$$B = 0, \tag{163}$$

and

$$\lambda_n = n, \; n = 0, 1, 2, 3, \cdots. \tag{164}$$

Using the results we obtained for λ_n, we write

$$
\begin{aligned}
F_n(x) &= \cos(nx), \\
T_n(t) &= e^{-n^2 t}, \; n = 0, 1, 2, \cdots.
\end{aligned}
\tag{165}
$$

Using the superposition principle, the general solution is given by

$$u(x,t) = \sum_{n=0}^{\infty} A_n e^{-n^2 t} \cos(nx). \tag{166}$$

To determine the constants A_n we use the initial condition to find

$$\sum_{n=0}^{\infty} A_n \cos(nx) = x. \tag{167}$$

The arbitrary constants A_n are determined by using the Fourier method, therefore we find

$$
\begin{aligned}
A_0 &= \frac{1}{\pi} \int_0^\pi x\,dx = \frac{\pi}{2} \\
A_n &= \frac{2}{\pi} \int_0^\pi x\cos(nx)\,dx = \frac{2}{\pi}\left(\frac{1}{n^2}\cos(n\pi) - \frac{1}{n^2}\right), \; n = 1, 2, \cdots.
\end{aligned}
\tag{168}
$$

so that

$$
A_n = \begin{cases} 0 & \text{if } n \text{ is even, } n \neq 0, \\[2mm] -\frac{4}{\pi n^2} & \text{if } n \text{ is odd.} \end{cases}
\tag{169}
$$

Based on these results for the constants A_n, the particular solution is given by

$$u(x,t) = \frac{\pi}{2} - \frac{4}{\pi} \sum_{m=0}^{\infty} \frac{1}{(2m+1)^2} e^{-(2m+1)^2 t} \cos(2m+1)x. \tag{170}$$

Exercises 3.3.1

Solve the following initial-boundary value problems by the method of separation of variables:

1. $u_t = u_{xx}$, $0 < x < \pi$, $t > 0$
   $u(0,t) = 0$, $u(\pi,t) = 0$
   $u(x,0) = \sin x + 2\sin(3x)$

2. $u_t = u_{xx}$, $0 < x < 1$, $t > 0$
   $u(0,t) = 0$, $u(1,t) = 0$
   $u(x,0) = \sin(\pi x) + \sin(2\pi x)$

3. $u_t = 4u_{xx}$, $0 < x < \pi$, $t > 0$
   $u(0,t) = 0$, $u(\pi,t) = 0$
   $u(x,0) = \sin(2x)$

4. $u_t = 2u_{xx}$, $0 < x < 1$, $t > 0$
   $u(0,t) = 0$, $u(1,t) = 0$
   $u(x,0) = \sin(\pi x)$

5. $u_t = u_{xx}$, $0 < x < \pi$, $t > 0$
   $u_x(0,t) = 0$, $u_x(\pi,t) = 0$
   $u(x,0) = 1 + \cos x$

6. $u_t = 2u_{xx}$, $0 < x < \pi$, $t > 0$
   $u_x(0,t) = 0$, $u_x(\pi,t) = 0$
   $u(x,0) = 3 + 4\cos x$

7. $u_t = 3u_{xx}$, $0 < x < \pi$, $t > 0$
   $u_x(0,t) = 0$, $u_x(\pi,t) = 0$
   $u(x,0) = 1 + \cos x + \cos(2x)$

8. $u_t = 4u_{xx}$, $0 < x < 1$, $t > 0$
   $u_x(0,t) = 0$, $u_x(1,t) = 0$
   $u(x,0) = 2 + 2\cos(2\pi x)$

9. $u_t = 4u_{xx}$, $0 < x < \pi$, $t > 0$
   $u(0,t) = 0$, $u(\pi,t) = 0$
   $u(x,0) = 2$

10. $u_t = 2u_{xx}$, $0 < x < \pi$, $t > 0$
    $u(0,t) = 0$, $u(\pi,t) = 0$
    $u(x,0) = 3$

11. $u_t = 3u_{xx}$, $0 < x < \pi$, $t > 0$
    $u(0,t) = 0$, $u(\pi,t) = 0$
    $u(x,0) = 2x$

12. $u_t = u_{xx}$, $0 < x < \pi$, $t > 0$
    $u(0,t) = 0$, $u(\pi,t) = 0$
    $u(x,0) = 1 + x$

3.3.2 Inhomogeneous Boundary Conditions

In this section we will consider the case where the ends of a rod are kept at constant temperatures different from zero. It is well known that the method of separation of variables is applicable if the equation and the boundary conditions are linear and homogeneous. Consequently, a transformation formula is needed that will enable us to convert the inhomogeneous boundary conditions to homogeneous boundary conditions. This is necessary in order to apply the method of separation of variables in a parallel way to that used above.

We begin our analysis by considering the initial-boundary value problem

$$
\begin{aligned}
\text{PDE} \qquad u_t &= u_{xx}, \ 0 < x < L, \ t > 0 \\[2mm]
\text{B.C} \qquad u(0,t) &= \alpha, \ t \geq 0 \\[2mm]
u(L,t) &= \beta, \ t \geq 0 \\[2mm]
\text{I.C} \qquad u(x,0) &= f(x)
\end{aligned}
\tag{171}
$$

To convert the boundary conditions from inhomogeneous to homogeneous we simply use the following transformation formula

$$
u(x,t) = \left(\alpha + \frac{x}{L}(\beta - \alpha)\right) + v(x,t).
\tag{172}
$$

This means that $u(x,t)$ consists of a **steady-state solution**, that does not depend on time, defined by

$$
w(x) = \alpha + \frac{\beta}{L}(x - \alpha),
\tag{173}
$$

that satisfies the boundary conditions, and a **transient solution** given by $v(x,t)$. We can easily show that $v(x,t)$ will be governed by the initial-boundary value problem

$$
\begin{aligned}
\text{PDE} \qquad v_t &= v_{xx}, \ 0 < x < L, \ t > 0 \\[2mm]
\text{B.C} \qquad v(0,t) &= 0, \ t \geq 0 \\[2mm]
v(L,t) &= 0, \ t \geq 0 \\[2mm]
\text{I.C} \qquad v(x,0) &= f(x) - \left(\alpha + \frac{x}{L}(\beta - \alpha)\right)
\end{aligned}
\tag{174}
$$

Consequently, the method of separation of variables can be used in a similar way to that used in the previous section. Recall that Adomian decomposition method can be implemented directly. To get a better understanding of converting inhomogeneous boundary conditions to homogeneous boundary conditions, the following illustrative example will be discussed.

Example 5. Solve the following initial- boundary value problem

$$
\begin{aligned}
\text{PDE} \qquad u_t &= u_{xx}, \ 0 < x < 1, \ t > 0 \\
\text{B.C} \qquad u(0,t) &= 1, \ t \geq 0 \\
u(L,t) &= 2, \ t \geq 0 \\
\text{I.C} \qquad u(x,0) &= 1 + x + 2\sin(\pi x)
\end{aligned}
\tag{175}
$$

Solution.

Using the transformation (172) we obtain

$$
u(x,t) = (1+x) + v(x,t).
\tag{176}
$$

In view of (176), Eq. (175) is transformed into

$$
\begin{aligned}
\text{PDE} \qquad v_t &= v_{xx}, \ 0 < x < 1, \ t > 0 \\
\text{B.C} \qquad v(0,t) &= 0, \ t \geq 0 \\
v(1,t) &= 0, \ t \geq 0 \\
\text{I.C} \qquad v(x,0) &= 2\sin(\pi x)
\end{aligned}
\tag{177}
$$

Assuming that

$$
v(x,t) = F(x)T(t),
\tag{178}
$$

and proceeding as before we obtain

$$
T(t) = Ce^{-\lambda^2 t},
\tag{179}
$$

and

$$
F(x) = A\cos(\lambda x) + B\sin(\lambda x),
\tag{180}
$$

where A, B, and C are constants. Using the boundary conditions gives

$$
v(x,t) = \sum_{n=1}^{\infty} B_n e^{-n^2\pi^2 t} \sin(n\pi x)
\tag{181}
$$

Using the initial condition in (177) and expanding (181) we obtain

$$
B_1 = 2, \ B_k = 0, \ k \geq 2.
\tag{182}
$$

This gives the solution for $v(x,t)$ by

$$
v(x,t) = 2e^{-\pi^2 t} \sin(\pi x),
\tag{183}
$$

so that the particular solution

$$u(x,t) = 1 + x + 2e^{-\pi^2 t}\sin(\pi x),\tag{184}$$

follows immediately.

At this point, it seems reasonable to use the Adomian decomposition method to solve the initial-boundary value problem of Ex. 5. The newly developed approach can be used to examine the performance of the decomposition method if compared with the classical method of separation of variables.

Applying the inverse operator L_t^{-1} to the operator form of (191) and using the initial condition we obtain

$$u(x,t) = 1 + x + 2\sin(\pi x) + L_t^{-1}\left(L_x u(x,t)\right).\tag{185}$$

Using the decomposition series (23) of $u(x,t)$ yields

$$\sum_{n=0}^{\infty} u_n(x,t) = 1 + x + 2\sin(\pi x) + L_t^{-1}\left(L_x\left(\sum_{n=0}^{\infty} u_n(x,t)\right)\right).\tag{186}$$

Using the recursive algorithm we obtain

$$
\begin{aligned}
u_0(x,t) &= 1 + x + 2\sin(\pi x),\\
u_1(x,t) &= L_t^{-1}(L_x u_0) = -2\pi^2 t \sin(\pi x),\\
u_2(x,t) &= L_t^{-1}(L_x u_1) = 2\pi^4 \frac{t^2}{2!} t \sin(\pi x),
\end{aligned}\tag{187}
$$

and so on. Consequently, the solution in a series form is given by

$$u(x,t) = 1 + x + 2\sin(\pi x)\left(1 - \pi^2 t + \pi^4 \frac{t^2}{2!} - \cdots\right),\tag{188}$$

and in a closed form

$$u(x,t) = 1 + x + 2\sin(\pi x)e^{-\pi^2 t}.\tag{189}$$

As can be verified from this comparison, the decomposition method approaches any problem directly without any need to transform the inhomogeneous boundary conditions to homogeneous conditions.

Exercises 3.3.2

In Exercises 1 – 4, use the method of separation of variables to solve the initial-boundary value problems:

1. $u_t = u_{xx}$, $0 < x < 1$, $t > 0$
   $u(0,t) = 1$, $u(1,t) = 3$
   $u(x,0) = 1 + 2x + 3\sin(\pi x)$

2. $u_t = u_{xx}$, $0 < x < \pi$, $t > 0$
   $u(0,t) = 1$, $u(\pi,t) = 1$
   $u(x,0) = 1 + 4\sin x$

3. $u_t = u_{xx}$, $0 < x < \pi$, $t > 0$
   $u(0,t) = 0$, $u(\pi,t) = \pi$
   $u(x,0) = x + \sin(2x)$

4. $u_t = u_{xx}$, $0 < x < \pi$, $t > 0$
   $u(0,t) = 4$, $u(\pi,t) = 4 - 4\pi$
   $u(x,0) = 4 - 4x + \sin(3x)$

5. Solve the following initial-boundary value problem by the decomposition method and by the separation of variables method:

   $u_t = u_{xx}$, $0 < x < \pi$, $t > 0$
   $u(0,t) = 2$, $u(\pi,t) = 2 + 3\pi$
   $u(x,0) = 2 + 3x + \sin x$

6. Solve the following initial-boundary value problem by the decomposition method and by the separation of variables method:

   $u_t = u_{xx}$, $0 < x < 1$, $t > 0$
   $u(0,t) = 1$, $u(1,t) = 3$
   $u(x,0) = 1 + 2x + \sin(2\pi x)$

3.3.3   Equations with Lateral Heat Loss

For a rod with a lateral heat loss, it can be proved that the heat flow is controlled by the homogeneous PDE

$$
\begin{aligned}
\text{PDE} \qquad u_t &= ku_{xx} - cu, 0 < x < L, t > 0 \\
\text{B.C} \qquad u(0,t) &= 0 \\
u(L,t) &= 0 \\
\text{I.C} \qquad u(x,0) &= f(x)
\end{aligned}
\tag{190}
$$

It is easily observed that this equation is not the standard heat equation we discussed so far. Instead, it includes the term $-cu(x,t)$ due to the lateral heat loss.

We will focus our attention on converting Eq. (190) to a standard heat equation. Thereafter, we can implement the separation of variables method in a straightforward way. This goal can be achieved by using the transformation formula

$$
u(x,t) = e^{-ct}w(x,t).
\tag{191}
$$

Accordingly, $w(x,t)$ will be governed by the IBVP

$$
\begin{array}{llll}
\text{PDE} & w_t & = & kw_{xx}, \ 0 < x < L, t > 0 \\
\text{B.C} & w(0,t) & = & 0 \\
& w(L,t) & = & 0 \\
\text{I.C} & w(x,0) & = & f(x)
\end{array}
\tag{192}
$$

where $w(x,t)$ can be easily obtained in a similar manner to the discussion stated above. The following example illustrates the use of the transformation formula (191).

Example 6. Solve the following initial boundary value problem

$$
\begin{array}{llll}
\text{PDE} & u_t & = & u_{xx} - u, 0 < x < 1, \ t > 0 \\
\text{B.C} & u(0,t) & = & 0 \\
& u(1,t) & = & 0 \\
\text{I.C} & u(x,0) & = & \sin(\pi x) + 2\sin(3\pi x)
\end{array}
\tag{193}
$$

Solution.

Using the transformation formula

$$
u(x,t) = e^{-t} w(x,t),
\tag{194}
$$

carries (193) into

$$
\begin{array}{llll}
\text{PDE} & w_t & = & w_{xx}, \ 0 < x < 1, t > 0, \\
\text{B.C} & w(0,t) & = & 0, \\
& w(1,t) & = & 0, \\
\text{I.C} & w(x,0) & = & \sin(\pi x) + 2\sin(3\pi x).
\end{array}
\tag{195}
$$

Setting

$$
w(x,t) = F(x)T(t),
\tag{196}
$$

and proceeding as before we obtain

$$
w(x,t) = \sum_{n=1}^{\infty} B_n e^{-n^2\pi^2 t} \sin(n\pi x),
\tag{197}
$$

obtained upon using the boundary conditions. To determine the arbitrary constants B_n, $n \geq 1$, we substitute $t = 0$ in (197) and use the initial condition to find

$$B_1 \sin(\pi x) + B_2 \sin(2\pi x) + B_3 \sin(3\pi x) + \cdots = \sin(\pi x) + 2\sin(3\pi x), \qquad (198)$$

which gives

$$B_1 = 1,\ B_3 = 2,\ B_k = 0,\ k \neq 1, 3. \qquad (199)$$

In view of (199), Eq. (197) becomes

$$w(x, t) = e^{-\pi^2 t} \sin(\pi x) + e^{-9\pi^2 t} \sin(3\pi x). \qquad (200)$$

Substituting (200) into (194) gives

$$u(x, t) = e^{-t} \left(e^{-\pi^2 t} \sin(\pi x) + e^{-9\pi^2 t} \sin(3\pi x) \right) \qquad (201)$$

For comparisons reasons, the Adomian decomposition method will be used to solve Example 6. Applying the operator L_t^{-1} to both sides of (193) and using the initial condition we obtain

$$u(x, t) = \sin(\pi x) + 2\sin(3\pi x) + L_t^{-1}(L_x u - u). \qquad (202)$$

Using the decomposition series for $u(x, t)$ gives

$$\sum_{n=0}^{\infty} u_n(x, t) \;=\; \sin(\pi x) + 2\sin(3\pi x)$$

$$- L_t^{-1} \left(L_x \left(\sum_{n=0}^{\infty} u_n(x, t) \right) - \sum_{n=0}^{\infty} u_n(x, t) \right). \qquad (203)$$

Using the recursive algorithm as discussed before we obtain

$$
\begin{aligned}
u_0 &= \sin(\pi x) + 2\sin(3\pi x), \\
u_1 &= -(\pi^2 + 1)t \sin(\pi x) - 2(9\pi^2 + 1)t \sin(3\pi x), \\
u_2 &= (\pi^2 + 1)^2 \tfrac{t^2}{2!} \sin(\pi x) + 2(9\pi^2 + 1)^2 \tfrac{t^2}{2!} \sin(3\pi x).
\end{aligned}
\qquad (204)
$$

and so on. Based on this. the solution in a series for is given by

$$
\begin{aligned}
u(x, t) &= \sin(\pi x) \left(1 - (\pi^2 + 1)t + (\pi^2 + 1)^2 \frac{t^2}{2!} - \cdots \right) \\
&\quad 2\sin(3\pi x) \left(1 - (9\pi^2 + 1)t + \cdots \right),
\end{aligned}
\qquad (205)
$$

and in a closed form by

$$u(x,t) = \sin(\pi x)e^{-(\pi^2+1)t} + 2\sin(3\pi x)e^{-(9\pi^2+1)t}. \qquad (206)$$

Exercises 3.3.3

Use the method of separation of variables to solve the following homogeneous heat equations with lateral heat loss:

1. $u_t = u_{xx} - u, \; 0 < x < \pi, \; t > 0$
   $u(0,t) = 0, \; u(\pi,t) = 0$
   $u(x,0) = \sin x$

2. $u_t = u_{xx} - u, \; 0 < x < 1, \; t > 0$
   $u(0,t) = 0, \; u(1,t) = 0$
   $u(x,0) = \sin(2\pi x)$

3. $u_t = u_{xx} - 3u, \; 0 < x < \pi, \; t > 0$
   $u(0,t) = 0, \; u(\pi,t) = 0$
   $u(x,0) = \sin(x)$

4. $u_t = u_{xx} - 2u, \; 0 < x < 1, \; t > 0$
   $u(0,t) = 0, \; u(1,t) = 0$
   $u(x,0) = \sin(\pi x)$

5. $u_t = u_{xx} - u + 1, \; 0 < x < \pi, \; t > 0$
   $u(0,t) = 1, \; u(\pi,t) = 1$
   $u(x,0) = 1 + \sin x$

6. $u_t = u_{xx} - 2\pi^2 u + 6\pi^2, \; 0 < x < 1, \; t > 0$
   $u(0,t) = 3, \; u(1,t) = 3$
   $u(x,0) = 3 + \sin(\pi x)$

Chapter 4

Higher Dimensional Heat Flow

4.1 Introduction

This chapter is devoted to the study of the PDEs that control the heat flow in two and three dimensional spaces. The higher dimensional heat flow has been the subject of intensive analytical and numerical investigations. The work in this chapter will run in a parallel manner to the work used in Chapter 3. The study of higher dimensional heat equation stems mainly from Adomian decomposition method and the method of separation of variables. The two methods have been outlined in Chapters 2 and 3 and were implemented for the heat equation in one dimension.

The decomposition method has been used to obtain analytic and approximate solutions to a wide class of linear and nonlinear, differential and integral equations. It was found by many researchers, such as [4,6,9,120], that unlike other series solution methods, the decomposition method is easy to program in engineering problems [115], and provides immediate and visible solution terms without linearization or discretization. As stated in Chapter 2, the concept of rapid convergence of the method was addressed extensively in [34–37] among others. The main advantage of the decomposition method is that it can be applied directly to all types of differential equations with homogeneous or inhomogeneous boundary conditions. Another important advantage is that the method is capable of reducing the size of computational work.

To examine the performance of Adomian's method compared to existing techniques, the method of separation of variables will be implemented. The method of separation of variables provides the solution of a partial differential equation through reducing the equation to a system of ordinary differential equations. In addition, the method requires that the problem and the boundary conditions be linear and homogeneous, hence transformation formulae are usually used to meet this need.

4.2   The Adomian Decomposition Method

The Adomian decomposition method has been receiving much attention in recent years in the area of series solutions. A considerable research work has been invested recently in applying this method to a wide class of differential and integral equations. A useful attraction of this method is that it has proved to be a competitive alternative to the Taylor series method and other series techniques.

The decomposition method consists of decomposing the unknown function $u(x, t)$ into an infinite sum of components. The zeroth component $u_0(x, t)$ is identified by the terms arising from integrating the inhomogeneous term and the initial/boundary conditions. The successive terms are determined in a recursive manner. The method attacks inhomogeneous problems and homogeneous problems in a like manner, thus providing an easily computable technique.

4.2.1   Two Dimensional Heat Flow

The distribution of heat flow in a two dimensional space is governed by the following initial boundary value problem

$$
\begin{aligned}
\text{PDE} \qquad u_t &= \overline{k}(u_{xx} + u_{yy}), 0 < x < a, \, 0 < y < b, t > 0 \\[4pt]
\text{B.C } u(0, y, t) &= u(a, y, t) = 0 \\[4pt]
u(x, 0, t) &= u(x, b, t) = 0 \\[4pt]
\text{I.C } u(x, y, 0) &= f(x, y)
\end{aligned}
\tag{1}
$$

where $u \equiv u(x, y, t)$ is the temperature of any point located at the position (x, y) of a rectangular plate at any time t, and $\overline{k}$ is the thermal diffusivity.

As discussed before, the solution in the t space, the x space, or the y space will produce the same series solution [6,12]. However, the solution in the t space reduces the size of calculations compared with the other space solutions. For this reason the solution in the t direction will be followed in this chapter.

We first rewrite (1) in an operator form by

$$
L_t u(x, y, t) = \overline{k} \left(L_x u + L_y u \right),
\tag{2}
$$

where the differential operators $L_t, L_x,$ and L_y are defined by

$$
L_t = \frac{\partial}{\partial t}, \; L_x = \frac{\partial^2}{\partial x^2}, \; L_y = \frac{\partial^2}{\partial y^2},
\tag{3}
$$

so that the integral operator L_t^{-1} exists and given by

$$L_t^{-1}(.) = \int_0^t (.)dt. \tag{4}$$

Applying L_t^{-1} to both sides of (2) and using the initial condition leads to

$$u(x,y,t) = f(x,y) + \overline{k}\, L_t^{-1}\left(L_x u + L_y u\right). \tag{5}$$

The decomposition method defines the solution $u(x,y,t)$ as a series given by

$$u(x,y,t) = \sum_{n=0}^{\infty} u_n(x,y,t), \tag{6}$$

where the components $u_n(x,y,t), n \geq 0$ will be easily computed by using a recursive algorithm. Substituting (6) into both sides of (5) yields

$$\sum_{n=0}^{\infty} u_n = f(x,y) + \overline{k}\, L_t^{-1}\left(L_x \left(\sum_{n=0}^{\infty} u_n\right) + L_y \left(\sum_{n=0}^{\infty} u_n\right)\right). \tag{7}$$

The decomposition method suggests that the zeroth component $u_0(x,y,t)$ is identified as the terms arising from the initial/boundary conditions and from source terms. The remaining components of $u(x,y,t)$ can be determined in a recursive manner such that each component is determined by using the previous component. Accordingly, the components $u_n(x,y,t)$, $n \geq 0$ can be completely determined by following the recurrence relation

$$\begin{aligned} u_0(x,y,t) &= f(x,y), \\ u_{k+1}(x,y,t) &= \overline{k}\, L_t^{-1}\left(L_x u_k + L_y u_k\right), k \geq 0. \end{aligned} \tag{8}$$

As a result, the successive components are completely determined, and hence the solution in a series form is thus obtained. Recall that the components can be determined recursively as far as we like. For numerical purposes, the accuracy level can be improved significantly by increasing the number of components determined. As discussed earlier, the closed form solution may also be obtained.

To give a clear overview of the implementation of the decomposition method, we have chosen several examples, homogeneous and inhomogeneous, to illustrate the discussion given above.

Homogeneous Heat Equations

The Adomian decomposition method will be used to solve the following homogeneous heat equation in two dimensions with homogeneous or inhomogeneous boundary conditions.

Example 1. Use the Adomian decomposition method to solve the initial-boundary value problem

$$
\begin{array}{rll}
\text{PDE} \quad u_t & = & u_{xx} + u_{yy}, \ 0 < x, y < \pi, \ t > 0 \\
\text{B.C} \ u(0, y, t) & = & u(\pi, y, t) = 0 \\
u(x, 0, t) & = & u(x, \pi, t) = 0 \\
\text{I.C} \ u(x, y, 0) & = & \sin x \sin y
\end{array}
\tag{9}
$$

Solution.

We first write (9) in an operator form by

$$
L_t u = L_x u + L_y u.
\tag{10}
$$

Applying the inverse operator L_t^{-1} to (10) and using the initial condition we obtain

$$
u(x, y, t) = \sin x \sin y + L_t^{-1} \left(L_x u + L_y u \right).
\tag{11}
$$

The decomposition method defines the solution $u(x, y, t)$ as a series given by

$$
u(x, y, t) = \sum_{n=0}^{\infty} u_n(x, y, t),
\tag{12}
$$

where the components $u_n(x, y, t), n \geq 0$ are to be determined by using a recursive algorithm. Substituting (12) into both sides of (11) yields

$$
\sum_{n=0}^{\infty} u_n = \sin x \sin y + L_t^{-1} \left(L_x \left(\sum_{n=0}^{\infty} u_n \right) + L_y \left(\sum_{n=0}^{\infty} u_n \right) \right).
\tag{13}
$$

The zeroth component $u_0(x, y, t)$ is identified by all terms that are not included under L_t^{-1}. The components $u_n(x, y, t)$, $n \geq 0$ can be completely determined by following the recursive algorithm

$$
\begin{array}{rll}
u_0(x, y, t) & = & \sin x \sin y, \\
u_{k+1}(x, y, t) & = & L_t^{-1} \left(L_x u_k + L_y u_k \right), k \geq 0.
\end{array}
\tag{14}
$$

With u_0 defined as shown above, the first few terms of the decomposition (12) are given by

$$
\begin{array}{rll}
u_0(x, y, t) & = & \sin x \sin y, \\
u_1(x, y, t) & = & L_t^{-1} \left(L_x u_0 + L_y u_0 \right) = -2t \sin x \sin y, \\
u_2(x, y, t) & = & L_t^{-1} \left(L_x u_1 + L_y u_1 \right) = \frac{(2t)^2}{2!} \sin x \sin y, \\
u_3(x, y, t) & = & L_t^{-1} \left(L_x u_2 + L_y u_2 \right) = -\frac{(2t)^3}{3!} \sin x \sin y,
\end{array}
\tag{15}
$$

and so on. Combining (12) and (15), the solution in a series form is given by

$$u(x, y, t) = \sin x \sin y \left(1 - (2t) + \frac{(2t)^2}{2!} - \frac{(2t)^3}{3!} + \cdots\right),\tag{16}$$

and in a closed form by

$$u(x, y, t) = e^{-2t} \sin x \sin y.\tag{17}$$

Example 2. Use the Adomian decomposition method to solve the initial-boundary value problem with lateral heat loss

$$
\begin{aligned}
\text{PDE} \quad u_t &= u_{xx} + u_{yy} - u, \, 0 < x, y < \pi, \, t > 0\\
\text{B.C } u(0, y, t) &= u(\pi, y, t) = 0\\
u(x, 0, t) &= -u(x, \pi, t) = e^{-3t} \sin x\\
\text{I.C } u(x, y, 0) &= \sin x \cos y
\end{aligned}
\tag{18}
$$

Solution.

Applying the inverse operator L_t^{-1} to (18) gives

$$u(x, y, t) = \sin x \cos y + L_t^{-1}\left(L_x u + L_y u - u\right).\tag{19}$$

The decomposition method defines the solution $u(x, y, t)$ as a series given by

$$u(x, y, t) = \sum_{n=0}^{\infty} u_n(x, y, t).\tag{20}$$

Substituting (20) into both sides of (19) yields

$$\sum_{n=0}^{\infty} u_n = \sin x \cos y + L_t^{-1}\left(L_x \left(\sum_{n=0}^{\infty} u_n\right) + L_y \left(\sum_{n=0}^{\infty} u_n\right) - \sum_{n=0}^{\infty} u_n\right).\tag{21}$$

Proceeding as before we find

$$
\begin{aligned}
u_0(x, y, t) &= \sin x \cos y,\\
u_{k+1}(x, y, t) &= L_t^{-1}\left(L_x u_k + L_y u_k - u_k\right), k \geq 0.
\end{aligned}
\tag{22}
$$

It follows that

$$
\begin{aligned}
u_0(x, y, t) &= \sin x \cos y,\\
u_1(x, y, t) &= L_t^{-1}\left(L_x u_0 + L_y u_0 - u_0\right) = -3t \sin x \cos y,\\
u_2(x, y, t) &= L_t^{-1}\left(L_x u_1 + L_y u_1 - u_1\right) = \frac{(3t)^2}{2!} \sin x \cos y,\\
u_3(x, y, t) &= L_t^{-1}\left(L_x u_2 + L_y u_2 - u_2\right) = -\frac{(3t)^3}{3!} \sin x \cos y,
\end{aligned}
\tag{23}
$$

and so on. Combining (20) and (23), the solution in a series form is given by

$$u(x, y, t) = \sin x \cos y \left(1 - (3t) + \frac{(3t)^2}{2!} - \frac{(3t)^3}{3!} + \cdots \right), \qquad (24)$$

and in a closed form by

$$u(x, y, t) = e^{-3t} \sin x \cos y. \qquad (25)$$

Example 3. Use the Adomian decomposition method to solve the initial-boundary value problem

$$
\begin{aligned}
\text{PDE} \qquad u_t &= u_{xx} + u_{yy}, \ 0 < x, y < \pi, \ t > 0 \\[1mm]
\text{B.C } u(0, y, t) &= -u(\pi, y, t) = e^{-2t} \sin y \\[1mm]
u(x, 0, t) &= -u(x, \pi, t) = e^{-2t} \sin x \\[1mm]
\text{I.C } u(x, y, 0) &= \sin(x + y)
\end{aligned}
\qquad (26)
$$

Solution.

It is obvious that the boundary conditions are inhomogeneous. One major advantage of the decomposition method is that it handles any problem in a direct way without any need to transform the inhomogeneous conditions to homogeneous conditions.

Applying the operator L_t^{-1} to the operator form of (26) yields

$$u(x, y, t) = \sin(x + y) + L_t^{-1} \left(L_x u + L_y u \right). \qquad (27)$$

Substituting the decomposition series for $u(x, y, t)$ into (27) gives

$$\sum_{n=0}^{\infty} u_n = \sin(x + y) + L_t^{-1} \left(L_x \left(\sum_{n=0}^{\infty} u_n \right) + L_y \left(\sum_{n=0}^{\infty} u_n \right) \right). \qquad (28)$$

Proceeding as before, we set the recurrence relation

$$
\begin{aligned}
u_0(x, y, t) &= \sin(x + y), \\[1mm]
u_{k+1}(x, y, t) &= L_t^{-1} \left(L_x u_k + L_y u_k \right), k \geq 0.
\end{aligned}
\qquad (29)
$$

Using few terms of the decomposition gives

$$
\begin{aligned}
u_0(x, y, t) &= \sin(x + y), \\[1mm]
u_1(x, y, t) &= L_t^{-1} \left(L_x u_0 + L_y u_0 \right) = -2t \sin(x + y), \\[1mm]
u_2(x, y, t) &= L_t^{-1} \left(L_x u_1 + L_y u_1 \right) = \frac{(2t)^2}{2!} \sin(x + y), \\[1mm]
u_3(x, y, t) &= L_t^{-1} \left(L_x u_2 + L_y u_2 \right) = -\frac{(2t)^3}{3!} \sin(x + y),
\end{aligned}
\qquad (30)
$$

and so on. The solution in a series form is given by

$$u(x, y, t) = \sin(x + y) \left(1 - (2t) + \frac{(2t)^2}{2!} - \frac{(2t)^3}{3!} + \cdots \right), \tag{31}$$

and in a closed form by

$$u(x, y, t) = e^{-2t} \sin(x + y). \tag{32}$$

Inhomogeneous Heat Equations

It was defined before that inhomogeneous heat equation contains one or more terms that do not contain the dependent variable $u(x, y, t)$. A useful advantage of Adomian's method is that it handles homogeneous and inhomogeneous problems in a like manner. The decomposition method, that has been outlined before, will be applied to solve inhomogeneous heat flow equations given by the following illustrative examples.

Example 4. Use the Adomian decomposition method to solve the initial-boundary value problem

$$
\begin{array}{llll}
\text{PDE} & u_t & = & u_{xx} + u_{yy} + \sin y, \ 0 < x, y < \pi, \ t > 0 \\[2mm]
\text{B.C} & u(0, y, t) & = & u(\pi, y, t) = \sin y \\[2mm]
& u(x, 0, t) & = & u(x, \pi, t) = 0 \\[2mm]
\text{I.C} & u(x, y, 0) & = & \sin x \sin y + \sin y
\end{array}
\tag{33}
$$

Solution.

It is obvious that the given equation is an inhomogeneous equation. Unlike the method of separation of variables, the decomposition method handles any problem in a direct way without any need to transform the inhomogeneous equation to a related homogeneous equation.

Operating with L_t^{-1} to the operator form of (33) gives

$$u(x, y, t) = \sin x \sin y + \sin y + t \sin y + L_t^{-1} (L_x u + L_y u). \tag{34}$$

Using the decomposition series for $u(x, y, t)$ into (34) leads to

$$\sum_{n=0}^{\infty} u_n = \sin x \sin y + \sin y + t \sin y + L_t^{-1} \left(L_x \left(\sum_{n=0}^{\infty} u_n \right) + L_y \left(\sum_{n=0}^{\infty} u_n \right) \right). \tag{35}$$

It follows that the recursive relationship is given by

$$
\begin{array}{lll}
u_0(x, y, t) & = & \sin x \sin y + \sin y + t \sin y, \\[2mm]
u_{k+1}(x, y, t) & = & L_t^{-1} (L_x u_k + L_y u_k)), \ k \geq 0.
\end{array}
\tag{36}
$$

This gives

$$u_0(x, y, t) = \sin x \sin y + \sin y + t \sin y,$$

$$u_1(x, y, t) = L_t^{-1}(L_x u_0 + L_y u_0) = -2t \sin x \sin y - t \sin y - \frac{t^2}{2!} \sin y, \qquad (37)$$

$$u_2(x, y, t) = L_t^{-1}(L_x u_1 + L_y u_1) = \frac{(2t)^2}{2!} \sin x \sin y + \frac{t^2}{2!} \sin y + \frac{t^3}{3!} \sin y,$$

and so on. The solution in a series form is given by

$$u(x, y, t) = \sin y + \sin x \sin y \left(1 - (2t) + \frac{(2t)^2}{2!} - \frac{(2t)^3}{3!} + \cdots \right)$$

$$+ \left(t \sin y - t \sin y - \frac{t^2}{2!} \sin y + \frac{t^2}{2!} \sin y - \cdots \right), \qquad (38)$$

and in a closed form by

$$u(x, y, t) = \sin y + e^{-2t} \sin x \sin y, \qquad (39)$$

where other terms vanish in the limit.

Example 5. Use the Adomian decomposition method to solve the initial-boundary value problem

$$\begin{aligned}
\text{PDE} \qquad & u_t = u_{xx} + u_{yy} + 2 \cos x \cos y, \ 0 < x, y < \pi, \ t > 0 \\[4pt]
\text{B.C} \qquad & u(0, y, t) = -u(\pi, y, t) = (1 - e^{-2t}) \cos y \\[4pt]
& u(x, 0, t) = -u(x, \pi, t) = (1 - e^{-2t}) \cos x \\[4pt]
\text{I.C} \qquad & u(x, y, 0) = 0
\end{aligned} \qquad (40)$$

Solution.

The given partial differential equation and the boundary conditions are inhomogeneous. Our approach will be analogous to that employed in the previous examples.

Applying the inverse operator L_t^{-1} we obtain

$$u(x, y, t) = 2t \cos x \cos y + L_t^{-1}(L_x u + L_y u). \qquad (41)$$

Using the decomposition series for $u(x, y, t)$ gives

$$\sum_{n=0}^{\infty} u_n = 2t \cos x \cos y + L_t^{-1} \left(L_x \left(\sum_{n=0}^{\infty} u_n \right) + L_y \left(\sum_{n=0}^{\infty} u_n \right) \right). \qquad (42)$$

Proceeding as before, we set

$$
\begin{aligned}
u_0(x, y, t) &= 2t \cos x \cos y, \\
u_{k+1}(x, y, t) &= L_t^{-1}\left(L_x u_k + L_y u_k\right), k \geq 0.
\end{aligned}
\tag{43}
$$

Using few terms of the decomposition gives

$$
\begin{aligned}
u_0(x, y, t) &= 2t \cos x \cos y, \\
u_1(x, y, t) &= L_t^{-1}\left(L_x u_0 + L_y u_0\right) = -\frac{(2t)^2}{2!} \cos x \cos y, \\
u_2(x, y, t) &= L_t^{-1}\left(L_x u_1 + L_y u_1\right) = \frac{(2t)^3}{3!} \cos x \cos y,
\end{aligned}
\tag{44}
$$

and so on. The solution in a series form is given by

$$
u(x, y, t) = \cos x \cos y \left((2t) - \frac{(2t)^2}{2!} + \frac{(2t)^3}{3!} + \cdots \right),
\tag{45}
$$

and in a closed form by

$$
u(x, y, t) = (1 - e^{-2t}) \cos x \cos y.
\tag{46}
$$

Exercises 4.2.1

In Exercises 1 – 6, use the decomposition method to solve the homogeneous initial-boundary value problems:

1. $u_t = 2(u_{xx} + u_{yy})$, $0 < x, y < \pi$, $t > 0$
   $u(0, y, t) = u(\pi, y, t) = 0$
   $u(x, 0, t) = u(x, \pi, t) = 0$
   $u(x, y, 0) = \sin x \sin y$

2. $u_t = u_{xx} + u_{yy}$, $0 < x, y < \pi$, $t > 0$
   $u(0, y, t) = u(\pi, y, t) = 0$
   $u(x, 0, t) = u(x, \pi, t) = 0$
   $u(x, y, 0) = 2 \sin x \sin y$

3. $u_t = 2(u_{xx} + u_{yy})$, $0 < x, y < \pi$, $t > 0$
   $u(0, y, t) = -u(\pi, y, t) = e^{-4t} \cos y$
   $u(x, 0, t) = -u(x, \pi, t) = e^{-4t} \cos x$
   $u(x, y, 0) = \cos(x + y)$

4. $u_t = 3(u_{xx} + u_{yy})$, $0 < x, y < \pi$, $t > 0$
   $u(0, y, t) = -u(\pi, y, t) = -e^{-6t} \sin y$
   $u(x, 0, t) = -u(x, \pi, t) = e^{-6t} \sin x$

$u(x, y, 0) = \sin(x - y)$

5. $u_t = 2(u_{xx} + u_{yy}) - u, \ 0 < x, y < \pi, \ t > 0$
   $u(0, y, t) = u(\pi, y, t) = 0$
   $u(x, 0, t) = u(x, \pi, t) = 0$
   $u(x, y, 0) = \sin x \sin y$

6. $u_t = 3(u_{xx} + u_{yy}) - 2u, \ 0 < x, y < \pi, \ t > 0$
   $u(0, y, t) = -u(\pi, y, t) = e^{-8t} \sin y$
   $u(x, 0, t) = -u(x, \pi, t) = e^{-8t} \sin x$
   $u(x, y, 0) = \sin(x + y)$

In Exercises 7 – 12, use the decomposition method to solve the inhomogeneous initial-boundary value problems:

7. $u_t = 2(u_{xx} + u_{yy}) + 2 \sin x, \ 0 < x, y < \pi, \ t > 0$
   $u(0, y, t) = u(\pi, y, t) = 0$
   $u(x, 0, t) = u(x, \pi, t) = \sin x$
   $u(x, y, 0) = \sin x \sin y + \sin x$

8. $u_t = 3(u_{xx} + u_{yy}) + 3 \cos x, \ 0 < x, y < \pi, \ t > 0$
   $u(0, y, t) = -u(\pi, y, t) = 1$
   $u(x, 0, t) = u(x, \pi, t) = \cos x$
   $u(x, y, 0) = \sin x \sin y + \cos x$

9. $u_t = u_{xx} + u_{yy} + 2 \cos(x + y), \ 0 < x, y < \pi, \ t > 0$
   $u(0, y, t) = (e^{-2t} \sin y + \cos y)$
   $u(\pi, y, t) = -(e^{-2t} \sin y + \cos y)$
   $u(x, 0, t) = (e^{-2t} \sin x + \cos x)$
   $u(x, \pi, t) = -(e^{-2t} \sin x + \cos x)$
   $u(x, y, 0) = \sin(x + y) + \cos(x + y)$

10. $u_t = u_{xx} + u_{yy} = \sin x + \sin y, \ 0 < x, y < \pi, \ t > 0$
    $u(0, y, t) = u(\pi, y, t) = \sin y$
    $u(x, 0, t) = u(x, \pi, t) = \sin x$
    $u(x, y, 0) = \sin x(1 + \sin y) + \sin y$

11. $u_t = u_{xx} + u_{yy} - 2, \ 0 < x, y < \pi, \ t > 0$
    $u(0, y, t) = 0, \ u(\pi, y, t) = \pi^2$
    $u(x, 0, t) = u(x, \pi, t) = x^2$
    $u(x, y, 0) = x^2 + \sin x \sin y$

12. $u_t = u_{xx} + u_{yy} - 2, \ 0 < x, y < \pi, \ t > 0$
    $u(0, y, t) = u(\pi, y, t) = y^2$
    $u(x, 0, t) = 0, \ u(x, \pi, t) = \pi^2$
    $u(x, y, 0) = \sin x \sin y + y^2$

4.2.2 Three Dimensional Heat Flow

The distribution of heat flow in a three dimensional space is governed by the following initial boundary value problem

$$\text{PDE} \qquad u_t = \overline{k}(u_{xx} + u_{yy} + u_{zz}), \; t > 0$$

$$0 < x < a, 0 < y < b, 0 < z < c$$

$$\text{B.C} \quad u(0, y, z, t) = u(a, y, z, t) = 0$$

$$u(x, 0, z, t) = u(x, b, z, t) = 0 \tag{47}$$

$$u(x, y, 0, t) = u(x, y, c, t) = 0$$

$$\text{I.C} \quad u(x, y, z, 0) = f(x, y, z)$$

where $u \equiv u(x, y, z, t)$ is the temperature of any point located at the position (x, y, z) of a rectangular volume at any time t, and $\overline{k}$ is the thermal diffusivity.

We first rewrite (47) in an operator form by

$$L_t u = \overline{k}(L_x u + L_y u + L_z u), \tag{48}$$

where the differential operators L_x, L_y, and L_z are defined by

$$L_t = \frac{\partial}{\partial t}, L_x = \frac{\partial^2}{\partial x^2}, L_y = \frac{\partial^2}{\partial y^2}, L_z = \frac{\partial^2}{\partial z^2}, \tag{49}$$

so that the integral operator L_t^{-1} exists and given by

$$L_t^{-1}(.) = \int_0^t (.)dt. \tag{50}$$

Applying L_t^{-1} to both sides of (48) and using the initial condition lead to

$$u(x, y, t) = f(x, y, z) + \overline{k} \, L_t^{-1} \left(L_x u + L_y u + L_z u \right). \tag{51}$$

The decomposition method defines the solution $u(x, y, z, t)$ as a series given by

$$u(x, y, z, t) = \sum_{n=0}^{\infty} u_n(x, y, z, t). \tag{52}$$

Substituting (52) into both sides of (51) yields

$$\sum_{n=0}^{\infty} u_n = f(x, y, z)$$

$$+ \overline{k} \, L_t^{-1} \left(L_x \left(\sum_{n=0}^{\infty} u_n \right) + L_y \left(\sum_{n=0}^{\infty} u_n \right) + L_z \left(\sum_{n=0}^{\infty} u_n \right) \right). \tag{53}$$

The components $u_n(x, y, z, t)$, $n \geq 0$ can be completely determined by using the recursive relationship

$$
\begin{aligned}
u_0(x, y, z, t) &= f(x, y, z), \\
u_{k+1}(x, y, z, t) &= \overline{k} \, L_t^{-1} \left(L_x u_k + L_y u_k + L_z u_k \right), \; k \geq 0.
\end{aligned}
\tag{54}
$$

The components can be determined recursively as far as we like. Consequently, the components $u_n, n \geq 0$, are completely determined and the solution in a series form follows immediately.

Homogeneous Heat Equations

The decomposition method will be used to discuss the following homogeneous heat equations.

Example 6. Solve the following initial boundary value problem

$$
\begin{array}{lrl}
\text{PDE} & u_t &= u_{xx} + u_{yy} + u_{zz}, \; 0 < x, y, z < \pi, \, t > 0 \\[2mm]
\text{B.C} & u(0, y, z, t) &= u(\pi, y, z, t) = 0 \\[2mm]
& u(x, 0, z, t) &= u(x, \pi, z, t) = 0 \\[2mm]
& u(x, y, 0, t) &= u(x, y, \pi, t) = 0 \\[2mm]
\text{I.C} & u(x, y, z, 0) &= 2 \sin x \sin y \sin z
\end{array}
\tag{55}
$$

Solution.

Applying the inverse operator L_t^{-1} to the operator form of (55) gives

$$
u(x, y, z, t) = 2 \sin x \sin y \sin z + L_t^{-1} \left(L_x u + L_y u + L_z u \right).
\tag{56}
$$

Using the decomposition series

$$
u(x, y, z, t) = \sum_{n=0}^{\infty} u_n(x, y, z, t),
\tag{57}
$$

into (56) yields

$$
\sum_{n=0}^{\infty} u_n = 2 \sin x \sin y \sin z
$$

$$
+ L_t^{-1} \left(L_x \left(\sum_{n=0}^{\infty} u_n \right) + L_y \left(\sum_{n=0}^{\infty} u_n \right) + L_z \left(\sum_{n=0}^{\infty} u_n \right) \right).
\tag{58}
$$

The components $u_n(x, y, z, t)$, $n \geq 0$ can be determined by using the recurrence relation

$$u_0(x, y, z, t) = 2\sin x \sin y \sin z,$$

$$u_{k+1}(x, y, t) = L_t^{-1}(L_x u_k + L_y u_k + L_z u_k), k \geq 0. \tag{59}$$

It follows that the first few terms of the decomposition series of $u(x, y, z, t)$ are given by

$$u_0(x, y, z, t) = 2\sin x \sin y \sin z,$$

$$u_1(x, y, z, t) = L_t^{-1}(L_x u_0 + L_y u_0 + L_z u_0) = -2(3t)\sin x \sin y \sin z,$$

$$u_2(x, y, z, t) = L_t^{-1}(L_x u_1 + L_y u_1 + L_z u_1) = \frac{2(3t)^2}{2!}\sin x \sin y \sin z, \tag{60}$$

$$u_3(x, y, z, t) = L_t^{-1}(L_x u_2 + L_y u_2 + L_z u_2) = -\frac{2(3t)^3}{3!}\sin x \sin y \sin z,$$

and so on. As indicated before, further components can be easily computed to increase the level of accuracy.

Combining (57) and (60), the solution in a series form is given by

$$u(x, y, z, t) = 2\sin x \sin y \sin z \left(1 - (3t) + \frac{(3t)^2}{2!} - \frac{(3t)^3}{3!} + \cdots\right), \tag{61}$$

and in a closed form by

$$u(x, y, z, t) = 2e^{-3t}\sin x \sin y \sin z. \tag{62}$$

Example 7. Solve the following initial boundary value problem with lateral heat loss

$$\text{PDE} \qquad u_t = u_{xx} + u_{yy} + u_{zz} - 2u, \ 0 < x, y, z < \pi, \ t > 0$$

$$\text{B.C} \quad u(0, y, z, t) = u(\pi, y, z, t) = 0$$

$$u(x, 0, z, t) = u(x, \pi, z, t) = 0 \tag{63}$$

$$u(x, y, 0, t) = u(x, y, \pi, t) = 0$$

$$\text{I.C} \quad u(x, y, z, 0) = \sin x \sin y \sin z$$

Solution.

Operating with L_t^{-1} to (63) we obtain

$$u(x, y, z, t) = \sin x \sin y \sin z + L_t^{-1}(L_x u + L_y u + L_z u - 2u). \tag{64}$$

Proceeding as before we find

$$\sum_{n=0}^{\infty} u_n = \sin x \sin y \sin z$$

$$+L_t^{-1}\left(L_x\left(\sum_{n=0}^{\infty}u_n\right)+L_y\left(\sum_{n=0}^{\infty}u_n\right)+L_z\left(\sum_{n=0}^{\infty}u_n\right)-2\left(\sum_{n=0}^{\infty}u_n\right)\right). \tag{65}$$

Using the assumptions of the decomposition method yields

$$
\begin{aligned}
u_0(x,y,z,t) &= \sin x \sin y \sin z, \\
u_{k+1}(x,y,z,t) &= L_t^{-1}\left(L_x u_k + L_y u_k + L_z u_k - 2u_k\right),\ k \geq 0.
\end{aligned}
\tag{66}
$$

Consequently, we obtain

$$
\begin{aligned}
u_0 &= \sin x \sin y \sin z, \\
u_1 &= L_t^{-1}\left(L_x u_0 + L_y u_0 + L_z u_0 - 2u_0\right) = -5t\sin x \sin y \sin z, \\
u_2 &= L_t^{-1}\left(L_x u_1 + L_y u_1 + L_z u_1 - 2u_1\right) = \frac{(5t)^2}{2!}\sin x \cos y \cos z, \\
u_3 &= L_t^{-1}\left(L_x u_2 + L_y u_2 + L_z u_2 - 2u_2\right) = -\frac{(5t)^3}{3!}\sin x \sin y \sin z.
\end{aligned}
\tag{67}
$$

The solution in a series form is given by

$$u(x,y,z,t) = \sin x \sin y \sin z\left(1 - (5t) + \frac{(5t)^2}{2!} - \frac{(5t)^3}{3!} + \cdots\right), \tag{68}$$

and in a closed form by

$$u(x,y,z,t) = e^{-5t}\sin x \sin y \sin z, \tag{69}$$

obtained upon using the Taylor series for the function e^{-5t}. **Example 8.** Solve the following initial boundary value problem

$$
\begin{aligned}
\text{PDE} \qquad u_t &= 2(u_{xx} + u_{yy} + u_{zz}),\ 0 < x,y,z < \pi,\ t > 0 \\
\text{B.C} \quad u(0,y,z,t) &= u(\pi,y,z,t) = 0 \\
u(x,0,z,t) &= -u(x,\pi,z,t) = e^{-6t}\sin x \cos z \\
u(x,y,0,t) &= -u(x,y,\pi,t) = e^{-6t}\sin x \cos y \\
\text{I.C} \quad u(x,y,z,0) &= \sin x \cos y \cos z
\end{aligned}
\tag{70}
$$

Solution.

We first note that the boundary conditions are inhomogeneous. Following the previous discussion we obtain

$$u(x, y, z, t) = \sin x \cos y \cos z + 2L_t^{-1}\left(L_x u + L_y u + L_z u\right), \tag{71}$$

and hence we find

$$\sum_{n=0}^{\infty} u_n = \sin x \cos y \cos z$$

$$+ 2L_t^{-1}\left(L_x\left(\sum_{n=0}^{\infty} u_n\right) + L_y\left(\sum_{n=0}^{\infty} u_n\right) + L_z\left(\sum_{n=0}^{\infty} u_n\right)\right). \tag{72}$$

With u_0 defined as shown above, we set the relation

$$\begin{aligned}
u_0(x, y, z, t) &= \sin x \cos y \cos z, \\
u_{k+1}(x, y, z, t) &= 2L_t^{-1}\left(L_x u_k + L_y u_k + L_z u_k\right), \ k \geq 0.
\end{aligned} \tag{73}$$

Consequently, the first few components

$$\begin{aligned}
u_0(x, y, z, t) &= \sin x \cos y \cos z, \\
u_1(x, y, z, t) &= 2L_t^{-1}\left(L_x u_0 + L_y u_0 + L_z u_0\right) = -6t \sin x \cos y \cos z, \\
u_2(x, y, z, t) &= 2L_t^{-1}\left(L_x u_1 + L_y u_1 + L_z u_1\right) = \frac{(6t)^2}{2!} \sin x \cos y \cos z, \\
u_3(x, y, z, t) &= 2L_t^{-1}\left(L_x u_2 + L_y u_2 + L_z u_2\right) = -\frac{(6t)^3}{3!} \sin x \cos y \cos z,
\end{aligned} \tag{74}$$

are obtained. The solution in a series form

$$u(x, y, z, t) = \sin x \cos y \cos z \left(1 - (6t) + \frac{(6t)^2}{2!} - \frac{(6t)^3}{3!} + \cdots\right), \tag{75}$$

is readily obtained, and hence the exact solution

$$u(x, y, z, t) = e^{-6t} \sin x \cos y \cos z, \tag{76}$$

follows immediately.

Inhomogeneous Heat Equations

In the following, the Adomian decomposition method will be applied to inhomogeneous heat equations. The method will be implemented in a like manner to that used in homogeneous cases.

Example 9. Solve the following initial boundary value problem

$$\text{PDE} \qquad u_t \;=\; (u_{xx} + u_{yy} + u_{zz}) - 2,\; 0 < x, y, z < \pi,\; t > 0$$

$$\begin{aligned}
\text{B.C} \quad u(0, y, z, t) &= u(\pi, y, z, t) = z^2 \\
u(x, 0, z, t) &= u(x, \pi, z, t) = z^2 \\
u(x, y, 0, t) &= 0,\; u(x, y, \pi, t) = \pi^2 \\
\text{I.C} \quad u(x, y, z, 0) &= z^2 + \sin x \sin y \sin z
\end{aligned} \qquad (77)$$

Solution.

We first note that the PDE and the boundary conditions are inhomogeneous. Applying the inverse operator L_t^{-1} to (77) and using the initial condition we obtain

$$u(x, y, z, t) = -2t + z^2 + \sin x \sin y \sin z + L_t^{-1}\left(L_x u + L_y u + L_z u\right), \qquad (78)$$

and proceeding as before we find

$$\sum_{n=0}^{\infty} u_n = -2t + z^2 + \sin x \sin y \sin z$$

$$+ L_t^{-1}\left(L_x\left(\sum_{n=0}^{\infty} u_n\right) + L_y\left(\sum_{n=0}^{\infty} u_n\right) + L_z\left(\sum_{n=0}^{\infty} u_n\right)\right). \qquad (79)$$

We next set the recurrence relation

$$\begin{aligned}
u_0(x, y, z, t) &= -2t + z^2 + \sin x \sin y \sin z, \\
u_{k+1}(x, y, z, t) &= L_t^{-1}\left(L_x u_k + L_y u_k + L_z u_k\right),\; k \geq 0.
\end{aligned} \qquad (80)$$

The first few terms of the decomposition series are

$$\begin{aligned}
u_0(x, y, z, t) &= -2t + z^2 + \sin x \sin y \sin z, \\
u_1(x, y, z, t) &= -3t \sin x \sin y \sin z + 2t, \\
u_2(x, y, z, t) &= \frac{(3t)^2}{2!} \sin x \cos y \cos z, \\
u_3(x, y, z, t) &= -\frac{(3t)^3}{3!} \sin x \sin y \sin z.
\end{aligned} \qquad (81)$$

The solution in a series form is given by

$$u(x, y, z, t) = z^2 + \sin x \sin y \sin z \left(1 - (3t) + \frac{(3t)^2}{2!} - \frac{(3t)^3}{3!} + \cdots\right), \qquad (82)$$

and in a closed form by

$$u(x, y, z, t) = z^2 + e^{-3t} \sin x \sin y \sin z. \tag{83}$$

Example 10. Solve the following initial boundary value problem

$$
\begin{array}{lrl}
\text{PDE} & u_t & = (u_{xx} + u_{yy} + u_{zz}) + \sin z, \ 0 < x, y, z < \pi, \ t > 0 \\[4pt]
\text{B.C} \quad u(0, y, z, t) & = & \sin z + e^{-2t} \sin y \\[4pt]
u(\pi, y, z, t) & = & \sin z - e^{-2t} \sin y \\[4pt]
u(x, 0, z, t) & = & \sin z + e^{-2t} \sin x \\[4pt]
u(x, pi, z, t) & = & \sin z - e^{-2t} \sin x \\[4pt]
u(x, y, 0, t) & = & u(x, y, \pi, t) = e^{-2t} \sin(x + y) \\[4pt]
\text{I.C} \quad u(x, y, z, 0) & = & \sin(x + y) + \sin z
\end{array}
\tag{84}
$$

Solution.

It is clear that the PDE and the boundary conditions are inhomogeneous. Applying the inverse operator L_t^{-1} to (84) gives

$$u(x, y, z, t) = \sin(x + y) + \sin z + t \sin z + L_t^{-1} (L_x u + L_y u + L_z u), \tag{85}$$

and this in turn gives

$$\sum_{n=0}^{\infty} u_n = \sin(x + y) + \sin z + t \sin z$$

$$+ L_t^{-1} \left(L_x \left(\sum_{n=0}^{\infty} u_n \right) + L_y \left(\sum_{n=0}^{\infty} u_n \right) + L_z \left(\sum_{n=0}^{\infty} u_n \right) \right). \tag{86}$$

Accordingly, we set the recursive relationship

$$
\begin{array}{rl}
u_0(x, y, z, t) & = \sin(x + y) + \sin z + t \sin z, \\[6pt]
u_{k+1}(x, y, z, t) & = L_t^{-1} (L_x u_k + L_y u_k + L_z u_k), \ k \geq 0.
\end{array}
\tag{87}
$$

The first few terms of the decomposition are

$$
\begin{array}{rl}
u_0(x, y, z, t) & = \sin(x + y) + \sin z + t \sin z, \\[6pt]
u_1(x, y, z, t) & = -2t \sin(x + y) - t \sin z - \frac{t^2}{2!} \sin z, \\[6pt]
u_2(x, y, z, t) & = \frac{(2t)^2}{2!} \sin(x + y) + \frac{t^2}{2!} \sin z + \frac{t^3}{3!} \sin z.
\end{array}
\tag{88}
$$

The solution in a series form is given by

$$u(x, y, z, t) = \sin z + \sin(x + y) \left(1 - (2t) + \frac{(2t)^2}{2!} - \frac{(2t)^3}{3!} + \cdots \right)$$
$$+ \left(t \sin z - t \sin z - \frac{t^2}{2!} \sin z + \frac{t^2}{2!} \sin z + \cdots \right),$$

(89)

and in a closed form by

$$u(x, y, z, t) = \sin z + e^{-2t} \sin(x + y).$$

(90)

Exercises 4.2.2

In Exercises 1 – 4, use the decomposition method to solve the homogeneous initial-boundary value problems:

1. $u_t = 2(u_{xx} + u_{yy} + u_{zz})$, $0 < x, y, z < \pi$, $t > 0$
   $u(0, y, z, t) = u(\pi, y, z, t) = 0$
   $u(x, 0, z, t) = u(x, \pi, z, t) = 0$
   $u(x, y, 0, t) = u(x, y, \pi, t) = 0$
   $u(x, y, z, 0) = \sin x \sin y \sin z$

2. $u_t = u_{xx} + u_{yy} + u_{zz}$, $0 < x, y, z < \pi$, $t > 0$
   $u(0, y, z, t) = u(\pi, y, z, t) = 0$
   $u(x, 0, z, t) = u(x, \pi, z, t) = 0$
   $u(x, y, 0, t) = u(x, y, \pi, t) = 0$
   $u(x, y, z, 0) = 2 \sin x \sin y \sin z$

3. $u_t = u_{xx} + u_{yy} + u_{zz}$, $0 < x, y, z < \pi$, $t > 0$
   $u(0, y, z, t) = -u(\pi, y, z, t) = e^{-3t} \sin(y + z)$
   $u(x, 0, z, t) = -u(x, \pi, z, t) = e^{-3t} \sin(x + z)$
   $u(x, y, 0, t) = -u(x, y, \pi, t) = e^{-3t} \sin(x + y)$
   $u(x, y, z, 0) = \sin(x + y + z)$

4. $u_t = u_{xx} + u_{yy} + u_{zz} - u$, $0 < x, y, z < \pi$, $t > 0$
   $u(0, y, z, t) = u(\pi, y, z, t) = 0$
   $u(x, 0, z, t) = u(x, \pi, z, t) = 0$
   $u(x, y, 0, t) = u(x, y, \pi, t) = 0$
   $u(x, y, z, 0) = \sin x \sin y \sin z$

In Exercises 5 – 8, solve the inhomogeneous initial-boundary value problems:

5. $u_t = u_{xx} + u_{yy} + u_{zz} - 4$, $0 < x, y, z < \pi$, $t > 0$
   $u(0, y, z, t) = 0$, $u(\pi, y, z, t) = 2\pi^2$
   $u(x, 0, z, t) = u(x, \pi, z, t) = 2x^2$
   $u(x, y, 0, t) = u(x, y, \pi, t) = 2x^2$
   $u(x, y, z, 0) = 2x^2 + \sin x \sin y \sin z$

6. $u_t = u_{xx} + u_{yy} + u_{zz} - 2,\ 0 < x, y, z < \pi,\ t > 0$
   $u(0, y, z, t) = u(\pi, y, z, t) = y^2$
   $u(x, 0, z, t) = 0,\ u(x, \pi, z, t) = \pi^2$
   $u(x, y, 0, t) = u(x, y, \pi, t) = y^2$
   $u(x, y, z, 0) = y^2 + \sin x \sin y \sin z$

7. $u_t = u_{xx} + u_{yy} + u_{zz} + \sin x,\ 0 < x, y, z < \pi,\ t > 0$
   $u(0, y, z, t) = u(\pi, y, z, t) = e^{-2t} \sin(y + z)$
   $u(x, 0, z, t) = \sin x + e^{-2t} \sin z,\ u(x, \pi, z, t) = \sin x - e^{-2t} \sin z$
   $u(x, y, 0, t) = \sin x + e^{-2t} \sin y,\ u(x, y, \pi, t) = \sin x - e^{-2t} \sin y$
   $u(x, y, z, 0) = \sin x + \sin(y + z)$

8. $u_t = u_{xx} + u_{yy} + u_{zz} - 2,\ 0 < x, y, z < \pi,\ t > 0$
   $u(0, y, z, t) = e^{-3t}(\sin y + \sin z)$
   $u(\pi, y, z, t) = \pi^2 + e^{-3t}(\sin y + \sin z)$
   $u(x, 0, z, t) = u(x, \pi, z, t) = x^2 + e^{-3t}(\sin x + \sin z)$
   $u(x, y, 0, t) = u(x, y, \pi, t) = x^2 + e^{-3t}(\sin x + \sin y)$
   $u(x, y, z, 0) = x^2 + (\sin x + \sin y + \sin z)$

4.3 Method of Separation of Variables

In this section, the heat flow in a two dimensional space and a three dimensional space will be discussed by using the classical method of the separation of variables. As discussed in Chapter 3, this method replaces the partial differential equation by a system of ordinary differential equations that are usually easy to handle. The resulting ODEs are then solved independently. We then proceed as discussed in Chapter 3 and apply the boundary and the initial equations to determine the constants of integration. Unlike the Adomian decomposition method, it is well known that the method of separation of variables is commonly used for the case where the PDE and the boundary conditions are linear and homogeneous. For inhomogeneous equations, transformation formulas are used to convert the inhomogeneous equations to homogeneous equations.

4.3.1 Two Dimensional Heat Flow

The distribution of heat flow in a two dimensional space is governed by the following initial boundary value problem

$$
\begin{aligned}
\text{PDE} \quad u_t &= \bar{k}(u_{xx} + u_{yy}),\ 0 < x < a,\ 0 < y < b, t > 0 \\
\text{B.C}\ u(0, y, t) &= u(a, y, t) = 0 \\
u(x, 0, t) &= u(x, b, t) = 0 \\
\text{I.C}\ u(x, y, 0) &= f(x, y)
\end{aligned}
\tag{91}
$$

where $u \equiv u(x,y,t)$ defines the temperature of any point at the position (x,y) of a rectangular plate at any time t, and $\bar{k}$ is the thermal diffusivity.

The method of separation of variables is based on an assumption that the solution $u(x,y,t)$ can be expressed as the product of distinct functions $F(x), G(y)$, and $T(t)$, such that each function depends on one variable only. Based on this assumption, we first set

$$u(x,y,t) = F(x)G(y)T(t). \tag{92}$$

Differentiating both sides of (92) with respect to t and twice with respect to x and y respectively, we obtain

$$
\begin{aligned}
u_t &= F(x)G(y)T'(t), \\
u_{xx} &= F''(x)G(y)T(t), \\
u_{yy} &= F(x)G''(y)T(t).
\end{aligned}
\tag{93}
$$

Substituting (93) into (91) leads to

$$F(x)G(y)T'(t) = \bar{k}\left(F''(x)G(y)T(t) + F(x)G''(y)T(t)\right). \tag{94}$$

Dividing both sides of (94) by $\bar{k}F(x)G(y)T(t)$ yields

$$\frac{T'(t)}{\bar{k}T(t)} = \frac{F''(x)}{F(x)} + \frac{G''(y)}{G(y)}. \tag{95}$$

It is obvious that the left hand side depends only on t and the right hand side depends only on x and y. This means that the equality holds only if both sides are equal to the same constant. Assuming that the right hand side is a constant, it is valid to assume that it is the sum of two constants. Therefore, we set

$$\frac{F''(x)}{F(x)} = -\lambda^2. \tag{96}$$

and

$$\frac{G''(y)}{G(y)} = -\mu^2. \tag{97}$$

Consequently, we find

$$F''(x) + \lambda^2 F(x) = 0. \tag{98}$$

and

$$G''(y) + \mu^2 G(y) = 0. \tag{99}$$

This means that the left hand side of (95) is equal to $(\lambda^2 - \mu^2)$. Accordingly, we obtain

$$\frac{T'(t)}{\bar{k}T(t)} = -(\lambda^2 + \mu^2). \tag{100}$$

or equivalently

$$T'(t) + \overline{k}(\lambda^2 + \mu^2)T(t) = 0. \tag{101}$$

The selection of $-(\lambda^2 + \mu^2)$ is the only selection that will provide nontrivial solutions. Besides, this selection is made in accordance with the natural fact that the factor $T(t)$, and hence the temperature $u(x, y, t)$, must vanish as $t \to \infty$. From physics, we know that the temperature component $T(t)$ follows the exponential decay phenomena.

It is interesting to note that the partial differential equation (91) has been transformed to three ordinary differential equations, two second order ODEs given by (98) and (99), and a first order ODE given by (101).

The solution of (101) is given by

$$T(t) = Ce^{-\overline{k}(\lambda^2 + \mu^2)t}, \tag{102}$$

where C is a constant. The result (102) explains the fact that $T(t)$ must follow the exponential decay of heat flow. If accidently we selected the constant to equal $(\lambda^2 + \mu^2)$, this will result in an exponential growth of the Temperature factor $T(t)$. In this case, $T(t) \to \infty$ and consequently $u(x, y, t) \to \infty$ as $t \to \infty$. This contradicts the natural behavior of the heat flow.

The second order differential equations (98) and (99) give the solutions

$$F(x) = A\cos(\lambda x) + B\sin(\lambda x), \tag{103}$$

and

$$G(y) = \alpha\cos(\mu y) + \beta\sin(\mu y), \tag{104}$$

where A, B, α, and β are constants that will be determined.

To determine the constants A and B, we use the boundary conditions at $x = 0$ and at $x = a$ to find that

$$\begin{aligned} F(0)G(y)T(t) &= 0, \\ F(a)G(y)T(t) &= 0, \end{aligned} \tag{105}$$

which gives

$$\begin{aligned} F(0) &= 0, \\ F(a) &= 0. \end{aligned} \tag{106}$$

Substituting (106) into (103) gives

$$A = 0, \tag{107}$$

and

$$\lambda_n = \frac{n\pi}{a}, \quad n = 1, 2, 3, \cdots. \tag{108}$$

It is important to note here that we exclude $n = 0$ and $B = 0$ because each will lead to the trivial solution $u(x, y, t) = 0$. Using the results obtained for the constants A and λ_n, we therefore write the functions

$$F_n(x) = B_n \sin(\frac{n\pi}{a}x), \; n = 1, 2, 3, \cdots. \tag{109}$$

In a parallel manner, we use the second boundary condition at $y = 0$ and at $y = b$ into (104) to find that

$$\alpha = 0, \tag{110}$$

and

$$\lambda_m = \frac{m\pi}{b}, m = 1, 2, 3, \cdots. \tag{111}$$

We exclude $m = 0$ and $\beta = 0$, because each will lead to the trivial solution as indicated before. Consequently, $G(y)$ can be the functions

$$G_m(y) = \beta_m \frac{m\pi}{b}, \; m = 1, 2, 3, \cdots. \tag{112}$$

Based on the infinite number of values for λ_n and μ_m, then $T(t)$ of (102) takes the functions

$$T_{nm} = \overline{C}_{nm} e^{-\overline{k}(\frac{n^2}{a^2} + \frac{m^2}{b^2})\pi^2 t} \tag{113}$$

Ignoring the constants B_n, β_m, and $\overline{C}_{nm}$, we conclude that the functions, that form the set of fundamental solutions,

$$
\begin{aligned}
u_{nm} &= F_n(x)G_m(y)T_{nm}(t) \\
&= \sin(\frac{n\pi}{a}x) \sin(\frac{m\pi}{b}y)e^{-\overline{k}(\frac{n^2}{a^2} + \frac{m^2}{b^2})\pi^2 t}, \; n, m = 1, 2, \cdots,
\end{aligned} \tag{114}
$$

satisfy (91) and the boundary conditions.

Using the superposition principle, we obtain

$$u(x, y, t) = \sum_{n=1}^{\infty} \sum_{m=1}^{\infty} C_{nm} e^{-\overline{k}(\frac{n^2}{a^2} + \frac{m^2}{b^2})\pi^2 t} \sin(\frac{n\pi}{a}x) \sin(\frac{m\pi}{b}y), \tag{115}$$

where the arbitrary constants C_{nm} are as yet undetermined.

To determine the constants C_{nm}, we use the given initial condition to find

$$\sum_{n=1}^{\infty} \sum_{m=1}^{\infty} C_{nm} \sin(\frac{n\pi}{a}x) \sin(\frac{m\pi}{b}y) = f(x, y). \tag{116}$$

The constants C_{nm} are completely determined by using a double Fourier coefficients where we find

$$C_{nm} = \frac{4}{ab} \int_0^b \int_0^a f(x, y) \sin(\frac{n\pi}{a}x) \sin(\frac{m\pi}{b}y) dx \, dy. \tag{117}$$

Having determined the constants C_{nm}, the solution given by (115) is completely determined.

It is interesting to point out that the constants C_{nm} can also be determined by equating the coefficients of both sides if the initial condition is defined explicitly in terms of $\sin(\gamma x) \sin(\delta y)$, where γ and δ are constants. This will reduce the massive size of calculations required by the computational work of the double Fourier series.

The method will be illustrated by discussing the following examples.

Example 1. Use the method of separation of variables to solve the initial-boundary value problem:

$$
\begin{aligned}
\text{PDE} \qquad u_t &= u_{xx} + u_{yy},\ 0 < x, y < \pi,\ t > 0 \\
\text{B.C} \quad u(0, y, t) &= u(\pi, y, t) = 0 \\
u(x, 0, t) &= u(x, \pi, t) = 0 \\
\text{I.C} \quad u(x, y, 0) &= 2 \sin x \sin y
\end{aligned}
\tag{118}
$$

Solution.

As discussed before, we first set

$$
u(x, y, t) = F(x)G(y)T(t).
\tag{119}
$$

Proceeding as before, we obtain

$$
F''(x) + \lambda^2 F(x) = 0,
\tag{120}
$$

$$
G''(y) + \mu^2 G(y) = 0.
\tag{121}
$$

and

$$
T'(t) + (\lambda^2 + \mu^2)T(t) = 0,
\tag{122}
$$

where λ and μ are constants.

The second order ordinary differential equations (120) and (121) give the solutions

$$
F(x) = A\cos(\lambda x) + B\sin(\lambda x),
\tag{123}
$$

and

$$
G(y) = \alpha\cos(\mu y) + \beta\sin(\mu y),
\tag{124}
$$

respectively, where A, B, α and β are constants. Using the boundary conditions of (118) into (123) gives

$$
A = 0,
\tag{125}
$$

and

$$
\lambda_n = n,\ n = 1, 2, 3, \cdots,
\tag{126}
$$

so that

$$F_n(x) = B_n \sin(nx), \; n = 1, 2, 3, \cdots. \tag{127}$$

Similarly, using the boundary conditions of (118) into (124) gives

$$\alpha = 0, \tag{128}$$

and

$$\mu_m = m, \; m = 1, 2, 3, \cdots. \tag{129}$$

so that

$$G_m(y) = \beta_m \sin(my), \; m = 1, 2, 3, \cdots. \tag{130}$$

The solution of the first order differential equation (122) is given by

$$T_{nm}(t) = \overline{C}_{nm} e^{-(\lambda^2 + \mu^2)t}, \tag{131}$$

and by substituting λ and μ we obtain

$$T_{nm}(t) = \overline{C}_{nm} e^{-(n^2 + m^2)t}. \tag{132}$$

Combining (127), (130) and (132) and using the superposition principle, the general solution of the problem is given by the double series

$$u(x, y, t) = \sum_{n=1}^{\infty} \sum_{m=1}^{\infty} C_{nm} e^{-(n^2 + m^2)t} \sin(nx) \sin(my). \tag{133}$$

To determine the constants C_{nm}, we use the given initial condition and expand the double series to find

$$C_{11} \sin(x) \sin(y) + C_{12} \sin(x) \sin(2y) + \cdots = 2 \sin(x) \sin(y). \tag{134}$$

Equating the coefficients on both sides yields

$$C_{11} = 2, \; C_{ij} = 0, i \neq 1, j \neq 1, n = m = 1. \tag{135}$$

Accordingly, the particular solution is given by

$$u(x, y, t) = 2e^{-2t} \sin x \sin y. \tag{136}$$

Example 2. Use the method of separation of variables to solve the initial-boundary value problem:

$$
\begin{array}{rll}
\text{PDE} & u_t = & u_{xx} + u_{yy}, \; 0 < x, y < \pi, \, t > 0 \\[2mm]
\text{B.C} \quad u(0, y, t) = & u(\pi, y, t) = 0 \\[2mm]
& u(x, 0, t) = & u(x, \pi, t) = 0 \\[2mm]
\text{I.C} \quad u(x, y, 0) = & \sin x \sin y + 2 \sin x \sin(2y)
\end{array}
\tag{137}
$$

Solution.

As discussed before, we set

$$u(x, y, t) = F(x)G(y)T(t). \tag{138}$$

Proceeding as before, we obtain

$$\frac{T'(t)}{T(t)} = \frac{F''(x)}{F(x)} + \frac{G''(y)}{G(y)}. \tag{139}$$

Proceeding as before, we obtain

$$F''(x) + \lambda^2 F(x) = 0, \tag{140}$$

$$G''(y) + \mu^2 G(y) = 0, \tag{141}$$

and

$$T'(t) + (\lambda^2 + \mu^2)T(t) = 0. \tag{142}$$

Solving (140) and (141) and using the boundary conditions leads to

$$F_n(x) = B_n \sin(nx), \ \lambda_n = n, \ n = 1, 2, 3, \cdots, \tag{143}$$

and

$$G_m(y) = \beta_m \sin(my), \ \mu_m = m, \ m = 1, 2, 3, \cdots, \tag{144}$$

respectively. The solution of the first order differential equation (142) is given by

$$T(t) = Ce^{-(\lambda^2+\mu^2)t}, \tag{145}$$

and by substituting λ and μ we obtain

$$T_{nm}(t) = \overline{C}_{nm} e^{-(n^2+m^2)t}. \tag{146}$$

Combining (143), (144), and (146) and using the superposition principle, the general solution of the problem is given by the double series

$$u(x, y, t) = \sum_{n=1}^{\infty} \sum_{m=1}^{\infty} C_{nm} e^{-(n^2+m^2)t} \sin(nx) \sin(my). \tag{147}$$

To determine the constants C_{nm}, we use the given initial condition and expand the double series to obtain

$$C_{11} \sin x \sin y + C_{21} \sin x \sin(2y) + \cdots = \sin x \sin y + 2 \sin x \sin(2y). \tag{148}$$

Equating the coefficients on both sides yields

$$C_{11} = 1, \ \text{for } n = 1, m = 1, \tag{149}$$

and
$$C_{12} = 2, \text{ for } n = 1, m = 2, \tag{150}$$

and other coefficients are zeros. Accordingly, the particular solution is given by

$$u(x, y, t) = e^{-2t} \sin x \sin y + 2e^{-5t} \sin x \sin(2y), \tag{151}$$

obtained by substituting (149) and (150) into (147).

Example 3. Use the method of separation of variables to solve the initial-boundary value problem with mixed boundary conditions

$$
\begin{array}{rlrl}
\text{PDE} & u_t & = & 3(u_{xx} + u_{yy}), \, 0 < x, y < \pi, \, t > 0 \\
\text{B.C} & u_x(0, y, t) & = & u_x(\pi, y, t) = 0 \\
& u(x, 0, t) & = & u(x, \pi, t) = 0 \\
\text{I.C} & u(x, y, 0) & = & \sin y + \cos x \sin y
\end{array}
\tag{152}
$$

Solution.

Proceeding as before we obtain

$$F''(x) + \lambda^2 F(x) = 0, \tag{153}$$

$$G''(y) + \mu^2 G(y) = 0, \tag{154}$$

and

$$T'(t) + 3(\lambda^2 + \mu^2)T(t) = 0. \tag{155}$$

Solving (153) we find
$$F(x) = A\cos(\lambda x) + B\sin(\lambda x). \tag{156}$$

It is important to note that the boundary conditions

$$u_x(0, y, t) = u_x(\pi, y, t) = 0, \tag{157}$$

implies that
$$F'(0) = 0, F'(\pi) = 0. \tag{158}$$

Using (158) into (156) gives

$$B = 0, \, \lambda = n, \, n = 0, 1, 2, \cdots, \tag{159}$$

so that $\lambda = 0$ is included because it does not provide the trivial solution. Consequently, we find
$$F_n(x) = A_n \cos(nx), \, n = 0, 1, 2, \cdots. \tag{160}$$

Solving (154) and using the proper boundary conditions we obtain

$$G_m(y) = \beta_m \sin(my), \ m = 1, 2, 3, \cdots. \tag{161}$$

The solution of the first order differential equation (155) is given by

$$T(t) = Ce^{-3(\lambda^2 + \mu^2)t}, \tag{162}$$

and by substituting λ and μ we obtain

$$T_{nm}(t) = \overline{C}_{nm} e^{-3(n^2 + m^2)t}. \tag{163}$$

Combining (160), (161), and (163) and using the superposition principle, the general solution of the problem is given by the double series

$$u(x, y, t) = \sum_{n=0}^{\infty} \sum_{m=1}^{\infty} C_{nm} e^{-3(n^2 + m^2)t} \cos(nx) \sin(my). \tag{164}$$

We then expand the double series (164) and use the given initial condition to find

$$C_{01} \sin y + C_{11} \cos x \sin y + \cdots = \sin y + \cos x \sin y. \tag{165}$$

Equating the coefficients on both sides yields

$$C_{01} = 1, \ \text{for } n = 0, m = 1, \tag{166}$$

and

$$C_{11} = 1, \ \text{for } n = 1, m = 1. \tag{167}$$

In addition, other coefficients are zeros. Accordingly, the particular solution is given by

$$u(x, y, t) = e^{-3t} \sin y + e^{-6t} \cos x \sin y. \tag{168}$$

Example 4. Use the method of separation of variables to solve the initial-boundary value problem with Neumann boundary conditions

$$
\begin{aligned}
\text{PDE} \quad u_t &= 2(u_{xx} + u_{yy}), \ 0 < x, y < \pi, \ t > 0 \\
\text{B.C} \quad u_x(0, y, t) &= u_x(\pi, y, t) = 0 \\
u_y(x, 0, t) &= u_y(x, \pi, t) = 0 \\
\text{I.C} \quad u(x, y, 0) &= 1 + \cos x \cos y
\end{aligned} \tag{169}
$$

Solution.

We first set

$$u(x, y, t) = F(x)G(y)T(t), \tag{170}$$

to obtain

$$F''(x) + \lambda^2 F(x) = 0, \tag{171}$$

$$G''(y) + \mu^2 G(y) = 0, \tag{172}$$

and

$$T'(t) + 2(\lambda^2 + \mu^2)T(t) = 0. \tag{173}$$

Solving (171) and (172) and using the boundary conditions we obtain

$$F_n(x) = A_n \cos(nx), \ \lambda_n = n, \ n = 0, 1, 2, \cdots, \tag{174}$$

and

$$G_m(y) = \beta_m \cos(my), \ \mu_m = m, \ m = 0, 1, 2, \cdots, \tag{175}$$

respectively. The solution of the equation (173) is therefore given by

$$T_{nm}(t) = \overline{C}_{nm} e^{-2(n^2+m^2)t}. \tag{176}$$

Combining the results obtained above and using the superposition principle lead to the general solution of the problem given by

$$u(x, y, t) = \sum_{n=0}^{\infty} \sum_{m=0}^{\infty} C_{nm} e^{-2(n^2+m^2)t} \cos(nx) \cos(my). \tag{177}$$

To determine the constants C_{nm}, we expand the double series (177) and we use the given initial condition to obtain

$$C_{00} + C_{11} \cos x \cos y + \cdots = 1 + \cos x \cos y. \tag{178}$$

Equating the coefficients of like terms on both sides yields

$$C_{00} = 1, \ \text{for} \ n = 0, m = 0, \tag{179}$$

and

$$C_{11} = 1, \ \text{for} \ n = 1, m = 1, \tag{180}$$

and other coefficients are zeros. Accordingly, the particular solution is given by

$$u(x, y, t) = 1 + e^{-4t} \cos x \cos y \tag{181}$$

Exercises 4.3.1

Use the method of separation of variables in the following initial-boundary value problems:

1. $u_t = u_{xx} + u_{yy}, \ 0 < x, y < \pi, \ t > 0$
   $u(0, y, t) = u(\pi, y, t) = 0$
   $u(x, 0, t) = u(x, \pi, t) = 0$
   $u(x, y, 0) = \sin(2x) \sin(3y)$

2. $u_t = 3(u_{xx} + u_{yy}),\ 0 < x, y < \pi,\ t > 0$
   $u(0, y, t) = u(\pi, y, t) = 0$
   $u(x, 0, t) = u(x, \pi, t) = 0$
   $u(x, y, 0) = \sin x \sin y + \sin(2x) \sin(2y)$

3. $u_t = 4(u_{xx} + u_{yy}),\ 0 < x, y < \pi,\ t > 0$
   $u(0, y, t) = u(\pi, y, t) = 0$
   $u(x, 0, t) = u(x, \pi, t) = 0$
   $u(x, y, 0) = \sin x \sin y + \sin x \sin(2y) + \sin(2x) \sin y$

4. $u_t = u_{xx} + u_{yy},\ 0 < x, y < \pi,\ t > 0$
   $u_x(0, y, t) = u_x(\pi, y, t) = 0$
   $u(x, 0, t) = u(x, \pi, t) = 0$
   $u(x, y, 0) = \cos x \sin y$

5. $u_t = u_{xx} + u_{yy},\ 0 < x, y < \pi,\ t > 0$
   $u(0, y, t) = u(\pi, y, t) = 0$
   $u_y(x, 0, t) = u_y(x, \pi, t) = 0$
   $u(x, y, 0) = \sin x \cos y$

6. $u_t = u_{xx} + u_{yy},\ 0 < x, y < \pi,\ t > 0$
   $u_x(0, y, t) = u_x(\pi, y, t) = 0$
   $u(x, 0, t) = u(x, \pi, t) = 0$
   $u(x, y, 0) = \cos x \sin y + \cos(2x) \sin(2y)$

7. $u_t = 2(u_{xx} + u_{yy}),\ 0 < x, y < \pi,\ t > 0$
   $u(0, y, t) = u(\pi, y, t) = 0$
   $u_y(x, 0, t) = u_y(x, \pi, t) = 0$
   $u(x, y, 0) = \sin x \cos y + \sin(2x) \cos(2y)$

8. $u_t = u_{xx} + u_{yy},\ 0 < x, y < \pi,\ t > 0$
   $u_x(0, y, t) = u_x(\pi, y, t) = 0$
   $u_y(x, 0, t) = u_y(x, \pi, t) = 0$
   $u(x, y, 0) = \cos(2x) \cos(3y)$

9. $u_t = u_{xx} + u_{yy},\ 0 < x, y < \pi,\ t > 0$
   $u_x(0, y, t) = u_x(\pi, y, t) = 0$
   $u_y(x, 0, t) = u_y(x, \pi, t) = 0$
   $u(x, y, 0) = 1 + \cos x \cos(2y)$

10. $u_t = 4(u_{xx} + u_{yy}),\ 0 < x, y < \pi,\ t > 0$
    $u_x(0, y, t) = u_x(\pi, y, t) = 0$
    $u_y(x, 0, t) = u_y(x, \pi, t) = 0$
    $u(x, y, 0) = 4 + \cos(2x) \cos(2y)$

4.3.2   Three Dimensional Heat Flow

The distribution of heat flow in a three dimensional space is governed by the initial boundary value problem:

$$\text{PDE} \qquad u_t \;=\; \overline{k}(u_{xx} + u_{yy} + u_{zz})$$

$$0 < x < a,\; 0 < y < b, 0 < z < c$$

$$
\begin{aligned}
\text{B.C} \quad u(0, y, z, t) &= u(a, y, z, t) = 0 \\[4pt]
u(x, 0, z, t) &= u(x, b, z, t) = 0 \\[4pt]
u(x, y, 0, t) &= u(x, y, c, t) = 0 \\[4pt]
\text{I.C} \qquad u(x, y, z, 0) &= f(x, y, z)
\end{aligned}
\tag{182}
$$

where $u \equiv u(x, y, z, t)$ defines the temperature of any point at the position (x, y, z) of a rectangular volume at any time t, $\overline{k}$ is the thermal diffusivity. In a parallel manner to the previous discussion, the separation of variables method assumes that the solution $u(x, y, z, t)$ consists of the product of four distinct functions each depends on one variable only. Accordingly, we set

$$u(x, y, z, t) = F(x)G(y)H(z)T(t). \tag{183}$$

As discussed before, differentiating (183) once with respect to t, and twice with respect to x, y, and z, substituting into (182), and by dividing both sides by $\overline{k}F(x)G(y)H(z)T(t)$ we obtain

$$\frac{T^{'}(t)}{\overline{k}T(t)} = \frac{F^{''}(x)}{F(x)} + \frac{G^{''}(y)}{G(y)} + \frac{H^{''}(z)}{H(z)}. \tag{184}$$

It is obvious that the equality in (184) holds only if both sides are equal to the same constant. This allows us to set

$$
\begin{aligned}
F^{''}(x) + \lambda^2 F(x) &= 0, & (185)\\[4pt]
G^{''}(y) + \mu^2 G(y) &= 0, & (186)\\[4pt]
H^{''}(z) + \nu^2 H(z) &= 0, & (187)\\[4pt]
T^{'}(t) + \overline{k}(\lambda^2 + \mu^2 + \nu^2)T(t) &= 0. & (188)
\end{aligned}
$$

where λ, μ, and ν are constants. By solving the second order normal forms (185) - (187), we obtain the following solutions

$$
\begin{aligned}
F(x) &= A\cos(\lambda x) + B\sin(\lambda x). & (189)\\[4pt]
G(y) &= \alpha\cos(\mu y) + \beta\sin(\mu y), & (190)\\[4pt]
H(z) &= \gamma\cos(\nu z) + \delta\sin(\nu z). & (191)
\end{aligned}
$$

respectively, where $A, B, \alpha, \beta, \gamma$, and δ are constants. Using the boundary conditions in a similar way as discussed before we find

$$A = 0, \quad \lambda = \frac{n\pi}{a}, \ n = 1, 2, 3, \cdots, \tag{192}$$

$$\alpha = 0 \quad \mu = \frac{m\pi}{b}, \ m = 1, 2, 3, \cdots, \tag{193}$$

$$\gamma = 0, \quad \nu = \frac{r\pi}{c}, \ r = 1, 2, 3, \cdots, \tag{194}$$

so that

$$F_n(x) = B_n \sin(\frac{n\pi}{a}x), \ n = 1, 2, 3, \cdots, \tag{195}$$

$$G_m(y) = \beta_m \sin(\frac{m\pi}{b}y), \ m = 1, 2, 3, \cdots, \tag{196}$$

$$H_r(z) = \delta_r \sin(\frac{r\pi}{c}z), \ r = 1, 2, 3, \cdots. \tag{197}$$

The solution of (188) is therefore given by

$$T_{nmr}(t) = \overline{C}_{nmr} e^{-\overline{k}(\frac{n^2}{a^2} + \frac{m^2}{b^2} + \frac{r^2}{c^2})\pi^2 t}. \tag{198}$$

Consequently, we can formulate the general solution of (182) by using the superposition principle, therefore we find

$$u = \sum_{r=1}^{\infty} \sum_{m=1}^{\infty} \sum_{n=1}^{\infty} C_{nmr} e^{-\overline{k}(\frac{n^2}{a^2} + \frac{m^2}{b^2} + \frac{r^2}{c^2})\pi^2 t} \sin(\frac{n\pi}{a}x) \sin(\frac{m\pi}{b}y) \sin(\frac{r\pi}{c}z). \tag{199}$$

It remains now to determine the constants C_{nmr}. Using the initial condition given in (182), the coefficients C_{nmr} are given by

$$C_{nmr} = \frac{8}{abc} \int_0^c \int_0^b \int_0^a f(x, y, z) \sin\frac{n\pi}{a}x \sin\frac{m\pi}{b}y \sin\frac{r\pi}{c}z \, dx \, dy \, dz. \tag{200}$$

The constants C_{nmr} can also be determined by equating the coefficients on both sides if the initial condition is given in terms of trigonometric functions identical to those included in $u(x, y, z, t)$. This technique reduces the massive size of calculations usually required by using the triple Fourier coefficients.

The following examples will illustrate the method presented above.

Example 5. Solve the initial-boundary value problem

$$\text{PDE} \quad u_t = u_{xx} + u_{yy} + u_{zz}, \ 0 < x, y, z < \pi, \ t > 0$$

$$\text{B.C} \quad u(0, y, z, t) = u(\pi, y, z, t) = 0$$

$$u(x, 0, z, t) = u(x, \pi, z, t) = 0 \tag{201}$$

$$u(x, y, 0, t) = u(x, y, \pi, t) = 0$$

$$\text{I.C} \quad u(x, y, z, 0) = 3 \sin x \sin y \sin z$$

Solution.

Proceeding as before, we set

$$u(x, y, z, t) = F(x)G(y)H(z)T(t). \tag{202}$$

Substituting (202) into (201) and following the discussions above we find

$$
\begin{aligned}
F_n(x) &= B_n \sin(nx), \ \lambda_n = n, \ n = 1, 2, 3, \cdots, & (203) \\
G_m(y) &= \beta_m \sin(my), \ \mu_m = m, \ m = 1, 2, 3, \cdots, & (204) \\
H_r(z) &= \delta_r \sin(rz), \ \nu_r = r, \ r = 1, 2, 3, \cdots, & (205) \\
T(t) &= Ce^{-(\lambda^2 + \mu^2 + \nu^2)t}. & (206)
\end{aligned}
$$

Consequently, we can formulate the general solution is given by

$$u = \sum_{r=1}^{\infty} \sum_{m=1}^{\infty} \sum_{n=1}^{\infty} C_{nmr} \, e^{-(n^2 + m^2 + r^2)t} \sin(nx) \sin(my) \sin(rz). \tag{207}$$

To determine the constants C_{nmr}, we use the initial condition and expand (207) to find

$$C_{111} \sin x \sin y \sin z + \cdots = 3 \sin x \sin y \sin z \tag{208}$$

Equation the coefficients of like terms in both sides we obtain

$$
\begin{aligned}
C_{111} &= 3, \text{ for } n = 1, m = 1, r = 1, \\
C_{ijk} &= 0, \text{ for } i \neq 1, j \neq 1, k \neq 1.
\end{aligned}
\tag{209}
$$

Consequently, the particular solution is given by

$$u(x, y, z, t) = 3e^{-3t} \sin x \sin y \sin z. \tag{210}$$

In the next example, the Neumann boundary conditions in the spatial domain are used.

Example 6. Solve the initial-boundary value problem with Neumann boundary conditions

$$
\begin{aligned}
\text{PDE} \quad u_t &= u_{xx} + u_{yy} + u_{zz}, \ 0 < x, y, z < \pi, \ t > 0 \\
\text{B.C } u_x(0, y, z, t) &= u_x(\pi, y, z, t) = 0 \\
u_y(x, 0, z, t) &= u_y(x, \pi, z, t) = 0 \\
u_z(x, y, 0, t) &= u_z(x, y, \pi, t) = 0 \\
\text{I.C } u(x, y, z, 0) &= 4 + \cos x \cos(2y) \cos(3z)
\end{aligned}
\tag{211}
$$

Solution.

Proceeding as before, we set

$$u(x, y, z, t) = F(x)G(y)H(z)T(t). \tag{212}$$

Substituting in (211), using the boundary conditions, and following the discussions above we find

$$
\begin{aligned}
F_n(x) &= A_n \cos(nx), \ \lambda_n = n, \ n = 0, 1, 2, 3, \cdots, & (213) \\
G_m(y) &= \alpha_m \cos(my), \ \mu_m = m, \ m = 0, 1, 2, 3, \cdots, & (214) \\
H_r(z) &= \gamma_r \cos(rz), \ \nu_r = r, \ r = 0, 1, 2, 3, \cdots, & (215) \\
T_{nmr}(t) &= \overline{C}_{nmr} e^{-\overline{k}(n^2 + m^2 + r^2)t}. & (216)
\end{aligned}
$$

Consequently, we can formulate the general solution expressed by

$$u = \sum_{r=0}^{\infty} \sum_{m=0}^{\infty} \sum_{n=0}^{\infty} C_{nmr} \, e^{-(n^2 + m^2 + r^2)t} \cos(nx) \cos(my) \cos(rz). \tag{217}$$

To determine the constants C_{nmr}, we using the initial condition and expand (217) to find

$$C_{000} + C_{123} \cos(x) \cos(2y) \cos(3z) + \cdots = 4 + \cos(x) \cos(2y) \cos(3z). \tag{218}$$

Equating the coefficients of like terms in both sides we obtain

$$
\begin{aligned}
C_{000} &= 4, \text{ for } n = 0, m = 0, r = 0, \\
C_{123} &= 1, \text{ for } n = 1, m = 2, r = 3,
\end{aligned} \tag{219}
$$

where other coefficients vanish. The result (219) can also be obtained by using the Fourier coefficients defined before.

Consequently, the particular solution is given by

$$u(x, y, z, t) = 4 + e^{-14t} \cos(x) \cos(2y) \cos(3z), \tag{220}$$

obtained upon combining (219) and (217).

Example 7. Solve the initial-boundary value problem

$$
\begin{aligned}
\text{PDE} \quad u_t &= u_{xx} + u_{yy} + u_{zz}, \ 0 < x, y, z < \pi, \ t > 0 \\
\text{B.C} \quad u(0, y, z, t) &= u(\pi, y, z, t) = 0 \\
u_y(x, 0, z, t) &= u_y(x, \pi, z, t) = 0 \\
u_z(x, y, 0, t) &= u_z(x, y, \pi, t) = 0 \\
\text{I.C} \quad u(x, y, z, 0) &= 2 \sin x \cos y \cos z
\end{aligned} \tag{221}
$$

Solution.

We first set

$$u(x, y, z, t) = F(x)G(y)H(z)T(t). \tag{222}$$

Substituting (222) in (221), and using the boundary conditions we obtain

$$
\begin{aligned}
F_n(x) &= B_n \sin(nx), \ \lambda_n = n, \ n = 1, 2, 3, \cdots, & (223) \\
G_m(y) &= \alpha_m \cos(my), \ \mu_m = m, \ m = 0, 1, 2, 3, \cdots, & (224) \\
H_r(z) &= \gamma_r \cos(rz), \nu_r = r, \ r = 0, 1, 2, 3, \cdots, & (225) \\
T_{nmr}(t) &= \overline{C}_{nmr} e^{-(n^2+m^2+r^2)t}. & (226)
\end{aligned}
$$

Consequently, the general solution is given by

$$u = \sum_{r=0}^{\infty} \sum_{m=0}^{\infty} \sum_{n=1}^{\infty} C_{nmr} \, e^{-(n^2+m^2+r^2)t} \sin(nx) \cos(my) \cos(rz). \tag{227}$$

To determine the constants C_{nmr}, we use the initial condition and expand (227) to find

$$C_{100} + C_{111} \sin(x) \cos(y) \cos(z) + \cdots = 2 \sin(x) \cos(y) \cos(z) \tag{228}$$

Equation the coefficients of like terms in both sides we obtain

$$C_{111} = 2, \text{ for } n = 1, m = 1, r = 1, \tag{229}$$

where other coefficients vanish. Consequently, the particular solution is given by

$$u(x, y, z, t) = 2e^{-3t} \sin x \, \cos y \, \cos z. \tag{230}$$

Exercises 4.3.2

Use the method of separation of variables in the following initial-boundary value problems:

1. $u_t = u_{xx} + u_{yy} + u_{zz}, \ 0 < x, y, z < \pi$
   $u(0, y, z, t) = u(\pi, y, z, t) = 0$
   $u(x, 0, z, t) = u(x, \pi, z, t) = 0$
   $u(x, y, 0, t) = u(x, y, \pi, 0) = 0$
   $u(x, y, z, 0) = \sin(2x) \sin(3y) \sin(4z)$

2. $u_t = u_{xx} + u_{yy} + u_{zz}, \ 0 < x, y, z < \pi$
   $u(0, y, z, t) = u(\pi, y, z, t) = 0$
   $u(x, 0, z, t) = u(x, \pi, z, t) = 0$
   $u(x, y, 0, t) = u(x, y, \pi, 0) = 0$
   $u(x, y, z, 0) = \sin x \sin y \sin z + \sin(2x) \sin(2y) \sin(2z)$

3. $u_t = u_{xx} + u_{yy} + u_{zz}, \ 0 < x, y, z < \pi$
   $u(0, y, z, t) = u(\pi, y, z, t) = 0$

$$u(x, 0, z, t) = u(x, \pi, z, t) = 0$$
$$u(x, y, 0, t) = u(x, y, \pi, 0) = 0$$
$$u(x, y, z, 0) = \sin x \sin y \sin(2z) + \sin x \sin(2y) \sin(3z)$$

4. $u_t = u_{xx} + u_{yy} + u_{zz}$, $0 < x, y, z < \pi$
$$u_x(0, y, z, t) = u_x(\pi, y, z, t) = 0$$
$$u(x, 0, z, t) = u(x, \pi, z, t) = 0$$
$$u(x, y, 0, t) = u(x, y, \pi, 0) = 0$$
$$u(x, y, z, 0) = \cos x \sin y \sin z$$

5. $u_t = u_{xx} + u_{yy} + u_{zz}$, $0 < x, y, z < \pi$
$$u(0, y, z, t) = u(\pi, y, z, t) = 0$$
$$u_y(x, 0, z, t) = u_y(x, \pi, z, t) = 0$$
$$u_z(x, y, 0, t) = u_z(x, y, \pi, 0) = 0$$
$$u(x, y, z, 0) = \sin x \cos y \cos z$$

6. $u_t = u_{xx} + u_{yy} + u_{zz}$, $0 < x, y, z < \pi$
$$u_x(0, y, z, t) = u_x(\pi, y, z, t) = 0$$
$$u(x, 0, z, t) = u(x, \pi, z, t) = 0$$
$$u_z(x, y, 0, t) = u_z(x, y, \pi, 0) = 0$$
$$u(x, y, z, 0) = \cos x \sin y \cos z$$

7. $u_t = u_{xx} + u_{yy} + u_{zz}$, $0 < x, y, z < \pi$
$$u(0, y, z, t) = u(\pi, y, z, t) = 0$$
$$u_y(x, 0, z, t) = u_y(x, \pi, z, t) = 0$$
$$u(x, y, 0, t) = u(x, y, \pi, 0) = 0$$
$$u(x, y, z, 0) = \sin x \cos y \sin z$$

8. $u_t = u_{xx} + u_{yy} + u_{zz}$, $0 < x, y, z < \pi$
$$u_x(0, y, z, t) = u_x(\pi, y, z, t) = 0$$
$$u_y(x, 0, z, t) = u_y(x, \pi, z, t) = 0$$
$$u_z(x, y, 0, t) = u_z(x, y, \pi, 0) = 0$$
$$u(x, y, z, 0) = 2 + 3 \cos x \cos(2y) \cos z$$

9. $u_t = u_{xx} + u_{yy} + u_{zz}$, $0 < x, y, z < \pi$
$$u_x(0, y, z, t) = u_x(\pi, y, z, t) = 0$$
$$u_y(x, 0, z, t) = u_y(x, \pi, z, t) = 0$$
$$u_z(x, y, 0, t) = u_z(x, y, \pi, 0) = 0$$
$$u(x, y, z, 0) = 1 + \cos x \cos y \cos z + \cos(2x) \cos(2y) \cos(2z)$$

10. $u_t = u_{xx} + u_{yy} + u_{zz}$, $0 < x, y, z < \pi$
$$u_x(0, y, z, t) = u_x(\pi, y, z, t) = 0$$
$$u_y(x, 0, z, t) = u_y(x, \pi, z, t) = 0$$
$$u_z(x, y, 0, t) = u_z(x, y, \pi, 0) = 0$$
$$u(x, y, z, 0) = 1 + 2 \cos x \cos y \cos z + 3 \cos(2x) \cos(3y) \cos(4z)$$

Chapter 5

One Dimensional Wave Equation

5.1 Introduction

In this chapter we will study the physical problem of the wave propagation. The wave equation usually describes water waves, the vibrations of a string or a membrane, the propagation of electromagnetic and sound waves, or the transmission of electric signals in a cable. The function $u(x, t)$ defines a small displacement of any point of a vibrating string at position x at time t. Unlike the heat equation, the wave equation contains the term u_{tt} that represents the vertical acceleration of a vibrating string at point x, which is due to the tension in the string.

The wave equation plays a significant role in various physical problems. The study of wave equation is needed in diverse areas of science and engineering.

The typical model that describes the wave equation, as will be discussed later, is an initial-boundary value problem valid in a bounded domain or an initial value problem valid in an unbounded domain. It is interesting to note here that two initial conditions should be prescribed, namely the initial displacement $u(x, 0) = f(x)$ and the initial velocity $u_t(x, 0) = g(x)$ that describe the initial displacement and the initial velocity at the starting time $t = 0$ respectively.

In a parallel manner to our approach applied to the heat equation in Chapters 3 and 4, our concern will be focused on solving the PDE in conjunction with the given conditions. The approach will be identical to that applied before, therefore the mathematical derivation of the wave equation will not be examined in this text.

In this chapter, we will apply the newly developed Adomian decomposition method and the traditional methods of separation of variables and D'Alembert method. Further, a particular solution of the wave equation will be established recalling that a general solution is of little use.

5.2   Adomian Decomposition Method

The Adomian decomposition method has been widely used with promising results in linear and nonlinear partial differential equations that describe wave propagations. The method has been presented in details in Chapters 2, 3, and 4 and the formal steps have been outlined and supported by several illustrative examples. The method introduces the solution of any equation in a series form with elegantly computed components. The method identifies the zeroth component u_0 by the terms that arise from the initial/boundary conditions and from integrating the source term if exists. The remaining components $u_n, n \geq 1$, are determined recursively as far as we like. The method will be illustrated by discussing the following typical wave model.

Without loss of generality, as a simple wave equation, we consider the following initial-boundary value problem:

$$
\begin{aligned}
\text{PDE} \qquad u_{tt} &= c^2 u_{xx}, \ 0 < x < L, \ t > 0 \\
\text{B.C} \qquad u(0,t) &= 0, \ u(L,t) = 0, \ t \geq 0 \qquad\qquad (1) \\
\text{I.C} \qquad u(x,0) &= f(x), \ u_t(x,0) = g(x)
\end{aligned}
$$

where $u = u(x,t)$ is the displacement of any point of the string at the position x and at time t, and c is a constant related to the elasticity of the material of the string. The given boundary conditions indicate that the end points of the vibrating string are fixed. It is obvious the IBVP (1), that governs the wave displacement, contains the term u_{tt}. Consequently, two initial conditions should be given. The initial conditions describe the initial displacement and the initial velocity of any point at the starting time $t = 0$.

We begin our analysis by rewriting (1) in an operator form by

$$
L_t u(x,t) = c^2 L_x u(x,t), \qquad\qquad (2)
$$

where the differential operators L_t and L_x are defined by

$$
L_t = \frac{\partial^2}{\partial t^2}, \ L_x = \frac{\partial^2}{\partial x^2}. \qquad\qquad (3)
$$

We assume that the integral operators L_t^{-1} and L_x^{-1} exist and may be regarded as two-fold indefinite integrals defined by

$$
L_t^{-1}(.) = \int_0^t \int_0^t (.) dt \, dt, \qquad\qquad (4)
$$

and

$$
L_x^{-1}(.) = \int_0^x \int_0^x (.) dx \, dx. \qquad\qquad (5)
$$

This means that

$$L_t^{-1} L_t u(x,t) = u(x,t) - tu_t(x,0) - u(x,0), \qquad (6)$$

and

$$L_x^{-1} L_x u(x,t) = u(x,t) - xu_x(0,t) - u(0,t). \qquad (7)$$

Recall that the solution can be obtained by using the inverse operator L_t^{-1} or the inverse operator L_x^{-1}. However, using the inverse operator L_t^{-1} requires the use of the initial conditions only, whereas operating wit L_x^{-1} imposes the use of initial and boundary conditions. For this reason, and to reduce the size of calculations, we will apply the decomposition method in the t direction. Applying L_t^{-1} to both sides of (2) and using the initial conditions we obtain

$$u(x,t) = f(x) + tg(x) + c^2 L_t^{-1}(L_x u(x,t)). \qquad (8)$$

The Adomian's method decomposes the displacement function $u(x,t)$ into a sum of an infinite components defined by the infinite series

$$u(x,t) = \sum_{n=0}^{\infty} u_n(x,t), \qquad (9)$$

where the components $u_n(x,t), n \geq 0$ will be easily calculated. Substituting (9) into both sides of (8) gives

$$\sum_{n=0}^{\infty} u_n(x,t) = f(x) + tg(x) + c^2 L_t^{-1} \left(L_x \left(\sum_{n=0}^{\infty} u_n(x,t) \right) \right), \qquad (10)$$

or by using few components

$$u_0 + u_1 + u_2 + \cdots = f(x) + tg(x) + c^2 L_t^{-1} \left(L_x \left(u_0 + u_1 + u_2 + \cdots \right) \right). \qquad (11)$$

The method suggests that the zeroth component $u_0(x,t)$ is identified by the terms that are not included under the integral sign in (10). The other components are determined by using the recursive relation

$$
\begin{aligned}
u_0(x,t) &= f(x) + tg(x), \\
u_{k+1}(x,t) &= c^2 L_t^{-1} \left(L_x \left(u_k(x,t) \right) \right), k \geq 0.
\end{aligned}
\qquad (12)
$$

In view of (12), the components $u_0(x,t), u_1(x,t), u_2(x,t), \cdots$ can be determined individually by

$$
\begin{aligned}
u_0(x,t) &= f(x) + tg(x), \\
u_1(x,t) &= c^2 L_t^{-1} L_x(u_0) = c^2 \left(\tfrac{t^2}{2!} f''(x) + \tfrac{t^3}{3!} g''(x) \right), \\
u_2(x,t) &= c^2 L_t^{-1} L_x(u_1) = c^4 \left(\tfrac{t^4}{4!} f^{(iv)}(x) + \tfrac{t^5}{5!} g^{(iv)}(x) \right), \\
u_3(x,t) &= c^2 L_t^{-1} L_x(u_2) = c^6 \left(\tfrac{t^6}{6!} f^{(vi)}(x) + \tfrac{t^7}{7!} g^{(vi)}(x) \right),
\end{aligned}
\qquad (13)
$$

and so on. It is obvious that the PDE (1) is reduced to solving simple integrals given in (13), where components can be determined easily as far as we like. The accuracy level can be enhanced significantly by determining more terms if a closed form solution is not obtained, where a truncated number of components is usually used for numerical purposes.

Having determined the components in (13), the solution of the partial differential equation (1) is obtained in a series form given by

$$u(x,t) = \sum_{n=0}^{\infty} c^{2n} \left(\frac{t^{2n}}{(2n)!} f^{(2n)}(x) + \frac{t^{2n+1}}{(2n+1)!} g^{(2n)}(x) \right), \tag{14}$$

obtained by substituting (13) into (9). It is important to note that the solution (14) can also be obtained by using the inverse operator L_x^{-1}. However, the solution in this way requires more work because the boundary condition $u_x(0,t)$ is not always available.

To give a clear overview of the decomposition method, we have selected homogeneous and inhomogeneous equations to illustrate the procedure discussed above.

5.2.1   Homogeneous Wave Equations

The Adomian decomposition method will be used to solve the following homogeneous equations.

Example 1. Use the Adomian decomposition method to solve the initial-boundary value problem

$$
\begin{array}{llll}
\text{PDE} & u_{tt} & = & u_{xx},\ 0 < x < \pi,\ t > 0 \\[2mm]
\text{B.C} & u(0,t) & = & 0,\ u(\pi,t) = 0,\ t \geq 0 \\[2mm]
\text{I.C} & u(x,0) & = & \sin x,\ u_t(x,0) = 0
\end{array}
\tag{15}
$$

Solution.

In an operator form, Eq. (15) can be written as

$$L_t u(x,t) = L_x u(x,t), \tag{16}$$

where the differential operators L_t and L_x are defined by

$$L_t = \frac{\partial^2}{\partial t^2},\ L_x = \frac{\partial^2}{\partial x^2}. \tag{17}$$

Accordingly, the inverse operator L_t^{-1} is a two-fold integral operator defined by

$$L_t^{-1}(.) = \int_0^t \int_0^t (.)\, dt\, dt, \tag{18}$$

so that

$$L_t^{-1} L_t u(x,t) = u(x,t) - t u_t(x,0) - u(x,0). \tag{19}$$

Applying L_t^{-1} to both sides of (16), noting (19), and using the initial conditions we find

$$u(x,t) = \sin x + L_t^{-1} \left(L_x u(x,t) \right). \tag{20}$$

The decomposition method defines the unknown function $u(x,t)$ by the series

$$u(x,t) = \sum_{n=0}^{\infty} u_n(x,t), \tag{21}$$

that carries (20) into

$$\sum_{n=0}^{\infty} u_n(x,t) = \sin x + L_t^{-1} \left(L_x \left(\sum_{n=0}^{\infty} u_n(x,t) \right) \right), \tag{22}$$

or equivalently

$$u_0 + u_1 + u_2 + \cdots = \sin x + L_t^{-1} \left(L_x \left(u_0 + u_1 + u_2 + \cdots \right) \right). \tag{23}$$

Following the discussions presented above we set the recursive relation

$$\begin{aligned}
u_0(x,t) &= \sin x, \\
u_{k+1}(x,t) &= L_t^{-1} \left(L_x \left(u_k \right) \right), k \geq 0,
\end{aligned} \tag{24}$$

and this in turn gives

$$\begin{aligned}
u_0(x,t) &= \sin x, \\
u_1(x,t) &= L_t^{-1} \left(L_x \left(u_0 \right) \right) = L_t^{-1} \left(-\sin x \right) = -\frac{1}{2!} t^2 \sin x, \\
u_2(x,t) &= L_t^{-1} \left(L_x \left(u_1 \right) \right) = L_t^{-1} \left(\frac{t^2}{2!} \sin x \right) = \frac{1}{4!} t^4 \sin x, \\
u_3(x,t) &= L_t^{-1} \left(L_x \left(u_2 \right) \right) = L_t^{-1} \left(-\frac{t^4}{4!} \sin x \right) = -\frac{1}{6!} t^6 \sin x,
\end{aligned} \tag{25}$$

and so on. Consequently, the solution $u(x,t)$ in a series form is given by

$$\begin{aligned}
u(x,t) &= u_0(x,t) + u_1(x,t) + u_2(x,t) + \cdots \\
&= \sin x \left(1 - \frac{1}{2!} t^2 + \frac{1}{4!} t^4 - \frac{1}{6!} t^6 + \cdots \right),
\end{aligned} \tag{26}$$

and in a closed form by

$$u(x,t) = \sin x \cos t, \tag{27}$$

obtained upon using the Taylor expansion of $\cos t$. It is clear that the particular solution (27) justifies the PDE, the boundary conditions and the initial conditions.

Example 2. Use the Adomian decomposition method to solve the initial-boundary value problem

$$
\begin{array}{llll}
\text{PDE} & u_{tt} & = & u_{xx}, \ 0 < x < \pi, \ t > 0 \\[2mm]
\text{B.C} & u(0,t) & = & 0, \ u(\pi,t) = 0 \\[2mm]
\text{I.C} & u(x,0) & = & 0, \ u_t(x,0) = \sin x
\end{array}
\tag{28}
$$

Solution.

Applying L_t^{-1} to both sides of the operator form (28) gives

$$
u(x,t) = t \sin x + L_t^{-1}\left(L_x(u(x,t))\right).
\tag{29}
$$

Substituting the decomposition series

$$
u(x,t) = \sum_{n-0}^{\infty} u_n(x,t),
\tag{30}
$$

into both sides of (29) yields

$$
\sum_{n=0}^{\infty} u_n(x,t) = t \sin x + L_t^{-1}\left(L_x\left(\sum_{n=0}^{\infty} u_n(x,t)\right)\right).
\tag{31}
$$

Proceeding as before we set

$$
\begin{array}{rcl}
u_0(x,0) & = & t \sin x, \\[2mm]
u_{k+1}(x,t) & = & L_t^{-1}\left(L_x(u_k(x,t))\right), \ k \geq 0,
\end{array}
\tag{32}
$$

hence we find

$$
\begin{array}{rcl}
u_0(x,t) & = & t \sin x, \\[2mm]
u_1(x,t) & = & L_t^{-1}\left(L_x\left(u_0\right)\right) = -\frac{1}{3!}t^3 \sin x, \\[2mm]
u_2(x,t) & = & L_t^{-1}\left(L_x\left(u_1\right)\right) = \frac{1}{5!}t^5 \sin x,
\end{array}
\tag{33}
$$

and so on. Consequently, the solution $u(x,t)$ in a series form is given by

$$
\begin{array}{rcl}
u(x,t) & = & u_0(x,t) + u_1(x,t) + u_2(x,t) + \cdots \\[2mm]
& = & \sin x \left(t - \dfrac{1}{3!}t^3 + \dfrac{1}{5!}t^5 - \cdots\right),
\end{array}
\tag{34}
$$

and in a closed form by

$$u(x,t) = \sin x \sin t. \tag{35}$$

Example 3. Use the Adomian decomposition method to solve the wave equation problem

$$\begin{array}{rcl}
\text{PDE} & u_{tt} & = & u_{xx}, \ \ 0 < x < \pi, \ t > 0 \\[2mm]
\text{B.C} & u(0,t) & = & 1 + \sin t, \ u(\pi,t) = 1 - \sin t \\[2mm]
\text{I.C} & u(x,0) & = & 1, \ u_t(x,0) = \cos x
\end{array} \tag{36}$$

Solution.

It is important to note that the boundary conditions are inhomogeneous. The method will be applied in a straightforward manner for all types of differential equations.

Applying L_t^{-1} to the operator form of (36) and using the initial conditions we find

$$u(x,t) = 1 + t \cos x + L_t^{-1} \left(L_x u(x,t) \right). \tag{37}$$

Substituting the decomposition series for $u(x,t)$ into both sides of (37) yields

$$\sum_{n=0}^{\infty} u_n(x,t) = 1 + t \cos x + L_t^{-1} \left(L_x \left(\sum_{n=0}^{\infty} u_n(x,t) \right) \right). \tag{38}$$

Proceeding as before we find

$$\begin{array}{rcl}
u_0(x,t) & = & 1 + t \cos x, \\[2mm]
u_1(x,t) & = & L_t^{-1} \left(L_x(u_0) \right) = -\frac{1}{3!} t^3 \cos x, \\[2mm]
u_2(x,t) & = & L_t^{-1} \left(L_x(u_1) \right) = \frac{1}{5!} t^5 \cos x,
\end{array} \tag{39}$$

and so on. It is clear that we can easily determine other components as far as we like.

In view of (39), the solution $u(x,t)$ in a series form is given by

$$\begin{array}{rcl}
u(x,t) & = & u_0(x,t) + u_1(x,t) + u_2(x,t) + \cdots \\[2mm]
& = & 1 + \cos x \left(t - \frac{1}{3!} t^3 + \frac{1}{5!} t^5 - \cdots \right),
\end{array} \tag{40}$$

and in a closed form by

$$u(x,t) = 1 + \cos x \sin t. \tag{41}$$

Example 4. Use the Adomian decomposition method to solve the initial-boundary value problem

$$
\begin{array}{llll}
\text{PDE} & u_{tt} & = & u_{xx} - 3u, \ 0 < x < \pi, \ t > 0 \\[2mm]
\text{B.C} & u(0,t) & = & \sin(2t), \ u(\pi,t) = -\sin(2t) \\[2mm]
\text{I.C} & u(x,0) & = & 0, \ u_t(x,0) = 2\cos x
\end{array}
\tag{42}
$$

Solution.

In this example, an additional term $-3u$ is involved. This term arises when each element of the string is subject to an additional force which is proportional to its displacement. Applying L_t^{-1} to the operator form of (42) and using the initial conditions we find

$$
u(x,t) = 2t\cos x + L_t^{-1}\left(L_x u(x,t) - 3u(x,t)\right).
\tag{43}
$$

Substituting the decomposition series for $u(x,t)$ into both sides of (43) yields

$$
\sum_{n=0}^{\infty} u_n = 2t\cos x + L_t^{-1}\left(L_x\left(\sum_{n=0}^{\infty} u_n\right) - 3\sum_{n=0}^{\infty} u_n\right).
\tag{44}
$$

The components $u_n(x,t), n \geq 0$ can be recursively determined as follows

$$
\begin{array}{lll}
u_0(x,t) & = & 2t\cos x, \\[2mm]
u_1(x,t) & = & -\frac{1}{3!}(2t)^3 \cos x, \\[2mm]
u_2(x,t) & = & \frac{1}{5!}(2t)^5 \cos x, \\[2mm]
u_3(x,t) & = & \frac{1}{7!}(2t)^7 \cos x,
\end{array}
\tag{45}
$$

and so on. Other components can be determined to improve the accuracy level if numerical approximations are required.

In view of (45), the solution $u(x,t)$ in a series form is given by

$$
\begin{aligned}
u(x,t) & = u_0(x,t) + u_1(x,t) + u_2(x,t) + \cdots \\
& = \cos x\left((2t) - \frac{1}{3!}(2t)^3 + \frac{1}{5!}(2t)^5 - \cdots\right),
\end{aligned}
\tag{46}
$$

and in a closed form by

$$
u(x,t) = \cos x \sin(2t),
\tag{47}
$$

obtained upon using the Taylor series of $\sin(2t)$.

Example 5. Use the Adomian decomposition method to solve the initial-boundary value problem

$$
\begin{array}{lll}
\text{PDE} & u_{tt} = u_{xx}, & 0 < x < \pi,\ t > 0 \\[2mm]
\text{B.C} & u_x(0,t) = 1,\ u_x(\pi,t) = 1 & \qquad (48) \\[2mm]
\text{I.C} & u(x,0) = x,\ u_t(x,0) = \cos x &
\end{array}
$$

Solution.

The given Neumann boundary conditions $u_x(0,t)$ and $u_x(\pi,t)$ are inhomogeneous. Operating with L_t^{-1} gives

$$
u(x,t) = x + t\cos x + L_t^{-1}\left(L_x u(x,t)\right), \tag{49}
$$

and proceeding as before we obtain

$$
\sum_{n=0}^{\infty} u_n(x,t) = x + t\cos x + L_t^{-1}\left(L_x\left(\sum_{n=0}^{\infty} u_n(x,t)\right)\right). \tag{50}
$$

The components $u_n(x,t), n \geq 0$ can be recursively determined as follows

$$
\begin{array}{lll}
u_0(x,t) &=& x + t\cos x, \\[2mm]
u_1(x,t) &=& L_t^{-1}\left(L_x\left(u_0\right)\right) = -\frac{1}{3!}t^3\cos x, \qquad (51) \\[2mm]
u_2(x,t) &=& L_t^{-1}\left(L_x\left(u_1\right)\right) = \frac{1}{5!}t^5\cos x.
\end{array}
$$

In view of (51), the series form for $u(x,t)$ is given by

$$
\begin{aligned}
u(x,t) &= u_0(x,t) + u_1(x,t) + u_2(x,t) + \cdots \\[2mm]
&= x + \cos x\left(t - \frac{1}{3!}t^3 + \frac{1}{5!}t^5 - \cdots\right), \qquad (52)
\end{aligned}
$$

and in a closed form by

$$
u(x,t) = x + \cos x \sin t. \tag{53}
$$

In the following example, we will discuss equations where the coefficient of u_{xx} is a function rather than a constant.

Example 6. Use the Adomian decomposition method to solve the initial-boundary value problem

$$
\begin{array}{lll}
\text{PDE} & u_{tt} = \frac{x^2}{2}u_{xx}, & 0 < x < 1,\ t > 0 \\[2mm]
\text{B.C} & u(0,t) = 0,\ u(1,t) = \sinh t & \qquad (54) \\[2mm]
\text{I.C} & u(x,0) = 0,\ u_t(x,0) = x^2 &
\end{array}
$$

Solution.

Applying L_t^{-1} to both sides of (54) gives

$$u(x,t) = tx^2 + L_t^{-1}\left(\frac{x^2}{2}L_x u(x,t)\right), \tag{55}$$

so that

$$\sum_{n=0}^{\infty} u_n(x,t) = tx^2 + L_t^{-1}\left(\frac{x^2}{2}L_x\left(\sum_{n=0}^{\infty} u_n(x,t)\right)\right). \tag{56}$$

The components $u_n(x,t), n \geq 0$ can be recursively determined as follows

$$
\begin{aligned}
u_0(x,t) &= tx^2, \\
u_1(x,t) &= L_t^{-1}\left(\frac{x^2}{2}L_x(u_0)\right) = \frac{x^2 t^3}{3!}, \\
u_2(x,t) &= L_t^{-1}\left(\frac{x^2}{2}L_x(u_1)\right) = \frac{x^2 t^5}{5!}, \\
u_3(x,t) &= L_t^{-1}\left(\frac{x^2}{2}L_x(u_2)\right) = \frac{x^2 t^7}{7!},
\end{aligned}
\tag{57}
$$

and so on. Other components can be determined to improve the accuracy level for numerical purposes.

In view of (57), the solution $u(x,t)$ in a series form is given by

$$
\begin{aligned}
u(x,t) &= u_0(x,t) + u_1(x,t) + u_2(x,t) + \cdots \\
&= x^2\left(t + \frac{1}{3!}t^3 + \frac{1}{5!}t^5 + \frac{1}{7!}t^7 + \cdots\right),
\end{aligned}
\tag{58}
$$

and in a closed form by

$$u(x,t) = x^2 \sinh t, \tag{59}$$

obtained upon using the Taylor series for $\sinh t$.

5.2.2   Inhomogeneous Wave Equations

As presented in Chapters 3 and 4, Adomian's method can handle easily any equation, homogeneous or inhomogeneous. This can be seen by discussing the following illustrative examples.

Example 7. Use Adomian decomposition method to solve the inhomogeneous PDE

$$
\begin{aligned}
\text{PDE} \qquad u_{tt} &= u_{xx} - 2, \ 0 < x < \pi, \ t > 0 \\[4pt]
\text{B.C} \qquad u(0,t) &= 0, \ u(\pi,t) = \pi^2, \ t \geq 0 \\[4pt]
\text{I.C} \qquad u(x,0) &= x^2, \ u_t(x,0) = \sin x
\end{aligned}
\tag{60}
$$

Solution.

In an operator form, Eq.(60) becomes

$$L_t u(x,t) = L_x u(x,t) - 2. \tag{61}$$

Operating with L_t^{-1} on both sides of (61) leads to

$$u(x,t) = x^2 + t \sin x - t^2 + L_t^{-1} \left(L_x u(x,t) \right), \tag{62}$$

and consequently we obtain

$$\sum_{n=0}^{\infty} u_n(x,t) = x^2 + t \sin x - t^2 + L_t^{-1} \left(L_x \left(\sum_{n=0}^{\infty} u_n(x,t) \right) \right). \tag{63}$$

It should be noted here that the zeroth component u_0 is assigned the terms that arise from integrating -2 and from using the initial conditions. The following recursive relation

$$\begin{aligned}
u_0(x,t) &= x^2 + t \sin x - t^2, \\
u_{k+1}(x,t) &= L_t^{-1} \left(L_x \left(u_k(x,t) \right) \right), k \geq 0,
\end{aligned} \tag{64}$$

should be used to determine the components of $u(x,t)$. Proceeding as before, we set

$$\begin{aligned}
u_0(x,t) &= x^2 + t \sin x - t^2, \\
u_1(x,t) &= L_t^{-1} \left(L_x \left(u_0 \right) \right) = t^2 - \tfrac{1}{3!} t^3 \sin x, \\
u_2(x,t) &= L_t^{-1} \left(L_x \left(u_1 \right) \right) = \tfrac{1}{5!} t^5 \sin x,
\end{aligned} \tag{65}$$

and so on. Consequently, the solution $u(x,t)$ in a series form is given by

$$\begin{aligned}
u(x,t) &= u_0(x,t) + u_1(x,t) + u_2(x,t) + \cdots \\
&= x^2 + \sin x \left(t - \frac{1}{3!} t^3 + \frac{1}{5!} t^5 - \cdots \right),
\end{aligned} \tag{66}$$

and in a closed form by

$$u(x,t) = x^2 + \sin x \sin t. \tag{67}$$

Example 8. Use the Adomian decomposition method to solve the inhomogeneous PDE

$$\begin{aligned}
\text{PDE} \qquad u_{tt} &= u_{xx} + \sin x, \ 0 < x < \pi, \ t > 0 \\
\text{B.C} \qquad u(0,t) &= 0, \ u(\pi,t) = 0, \ t \geq 0 \\
\text{I.C} \qquad u(x,0) &= \sin x, \ u_t(x,0) = \sin x
\end{aligned} \tag{68}$$

Solution.

Following the discussion presented above we obtain

$$u(x,t) = \sin x + t\sin x + \frac{1}{2!}t^2 \sin x + L_t^{-1}\left(L_x u(x,t)\right). \tag{69}$$

Using the decomposition series for $u(x,t)$ we obtain

$$\sum_{n=0}^{\infty} u_n(x,t) = \sin x + t\sin x + \frac{1}{2!}t^2 \sin x + L_t^{-1}\left(L_x\left(\sum_{n=0}^{\infty} u_n(x,t)\right)\right). \tag{70}$$

Proceeding as before, we use the recursive algorithm

$$\begin{aligned}
u_0(x,t) &= \sin x + t\sin x + \tfrac{1}{2!}t^2 \sin x, \\
u_1(x,t) &= L_t^{-1}\left(L_x\left(u_0\right)\right) = -\tfrac{1}{2!}t^2 \sin x - \tfrac{1}{3!}t^3 \sin x - \tfrac{1}{4!}t^4 \sin x, \\
u_2(x,t) &= L_t^{-1}\left(L_x\left(u_1\right)\right) = \tfrac{1}{4!}t^4 \sin x + \tfrac{1}{5!}t^5 \sin x + \tfrac{1}{6!}t^6 \sin x,
\end{aligned} \tag{71}$$

and so on. In view of (71), the solution $u(x,t)$ in a series form is given by

$$\begin{aligned}
u(x,t) &= u_0(x,t) + u_1(x,t) + u_2(x,t) + \cdots \\
&= \sin x + \sin x \left(t - \frac{1}{3!}t^3 + \frac{1}{5!}t^5 - \cdots\right),
\end{aligned} \tag{72}$$

and in a closed form by

$$u(x,t) = \sin x + \sin x \sin t. \tag{73}$$

Example 9. Use the decomposition method to solve the initial-boundary value problem

$$\begin{array}{lll}
\text{PDE} & u_{tt} = u_{xx} + 6t + 2x, \ 0 < x < \pi, \ t > 0 \\[4pt]
\text{B.C} & u_x(0,t) = t^2 + \sin t, \ u_x(\pi,t) = t^2 - \sin t & (74) \\[4pt]
\text{I.C} & u(x,0) = 0, \ u_t(x,0) = \sin x
\end{array}$$

Solution.

Operating with L_t^{-1} on both sides of (74) yields

$$u(x,t) = t^3 + t^2 x + t\sin x + L_t^{-1}\left(L_x u(x,t)\right), \tag{75}$$

so that

$$\sum_{n=0}^{\infty} u_n(x,t) = t^3 + t^2 x + t\sin x + L_t^{-1}\left(L_x\left(\sum_{n=0}^{\infty} u_n(x,t)\right)\right). \tag{76}$$

Following our discussion above we find

$$
\begin{aligned}
u_0(x,t) &= t^3 + t^2 x + t \sin x, \\
u_1(x,t) &= L_t^{-1}\left(L_x\left(u_0\right)\right) = -\frac{1}{3!}t^3 \sin x, \\
u_2(x,t) &= L_t^{-1}\left(L_x\left(u_1\right)\right) = \frac{1}{5!}t^5 \sin x,
\end{aligned}
\tag{77}
$$

and so on. In view of (77), the solution $u(x,t)$ in a series form is given by

$$
\begin{aligned}
u(x,t) &= u_0(x,t) + u_1(x,t) + u_2(x,t) + \cdots \\
&= t^3 + t^2 x + \sin x \left(t - \frac{1}{3!}t^3 + \frac{1}{5!}t^5 - \cdots \right),
\end{aligned}
\tag{78}
$$

and in a closed form by

$$
u(x,t) = t^3 + t^2 x + \sin x \sin t.
\tag{79}
$$

Exercises   5.2.2

In Exercises 1 – 8, use the decomposition method to solve the following homogeneous partial differential equations:

1. $u_{tt} = 4u_{xx}$, $0 < x < \pi$, $t > 0$
   $u(0,t) = 0$, $u(\pi,t) = 0$, $t \geq 0$
   $u(x,0) = \sin(2x)$, $u_t(x,0) = 0$

2. $u_{tt} = u_{xx}$, $0 < x < \pi$, $t > 0$
   $u(0,t) = 0$, $u(\pi,t) = 0$, $t \geq 0$
   $u(x,0) = \sin x$, $u_t(x,0) = \sin x$

3. $u_{tt} = u_{xx}$, $0 < x < \pi$, $t > 0$
   $u(0,t) = 2 + \cos t$, $u(\pi,t) = 2 - \cos t$, $t \geq 0$
   $u(x,0) = 2 + \cos x$, $u_t(x,0) = 0$

4. $u_{tt} = u_{xx}$, $0 < x < \pi$, $t > 0$
   $u(0,t) = 1$, $u(\pi,t) = 1 + \pi$, $t \geq 0$
   $u(x,0) = 1 + x$, $u_t(x,0) = \sin x$

5. $u_{tt} = u_{xx} - 8u$, $0 < x < \pi$, $t > 0$
   $u(0,t) = 0$, $u(\pi,t) = 0$, $t \geq 0$
   $u(x,0) = \sin x$, $u_t(x,0) = 0$

6. $u_{tt} = u_{xx} - 3u$, $0 < x < \pi$, $t > 0$
   $u(0,t) = 0$, $u(\pi,t) = 0$, $t \geq 0$
   $u(x,0) = 0$, $u_t(x,0) = 2 \sin x$

7. $u_{tt} = u_{xx}$, $0 < x < \pi$, $t > 0$

$$u_x(0,t) = 0, \ u_x(\pi,t) = 0, \ t \geq 0$$
$$u(x,0) = \cos x, \ u_t(x,0) = 0$$

8. $u_{tt} = u_{xx}, \ 0 < x < \pi, \ t > 0$
$$u_x(0,t) = 1, \ u_x(\pi,t) = 1, \ t \geq 0$$
$$u(x,0) = x + \cos x, \ u_t(x,0) = 0$$

In Exercises 9 – 14, solve the inhomogeneous initial-boundary value problems:

9. $u_{tt} = u_{xx} + \cos x, \ 0 < x < \pi, \ t > 0$
$$u(0,t) = 1, \ u(\pi,t) = -1, \ t \geq 0$$
$$u(x,0) = \cos x, \ u_t(x,0) = \sin x$$

10. $u_{tt} = u_{xx} + \sin x, \ 0 < x < \pi, \ t > 0$
$$u(0,t) = 0, \ u(\pi,t) = 0, \ t \geq 0$$
$$u(x,0) = 2\sin x, \ u_t(x,0) = 0$$

11. $u_{tt} = u_{xx} - 3u + 3, \ 0 < x < \pi, \ t > 0$
$$u(0,t) = 1, \ u(\pi,t) = 1, \ t \geq 0$$
$$u(x,0) = 1, \ u_t(x,0) = 2\sin x$$

12. $u_{tt} = u_{xx} - 12x^2, \ 0 < x < \pi, \ t > 0$
$$u(0,t) = 0, \ u(\pi,t) = \pi^4, \ t \geq 0$$
$$u(x,0) = x^4 + \sin x, \ u_t(x,0) = 0$$

13. $u_{tt} = u_{xx} - 6x, \ 0 < x < \pi, \ t > 0$
$$u(0,t) = 0, \ u(\pi,t) = \pi^3, \ t \geq 0$$
$$u(x,0) = x^3, \ u_t(x,0) = \sin x$$

14. $u_{tt} = u_{xx} + \cos x, \ 0 < x < \pi, \ t > 0$
$$u(0,t) = 2, \ u(\pi,t) = 0, \ t \geq 0$$
$$u(x,0) = 1 + \cos x, \ u_t(x,0) = \sin x$$

In Exercises 15 – 20, solve the initial-boundary value problems:

15. $u_{tt} = u_{xx} - 4, \ 0 < x < \pi, \ t > 0$
$$u(0,t) = 0, \ u(\pi,t) = 2\pi^2, \ t \geq 0$$
$$u(x,0) = 2x^2 + \sin x, \ u_t(x,0) = 0$$

16. $u_{tt} = u_{xx} - 2, \ 0 < x < \pi, \ t > 0$
$$u(0,t) = \sin t, \ u(\pi,t) = \pi^2 - \sin t, \ t \geq 0$$
$$u(x,0) = x^2, \ u_t(x,0) = \cos x$$

17. $u_{tt} = u_{xx} + \sin x, \ 0 < x < \pi, \ t > 0$
$$u_x(0,t) = 1, \ u_x(\pi,t) = -1, \ t \geq 0$$
$$u(x,0) = \sin x, \ u_t(x,0) = \cos x$$

18. $u_{tt} = u_{xx} - 2, \ 0 < x < \pi, \ t > 0$
$$u_x(0,t) = 0, \ u_x(\pi,t) = 2\pi, \ t \geq 0$$

$$u(x,0) = x^2 + \cos x, \ u_t(x,0) = 0$$

19. $u_{tt} = u_{xx} + 12t^2 + 6xt, \ 0 < x < \pi, \ t > 0$
    $u_x(0,t) = t^3, \ u_x(\pi,t) = t^3, \ t \geq 0$
    $u(x,0) = 0, \ u_t(x,0) = \cos x$

20. $u_{tt} = u_{xx} - 6x + 2, \ 0 < x < \pi, \ t > 0$
    $u(0,t) = t^2, \ u(\pi,t) = t^2 + \pi^3, \ t \geq 0$
    $u(x,0) = x^3, \ u_t(x,0) = \sin x$

In Exercise 21   24, solve the partial differential equations where coefficient of u_{xx} is a function:

21. $u_{tt} = \frac{x^2}{2} u_{xx}, \ 0 < x < 1, \ t > 0$
    $u(0,t) = 0, \ u(1,t) = \cosh t, \ t \geq 0$
    $u(x,0) = x^2, \ u_t(x,0) = 0$

22. $u_{tt} = \frac{x^2}{2} u_{xx}, \ 0 < x < 1, \ t > 0$
    $u(0,t) = 0, \ u_x(1,t) = 2e^t, \ t \geq 0$
    $u(x,0) = x^2, \ u_t(x,0) = x^2$

23. $u_{tt} = \frac{x^2}{12} u_{xx}, \ 0 < x < 1, \ t > 0$
    $u(0,t) = 0, \ u_x(1,t) = 4\sinh t, \ t \geq 0$
    $u(x,0) = 0, \ u_t(x,0) = x^4$

24. $u_{tt} = \frac{x^2}{6} u_{xx}, \ 0 < x < 1, \ t > 0$
    $u(0,t) = 0, \ u_x(1,t) = 3\cosh t, \ t \geq 0$
    $u(x,0) = x^3, \ u_t(x,0) = 0$

5.2.3   Wave Equation In An Infinite Domain

The **initial value problem** of the one dimensional wave equation, where the domain of the space variable x is unbounded, will be discussed by using Adomian decomposition method. This type of equations describes the motion of a very long string that is considered not to have boundaries. Based on this, the wave motion is described by a PDE and initial conditions only, therefore, it is called initial value problem. It was discovered before that solutions of the wave equation behave quite differently than solutions of the heat equation. The solution $u(x,t)$ of the wave equation represents the displacement of the point x at time $t \geq 0$.

It is interesting to note that the classical method of separation of variables is not applicable for this type of problems because of the lack of boundary conditions. However, a classical method called D'Alembert solution is usually used. The D'Alembert solution will be discussed later.

In this section, the Adomian decomposition method will be used to handle the wave equation where the space of the variable x is unbounded. Recall that Adomian's method can easily handle problems with initial conditions only.

To achieve our goal, we consider the initial value problem:

$$\text{PDE} \qquad u_{tt} = c^2 u_{xx}, \quad -\infty < x < \infty, \ t > 0$$

$$\text{I.C} \qquad u(x,0) = f(x), \ u_t(x,0) = g(x)$$

(80)

The attention will be focused upon the disturbance occurred at the center of the very long string. The initial displacement $u(x,0)$ and the initial displacement $u_t(x,0)$ are prescribed by $f(x)$ and $g(x)$ respectively.

Applying the inverse operator L_t^{-1} to both sides of the operator form of (80) and using the initial conditions we find

$$u(x,t) = f(x) + tg(x) + c^2 L_t^{-1}(L_x u(x,t)).$$

(81)

Identifying the zeroth component $u_0(x,t)$ and proceeding as before we find

$$u_0(x,t) = f(x) + tg(x),$$

$$u_1(x,t) = c^2 L_t^{-1}(L_x u_0),$$

$$= f''(x)\frac{(ct)^2}{2!} + c^2 g''(x)\frac{t^3}{3!},$$

$$u_2(x,t) = c^2 L_t^{-1}(L_x u_1),$$

$$= f^{(iv)}(x)\frac{(ct)^4}{4!} + c^4 g^{(iv)}(x)\frac{t^5}{5!},$$

(82)

and so on. In view of (82), the solution $u(x,t)$ of (80) in a series form is given by

$$u(x,t) = \left(f(x) + f''(x)\frac{(ct)^2}{2!} + f^{(iv)}(x)\frac{(ct)^4}{4!} + \cdots \right)$$
$$+ \left(g(x)t + c^2 g''(x)\frac{t^3}{3!} + c^4 g^{(iv)}(x)\frac{t^5}{5!} + \cdots \right),$$

(83)

or equivalently

$$u(x,t) = \sum_{n=0}^{\infty} \left(\frac{(ct)^{2n}}{(2n)!} f^{(2n)}(x) + c^{2n} \frac{t^{2n+1}}{(2n+1)!} g^{(2n)}(x) \right),$$

(84)

An important conclusion can be made here in that the series solution (84) is easily obtained because it relies completely on differentiating the initial conditions $f(x)$ and $g(x)$ which is mostly an east task. However, as will be seen later, the D'Alembert solution requires integrating $g(x)$ which is not always easily integrable such as $g(x) = e^{-x^2}$. The approach we followed will be illustrated by discussing the following examples.

Example 10. Use the Adomian decomposition method to solve the initial value problem

$$
\begin{array}{lll}
\text{PDE} & u_{tt} = 16 \quad u_{xx}, \quad -\infty < x < \infty, \ t > 0 \\
\\
\text{I.C} & u(x,0) = \sin x, \ u_t(x,0) = 2
\end{array}
\tag{85}
$$

Solution.

Note that $c = 4$, $f(x) = \sin x$ and $g(x) = 2$. We can easily apply the inverse operator as used in other examples. However, for simplicity reasons, we will use the result (84) hence we set

$$
f^{(2n)}(x) = (-1)^n \sin x, \quad n = 0, 1, 2, \cdots,
\tag{86}
$$

and

$$
g^{(2n)}(x) = \begin{cases} 2 & \text{for} \quad n = 0 \\[2mm] 0 & \text{for} \quad n = 1, 2, \cdots \end{cases}
\tag{87}
$$

The solution in a series form is readily obtained by substituting (86) and (87) into (83) and given by

$$
u(x,t) = \sin x \left(1 - \frac{(4t)^2}{2!} + \frac{(4t)^4}{4!} + \cdots \right) + 2t,
\tag{88}
$$

and

$$
u(x,t) = \sin x \cos(4t) + 2t.
\tag{89}
$$

Example 11. Use the Adomian decomposition method to solve the initial value problem

$$
\begin{array}{lll}
\text{PDE} & u_{tt} = 4 \quad u_{xx}, \quad -\infty < x < \infty, \ t > 0 \\
\\
\text{I.C} & u(x,0) = \sin x, \ u_t(x,0) = 2 \cos x
\end{array}
\tag{90}
$$

Solution.

Note that $c = 2$, $f(x) = \sin x$ and $g(x) = 2 \cos x$. Proceeding as before, we set

$$
f^{(2n)}(x) = (-1)^n \sin x, \quad n = 0, 1, 2, \cdots,
\tag{91}
$$

and

$$
g^{(2n)}(x) = 2(-1)^n \cos x, \quad n = 0, 1, 2, \cdots.
\tag{92}
$$

The solution in a series form is readily obtained by substituting (92) and (91) into (83) and given by

$$
\begin{aligned}
u(x,t) \;=\;& \sin x \left(1 - \frac{(2t)^2}{2!} + \frac{(2t)^4}{4!} + \cdots\right) \\
& + \cos x \left((2t) - \frac{(2t)^3}{3!} + \frac{(2t)^5}{5!} - \cdots\right),
\end{aligned}
\tag{93}
$$

and in a closed form by

$$
\begin{aligned}
u(x,t) \;=\;& \sin x \cos(2t) + \cos x \sin(2t), \\
\;=\;& \sin(x + 2t).
\end{aligned}
\tag{94}
$$

Example 12. Use the decomposition method to solve the initial value problem

$$
\text{PDE} \qquad u_{tt} \;=\; u_{xx} + xt^2 + t^3, -\infty < x < \infty, t > 0
\tag{95}
$$

$$
\text{I.C} \qquad u(x,0) \;=\; 0,\; u_t(x,0) = \sin x
$$

Solution.

Note that the initial value problem is inhomogeneous. Operating with L_t^{-1} on both sides of (95) and using the initial conditions we obtain

$$
u(x,t) = xt^2 + t^3 + \sin x + L_t^{-1}\left(L_x u(x,t)\right).
\tag{96}
$$

Using the decomposition series for $u(x,t)$ we obtain

$$
\sum_{n=0}^{\infty} u_n(x,t) = xt^2 + t^3 + \sin x + t \sin x + L_t^{-1}\left(L_x \left(\sum_{n=0}^{\infty} u_n(x,t)\right)\right).
\tag{97}
$$

Proceeding as before, we use the recursive algorithm

$$
\begin{aligned}
u_0(x,t) \;&=\; xt^2 + t^3 + \sin x, \\
u_1(x,t) \;&=\; L_t^{-1}\left(L_x\left(u_0\right)\right) = -\frac{1}{3!}t^3 \sin x, \\
u_2(x,t) \;&=\; L_t^{-1}\left(L_x\left(u_2\right)\right) = \frac{1}{5!}t^5 \sin x,
\end{aligned}
\tag{98}
$$

and so on. In view of (98), the solution $u(x,t)$ in a series form is given by

$$
\begin{aligned}
u(x,t) \;&=\; u_0(x,t) + u_1(x,t) + u_2(x,t) + \cdots \\
\;&=\; xt^2 + t^3 + \sin x \left(t - \frac{1}{3!}t^3 + \frac{1}{5!}t^5 - \cdots\right).
\end{aligned}
\tag{99}
$$

and in a closed form by

$$u(x,t) = xt^2 + t^3 + \sin x \sin t. \tag{100}$$

Example 13. Use the decomposition method to solve the initial value problem

$$
\begin{array}{lll}
\text{PDE} & u_{tt} = & u_{xx} + e^{-t}, \ -\infty < x < \infty, \ t > 0 \\
\text{I.C} & u(x,0) = & 1, \ u_t(x,0) = -1 + \sin x
\end{array}
\tag{101}
$$

Solution.

Note that the initial value problem is inhomogeneous. Operating with L_t^{-1} on both sides of (101) gives

$$u(x,t) = t \sin x + e^{-t} + L_t^{-1}\left(L_x u(x,t)\right), \tag{102}$$

so that

$$\sum_{n=0}^{\infty} u_n(x,t) = t \sin x + e^{-t} + L_t^{-1}\left(L_x\left(\sum_{n=0}^{\infty} u_n(x,t)\right)\right). \tag{103}$$

Proceeding as before we find

$$
\begin{array}{lll}
u_0(x,t) & = & t \sin x + e^{-t}, \\
u_1(x,t) & = & -\frac{1}{3!}t^3 \sin x, \\
u_2(x,t) & = & \frac{1}{5!}t^5 \sin x,
\end{array}
\tag{104}
$$

and so on. In view of (104), the solution $u(x,t)$ in a series form is given by

$$
\begin{aligned}
u(x,t) & = u_0(x,t) + u_1(x,t) + u_2(x,t) + \cdots \\
& = e^{-t} + \sin x \left(t - \frac{1}{3!}t^3 + \frac{1}{5!}t^5 - \cdots\right),
\end{aligned}
\tag{105}
$$

and in a closed form by

$$u(x,t) = e^{-t} + \sin x \sin t. \tag{106}$$

Exercises   5.2.3

In Exercises 1 – 8, use the decomposition method to solve the following initial value problems:

1. $u_{tt} = u_{xx}, \ -\infty < x < \infty, \ t > 0$
   $u(x,0) = 0, \ u_t(x,0) = 4 + \sin x$

2. $u_{tt} = u_{xx},\ -\infty < x < \infty,\ t > 0$
   $u(x,0) = \sin x,\ u_t(x,0) = \cos x$

3. $u_{tt} = u_{xx},\ -\infty < x < \infty,\ t > 0$
   $u(x,0) = \cos x,\ u_t(x,0) = -\sin x$

4. $u_{tt} = u_{xx},\ -\infty < x < \infty,\ t > 0$
   $u(x,0) = \sin x,\ u_t(x,0) = -\cos x$

5. $u_{tt} = u_{xx},\ -\infty < x < \infty,\ t > 0$
   $u(x,0) = \sin x,\ u_t(x,0) = 6$

6. $u_{tt} = u_{xx} + 4x,\ -\infty < x < \infty,\ t > 0$
   $u(x,0) = 0,\ u_t(x,0) = 6$

7. $u_{tt} = u_{xx} + 4t,\ -\infty < x < \infty,\ t > 0$
   $u(x,0) = 0,\ u_t(x,0) = x^2 + e^x$

8. $u_{tt} = u_{xx} + xe^t,\ -\infty < x < \infty,\ t > 0$
   $u(x,0) = x,\ u_t(x,0) = x + \cos x$

9. $u_{tt} = u_{xx},\ -\infty < x < \infty,\ t > 0$
   $u(x,0) = x^2,\ u_t(x,0) = \sin x$

10. $u_{tt} = u_{xx} - \cos x,\ -\infty < x < \infty,\ t > 0$
    $u(x,0) = 0,\ u_t(x,0) = 1 + 2x$

11. $u_{tt} = u_{xx} - \sin x,\ -\infty < x < \infty,\ t > 0$
    $u(x,0) = x^2 - \sin x,\ u_t(x,0) = \sin x$

12. $u_{tt} = u_{xx} + \cos x,\ -\infty < x < \infty,\ t > 0$
    $u(x,0) = 2\cos x,\ u_t(x,0) = 0$

5.3   Method of Separation of Variables

In this section the homogeneous partial differential equation that describes the vibrations of a vibrating string will be discussed by using a well-known method called the *method of separation of variables*. The most important feature of the method of separation of variables is that it reduces the partial differential equation into a system of ordinary differential equations that can be easily handled.

5.3.1   Analysis of the Method

As discussed before in the heat equation, the method of separation of variables requires that the PDE and the boundary conditions be linear and homogeneous. For this reason, we begin our analysis by discussing the vibrations of a freely vibrating string with fixed ends at $x = 0$ and $x = L$, initial position $u(x,0) = f(x)$ and initial velocity $u_t(x,0) = g(x)$. The initial-boundary value problem that controls the vibrations of a string is given by

$$
\begin{array}{lrcl}
\text{PDE} & u_{tt} & = & c^2 u_{xx}, \ 0 < x < L, \ t > 0 \\[2mm]
\text{B.C} & u(0,t) & = & 0, \ u(L,t) = 0 \\[2mm]
\text{I.C} & u(x,0) & = & f(x), \ u_t(x,0) = g(x)
\end{array}
\tag{107}
$$

The wave function $u(x,t)$ is the displacement of any point of a vibrating string at position x at time t. The method of separation of variables consists of assuming that the displacement $u(x,t)$ is identified as the product of two distinct functions $F(x)$ and $T(t)$, where $F(x)$ depends on the space variable x and $T(t)$ depends on the time variable t. This assumption allows us to set

$$
u(x,t) = F(x)T(t), \tag{108}
$$

assuming that $F(x)$ and $T(t)$ are twice continuously differentiable. Differentiating both sides of (108) twice with respect to t and twice with respect to x we obtain

$$
\begin{aligned}
u_{tt}(x,t) &= F(x)T^{''}(t), \\[2mm]
u_{xx}(x,t) &= F^{''}(x)T(t).
\end{aligned}
\tag{109}
$$

Substituting (109) into (107) yields

$$
F(x)T^{''}(t) = c^2 F^{''}(x)T(t). \tag{110}
$$

Dividing both sides of (110) by $c^2 F(x)T(t)$ gives

$$
\frac{T^{''}(t)}{c^2 T(t)} = \frac{F^{''}(x)}{F(x)}. \tag{111}
$$

The left hand side of (111) depends only on t and the right hand side depends only on x. This means that the equality holds only if both sides are equal to the same constant. Therefore, we set

$$
\frac{T^{''}(t)}{c^2 T(t)} = \frac{F^{''}(x)}{F(x)} = -\lambda^2. \tag{112}
$$

The selection of $-\lambda^2$ in (112) is essential to obtain nontrivial solutions. However, we can easily show that selecting the constant to be zero or λ^2 will produce the trivial solution $u(x,t) = 0$.

The result (112) gives two distinct ordinary differential equations given by

$$
\begin{aligned}
F^{''}(x) + \lambda^2 F(x) &= 0, \\[2mm]
T^{''}(t) + c^2 \lambda^2 T(t) &= 0.
\end{aligned}
\tag{113}
$$

This means that the partial differential equation of (107) is reduced to the more familiar second order ordinary differential equations ODEs (113) where each equation relies only on one distinct variable.

To determine the function $F(x)$, we solve the second order linear ODE

$$F^{''}(x) + \lambda^2 F(x) = 0, \tag{114}$$

to find that

$$F(x) = A\cos(\lambda x) + B\sin(\lambda x), \tag{115}$$

where A and B are constants. To determine the constants A, B, and λ, we use the homogeneous boundary conditions

$$u(0,t) \quad = \quad 0, \quad u(L,t) \quad = \quad 0, \tag{116}$$

as given above by (107). Substituting (116) into the assumption (108) gives

$$F(0) \quad = \quad 0, \quad F(L) \quad = \quad 0. \tag{117}$$

Using (117) into (115) leads to

$$A = 0, \tag{118}$$

and

$$\sin(\lambda L) = 0. \tag{119}$$

We exclude $B = 0$ since it gives the trivial solution $u(x,t) = 0$. Accordingly, we find

$$\lambda_n = \frac{n\pi}{L}, \quad n = 1, 2, 3, \cdots. \tag{120}$$

It is important to note that $n = 0$ is excluded since it gives the trivial solution $u(x,t) = 0$. The function $F_n(x)$ associated with λ_n is

$$F_n(x) = \sin(\frac{n\pi}{L}x), n = 1, 2, 3, \cdots. \tag{121}$$

Consequently, the solution $T_n(t)$ associated with λ_n must satisfy

$$T_n^{''}(t) + c^2 \lambda_n^2 T_n(t) = 0. \tag{122}$$

The general solution of (122) is given by

$$\begin{aligned} T_n(t) \quad &= \quad C_n\cos(\lambda_n ct) + D_n\sin(\lambda_n ct), \\ &= \quad C_n\cos(\tfrac{n\pi c}{L}t) + D_n\sin(\tfrac{n\pi c}{L}t), n = 1, 2, 3, \cdots, \end{aligned} \tag{123}$$

where C_n and D_n are constants.

Combining the results (121) and (123) we obtain the infinite sequence of product functions

$$
\begin{aligned}
u_n(x,t) &= F_n(x)T_n(t), \\
&= \sin(\tfrac{n\pi}{L}x)\left(C_n\cos(\tfrac{n\pi c}{L}t) + D_n\sin(\tfrac{n\pi c}{L}t)\right), n = 1, 2, \cdots.
\end{aligned}
\tag{124}
$$

Recall that the superposition principle admits that a linear combination of the functions $u_n(x,t)$ also satisfies the given equation and the boundary conditions. Therefore, using this principle gives the general solution by

$$
u(x,t) = \sum_{n=1}^{\infty} \sin(\frac{n\pi}{L}x)\left(C_n\cos(\frac{n\pi c}{L}t) + D_n\sin(\frac{n\pi c}{L}t)\right),
\tag{125}
$$

where the arbitrary constants $C_n, D_n, n \geq 1$, are as yet undetermined. The derivative of (125) with respect to t is

$$
u_t(x,t) = \sum_{n=1}^{\infty} \sin(\frac{n\pi}{L}x)\left(-\frac{n\pi c}{L}C_n\sin(\frac{n\pi c}{L}t) + \frac{n\pi c}{L}D_n\cos(\frac{n\pi c}{L}t)\right).
\tag{126}
$$

To determine $C_n, n \geq 1$, we substitute $t = 0$ in (125) and by using the initial condition $u(x,0) = f(x)$, we obtain

$$
\sum_{n=1}^{\infty} C_n\sin(\frac{n\pi}{L}x) = f(x),
\tag{127}
$$

so that the constants C_n can be determined in this case by using Fourier coefficients given by the formula

$$
C_n = \frac{2}{L}\int_0^L f(x)\sin(\frac{n\pi}{L}x)\,dx.
\tag{128}
$$

To determine $D_n, n \geq 1$, we substitute $t = 0$ in (126) and by using the initial condition $u_t(x,0) = g(x)$ we obtain

$$
\sum_{n=1}^{\infty} \frac{n\pi c}{L}D_n\sin(\frac{n\pi}{L}x) = g(x),
\tag{129}
$$

so that

$$
D_n = \frac{2}{n\pi c}\int_0^L g(x)\sin(\frac{n\pi}{L}x)\,dx.
\tag{130}
$$

Having determined the constants C_n and D_n, the particular solution $u(x,t)$ follows immediately upon substituting (128) and (130) into (125).

It is to be noted that the use of the Fourier coefficients requires a considerable size of calculations. However, if the initial conditions $f(x)$ and $g(x)$ are given in terms of $\sin(nx)$ and $\cos(mx)$, it seems reasonable to expand the solution (125) and then equate the coefficients of like terms in both sides to determine the constants C_n and D_n.

To give a clear overview of the method of separation of variables, we have selected several examples of homogeneous PDEs with homogeneous boundary conditions to illustrate the discussion presented above.

Example 1. Use the method of separation of variables to solve the following initial-boundary value problem

$$\text{PDE} \qquad u_{tt} \;=\; u_{xx}, \; 0 < x < \pi, \; t > 0$$

$$\text{B.C} \qquad u(0,t) \;=\; 0, \; u(\pi,t) = 0 \tag{131}$$

$$\text{I.C} \qquad u(x,0) \;=\; \sin(2x), \; u_t(x,0) = 0$$

Solution.

We first set
$$u(x,t) = F(x)T(t). \tag{132}$$

Using (132) into (131) and proceeding as discussed before we find

$$F(x) = A\cos(\lambda x) + B\sin(\lambda x). \tag{133}$$

$$T(t) = C\cos(\lambda t) + B\sin(\lambda t). \tag{134}$$

Using the boundary conditions of (131) into (133) gives

$$A \;=\; 0,$$
$$\lambda_n \;=\; n, \; n = 1,2,3,\cdots. \tag{135}$$

Based on this, equations (133) and (134) become

$$F_n(x) \;=\; B_n\sin(nx), \; n = 1,2,\cdots,$$
$$T_n(t) \;=\; C_n\cos(nt) + D_n\sin(nt). \tag{136}$$

This gives the infinite sequence of product functions

$$u_n(x,t) = \sin(nx)\left(C_n\cos(nt) + D_n\sin(nt)\right). \tag{137}$$

Using the superposition principle we obtain

$$u(x,t) = \sum_{n=1}^{\infty}\sin(nx)\left(C_n\cos(nt) + D_n\sin(nt)\right). \tag{138}$$

and its derivative with respect to t is

$$u_t(x,t) = \sum_{n=1}^{\infty} \sin(nx)\left(-nC_n\sin(nt) + nD_n\cos(nt)\right). \tag{139}$$

To determine C_n, we use the initial condition $u(x,0) = \sin(2x)$ and substitute $t = 0$ in (138) to find

$$C_1\sin(x) + C_2\sin(2x) + C_3\sin(3x) + \cdots = \sin(2x). \tag{140}$$

Equating the coefficients of like terms of both sides gives

$$C_2 = 1,\ C_j = 0, j \neq 2. \tag{141}$$

To determine D_n, substitute $t = 0$ in (139), and use the initial condition $u_t(x,0) = 0$ to find

$$D_1\sin(x) + 2D_2\sin(2x) + 3D_3\sin(3x) + \cdots = 0, \tag{142}$$

so that

$$D_j = 0,\ j \geq 1. \tag{143}$$

Combining (138), (141), and (143), the particular solution is given by

$$u(x,t) = \sin(2x)\cos(2t). \tag{144}$$

It is obvious that the particular solution (144) is obtained by expanding (138) and (139), using the initial conditions, and by equating coefficients of like terms of both sides.

Example 2. Use the method of separation of variables to solve the following initial-boundary value problem

$$
\begin{array}{lrcl}
\text{PDE} & u_{tt} &=& u_{xx},\ 0 < x < \pi,\ t > 0 \\[4pt]
\text{B.C} & u(0,t) &=& 0,\ u(\pi,t) = 0 \\[4pt]
\text{I.C} & u(x,0) &=& \sin x,\ u_t(x,0) = 2\sin x
\end{array}
\tag{145}
$$

Solution.

We first set

$$u(x,t) = F(x)T(t). \tag{146}$$

Proceeding as before and substituting (146) into (145), the general solution is given by

$$u(x,t) = \sum_{n=1}^{\infty} \sin(nx)\left(C_n\cos(nt) + D_n\sin(nt)\right). \tag{147}$$

To determine C_n, we use the initial condition $u(x,0) = \sin x$ and replace t by zero in (147) to find

$$C_1 \sin(x) + C_2 \sin(2x) + \cdots = \sin(x). \tag{148}$$

Equating coefficients of like terms of both sides of (148) gives

$$C_1 = 1, \; C_j = 0, \; j \neq 1. \tag{149}$$

To determine D_n, we use the initial condition $u_t(x,0) = 2\sin x$ and replace t by zero in the derivative of (147) to find

$$D_1 \sin(x) + 2D_2 \sin(2x) + \cdots = 2\sin(x). \tag{150}$$

Equating coefficients of like terms of both sides of (150) gives

$$D_1 = 2, \; D_j = 0, \; j \neq 1. \tag{151}$$

Combining the results (147), (149) and (151), the particular solution is given by

$$u(x,t) = \sin x \cos t + 2\sin x \sin t. \tag{152}$$

Example 3. Use the method of separation of variables to solve the following initial-boundary value problem

$$
\begin{array}{lrl}
\text{PDE} & u_{tt} &= u_{xx}, \; 0 < x < \pi, \; t > 0 \\[4pt]
\text{B.C} & u_x(0,t) &= 0, \; u_x(\pi,t) = 0 \\[4pt]
\text{I.C} & u(x,0) &= 0, \; u_t(x,0) = \cos x
\end{array} \tag{153}
$$

Solution.

It is interesting to note that the boundary conditions are of the second kind defined by the derivatives $u_x(0,t) = 0$ and $u_x(\pi,t) = 0$. This means that the ends of the string are free and not fixed. We first set

$$u(x,t) = F(x)T(t). \tag{154}$$

Proceeding as before and substituting (154) into (153) and solving the resulting equations we obtain

$$F(x) = A\cos(\lambda x) + B\sin(\lambda x), \tag{155}$$

so that by using the boundary conditions we find

$$
\begin{array}{rcl}
B &=& 0, \; A \neq 0, \\[6pt]
\lambda_n &=& n, \; n = 0, 1, 2, 3, \cdots.
\end{array} \tag{156}
$$

Note that $\lambda_n = 0$ is considered because it did not give the trivial solution. Accordingly, we find an infinite number of solutions for $F_n(x)$ defined by

$$F_n(x) = A_n \cos(nx), n = 0, 1, 2, 3, \cdots. \tag{157}$$

Using the values obtained for λ_n, we obtain

$$T_n(t) = C_n \cos(nt) + D_n \sin(nt), \tag{158}$$

where C_n and D_n are as yet undetermined constants.

Using the superposition principle, the general solution is given by

$$u(x,t) = \sum_{n=0}^{\infty} \cos(nx) \left(C_n \cos(nt) + D_n \sin(nt) \right). \tag{159}$$

To determine C_n, we use the initial condition $u(x,0) = 0$ and replace t by zero in (159) to find

$$C_0 + C_1 \cos(x) + C_2 \cos(2x) + \cdots = 0, \tag{160}$$

and this gives

$$C_j = 0, j \geq 0. \tag{161}$$

To determine D_n, we use the initial condition $u_t(x,0) = \cos x$ and replace t by zero in the derivative of (159) with respect to t to find

$$D_1 \cos(x) + 2D_2 \cos(2x) + \cdots = \cos(x), \tag{162}$$

so that

$$D_1 = 1, D_j = 0, j \neq 1. \tag{163}$$

Combining the results (159), (161) and (163), the particular solution is given by

$$u(x,t) = \cos x \sin t, \tag{164}$$

obtained by substituting (161) and (163) into (159).

Example 4. Use the method of separation of variables to solve the following initial-boundary value problem

$$
\begin{array}{lll}
\text{PDE} & u_{tt} = u_{xx}, & 0 < x < \pi, \ t > 0 \\
\text{B.C} & u_x(0,t) = 0, \ u_x(\pi,t) = 0 & \\
\text{I.C} & u(x,0) = 1 + \cos x, \ u_t(x,0) = 0 &
\end{array} \tag{165}
$$

Solution.

Proceeding as before we obtain

$$F(x) = A \cos(\lambda x) + B \sin(\lambda x), \tag{166}$$

so that by using the boundary conditions we find

$$B = 0, A \neq 0,$$
$$\lambda_n = n, n = 0, 1, 2, 3, \cdots.$$
$$(167)$$

Consequently we find

$$F_n(x) = A_n \cos(nx), n = 0, 1, 2, 3, \cdots. \qquad (168)$$

Using the values obtained for λ_n, we obtain

$$T_n(t) = C_n \cos(nt) + D_n \sin(nt). \qquad (169)$$

Using the superposition principle gives the general solution by

$$u(x, t) = \sum_{n=0}^{\infty} \cos(nx) \left(C_n \cos(nt) + D_n \sin(nt) \right), \qquad (170)$$

so that

$$u_t(x, t) = \sum_{n=0}^{\infty} \cos(nx) \left(-nC_n \sin(nt) + nD_n \cos(nt) \right). \qquad (171)$$

To determine C_n, we replace t by zero in (170) to find

$$C_0 + C_1 \cos(x) + C_2 \cos(2x) + \cdots = 1 + \cos x, \qquad (172)$$

and this gives

$$C_0 = 1, \ C_1 = 1, \ C_j = 0, j \geq 2. \qquad (173)$$

We next use the initial condition $u_t(x, 0) = 0$ and replace t by zero in the derivative of (170) to find

$$D_1 \cos(x) + 2D_2 \cos(2x) + \cdots = 0, \qquad (174)$$

so that

$$D_j = 0, \ j \geq 1. \qquad (175)$$

Combining the results obtained above gives

$$u(x, t) = 1 + \cos x \cos t. \qquad (176)$$

Example 5. Use the method of separation of variables to solve the following initial-boundary value problem

$$
\begin{array}{llll}
\text{PDE} & u_{tt} & = & u_{xx}, \ 0 < x < \pi, \ t > 0 \\
\text{B.C} & u(0, t) & = & 0, \ u(\pi, t) = 0 \\
\text{I.C} & u(x, 0) & = & 1, \ u_t(x, 0) = 0
\end{array}
\qquad (177)
$$

Solution.

Following the analysis introduced before leads to

$$F_n(x) = A_n \sin(nx), n = 1, 2, 3, \cdots. \tag{178}$$

and

$$T_n(t) = C_n \cos(nt) + D_n \sin(nt), \tag{179}$$

which in turn gives

$$u(x,t) = \sum_{n=0}^{\infty} \sin(nx) \left(C_n \cos(nt) + D_n \sin(nt) \right). \tag{180}$$

To determine C_n, we substitute $t = 0$ into (180), and use the initial condition $u(x,0) = 1$ to find

$$\sum_{n=1}^{\infty} C_n \sin(nx) = 1. \tag{181}$$

The arbitrary constants C_n are determined by using the Fourier coefficients method, therefore we find

$$C_n = \frac{2}{\pi} \int_0^{\pi} \sin(nx)dx = \frac{2}{n\pi}(1 - \cos(n\pi)), \tag{182}$$

so that

$$C_n = 0 \text{ if } n \text{ is even}, \quad \frac{4}{n\pi} \text{ if } n \text{ is odd}. \tag{183}$$

This means that we can express C_n by

$$
\begin{aligned}
C_{2m} &= 0, \\
C_{2m+1} &= \frac{4}{(2m+1)\pi}, \ m = 0, 1, 2, \cdots.
\end{aligned}
\tag{184}
$$

To determine D_n, we substitute $t = 0$ into (180) to find

$$D_n = 0, n = 1, 2, 3, \cdots. \tag{185}$$

Combining the results obtained above, the particular solution is given by

$$u(x,t) = \frac{4}{\pi} \sum_{m=0}^{\infty} \frac{1}{(2m+1)} \sin(2m+1)x \, \cos(2m+1)t. \tag{186}$$

Exercises 5.3.1

In Exercises 1 – 6, where the ends of the string are fixed, solve the initial-boundary value problems by the method of separation of variables:

1. $u_{tt} = u_{xx}$, $0 < x < \pi$, $t > 0$
   $u(0, t) = 0$, $u(\pi, t) = 0$
   $u(x, 0) = 0$, $u_t(x, 0) = 3\sin(3x)$

2. $u_{tt} = u_{xx}$, $0 < x < \pi$, $t > 0$
   $u(0, t) = 0$, $u(\pi, t) = 0$
   $u(x, 0) = \sin x$, $u_t(x, 0) = 0$

3. $u_{tt} = u_{xx}$, $0 < x < \pi$, $t > 0$
   $u(0, t) = 0$, $u(\pi, t) = 0$
   $u(x, 0) = 0$, $u_t(x, 0) = 4\sin(2x)$

4. $u_{tt} = 4u_{xx}$, $0 < x < \pi$, $t > 0$
   $u(0, t) = 0$, $u(\pi, t) = 0$
   $u(x, 0) = \sin x$, $u_t(x, 0) = 0$

5. $u_{tt} = 4u_{xx}$, $0 < x < \pi$, $t > 0$
   $u(0, t) = 0$, $u(\pi, t) = 0$
   $u(x, 0) = \sin(2x)$, $u_t(x, 0) = 0$

6. $u_{tt} = 9u_{xx}$, $0 < x < \pi$, $t > 0$
   $u(0, t) = 0$, $u(\pi, t) = 0$
   $u(x, 0) = 0$, $u_t(x, 0) = 3\sin x$

In Exercises 7 – 10, where the ends of the string are free, solve the initial-boundary value problems by the method of separation of variables:

7. $u_{tt} = 9u_{xx}$, $0 < x < \pi$, $t > 0$
   $u_x(0, t) = 0$, $u_x(\pi, t) = 0$
   $u(x, 0) = 1$, $u_t(x, 0) = 3\cos x$

8. $u_{tt} = 4u_{xx}$, $0 < x < \pi$, $t > 0$
   $u_x(0, t) = 0$, $u_x(\pi, t) = 0$
   $u(x, 0) = 2 + \cos x$, $u_t(x, 0) = 0$

9. $u_{tt} = 9u_{xx}$, $0 < x < \pi$, $t > 0$
   $u_x(0, t) = 0$, $u_x(\pi, t) = 0$
   $u(x, 0) = 0$, $u_t(x, 0) = 3\cos x$

10. $u_{tt} = u_{xx}$, $0 < x < \pi$, $t > 0$
    $u_x(0, t) = 0$, $u_x(\pi, t) = 0$
    $u(x, 0) = \cos x$, $u_t(x, 0) = \cos x$

In Exercises 11 – 12, use the Fourier coefficients to solve the following initial-boundary value problems:

11. $u_{tt} = u_{xx}$, $0 < x < \pi$, $t > 0$
    $u(0, t) = 0$, $u(\pi, t) = 0$

$$u(x,0) = 0, u_t(x,0) = x$$

12. $u_{tt} = 4u_{xx}, \ 0 < x < \pi, \ t > 0$
   $u(0,t) = 0, \ u(\pi,t) = 0$
   $u(x,0) = 0, u_t(x,0) = x(1-x)$

5.3.2 Inhomogeneous Boundary Conditions

In this section we will consider the case where the boundary conditions of the vibrating string are inhomogeneous. It is well known that the method of separation of variables requires that the equation and the boundary conditions are linear and homogeneous. Therefore, transformation formulas should be used to convert the inhomogeneous boundary conditions to homogeneous boundary conditions.

In this section we will discuss wave equations where Dirichlet boundary conditions and Neumann boundary conditions are not homogeneous. It is normal to seek transformation formulas to convert these inhomogeneous conditions to homogeneous conditions.

Dirichlet Boundary Conditions

In this first type of boundary conditions, the displacements $u(0,t) = \alpha$ and $u(L,t) = \beta$ of a vibrating string of length L are given. We begin our analysis by considering the initial-boundary value problem

$$\begin{aligned}
\text{PDE} \qquad u_{tt} &= c^2 u_{xx}, \ 0 < x < L, \ t > 0 \\
\text{B.C} \qquad u(0,t) &= \alpha, \ u(L,t) = \beta, \ t \geq 0 \\
\text{I.C} \qquad u(x,0) &= f(x), \ u_t(x,0) = g(x)
\end{aligned} \qquad (187)$$

To convert the inhomogeneous boundary conditions of (187) to homogeneous boundary conditions, we simply use the conversion formula that we used before in Section 3.3.2. In other words, the following transformation formula

$$u(x,t) = \left(\alpha + \frac{x}{L}(\beta - \alpha)\right) + v(x,t), \qquad (188)$$

should be used to achieve this goal.

Substituting (188) into (187) shows that $v(x,t)$ is governed by the initial-boundary value problem

$$\begin{aligned}
\text{PDE} \qquad v_{tt} &= c^2 v_{xx}, \ 0 < x < L, \ t > 0 \\
\text{B.C} \qquad v(0,t) &= 0, \ v(L,t) = 0 \\
\text{I.C} \qquad v(x,0) &= f(x) - \left(\alpha + \frac{x}{L}(\beta - \alpha)\right), \ v_t(L,t) = g(x)
\end{aligned} \qquad (189)$$

In view of (189), the method of separation of variables can be easily used in (189) as discussed before. Having determined $v(x, t)$ of (189), the wave function $u(x, t)$ of (187) follows immediately upon substituting $v(x, t)$ into (188).

To get a better understanding of the implementation of the transformation formula (188), we will discuss the following illustrative examples.

Example 6. Solve the following initial-boundary value problem

$$
\begin{array}{llll}
\text{PDE} & u_{tt} & = & u_{xx}, \ 0 < x < 1, \ t > 0 \\[2mm]
\text{B.C} & u(0, t) & = & 1, \ u(1, t) = 2, \ t \geq 0 \\[2mm]
\text{I.C} & u(x, 0) & = & 1 + x, \ u_t(x, 0) = \pi \sin(\pi x)
\end{array}
\tag{190}
$$

Solution.

Using the transformation formula (188) we obtain

$$
u(x, t) = (1 + x) + v(x, t),
\tag{191}
$$

that carries (190) into

$$
\begin{array}{llll}
\text{PDE} & v_{tt} & = & v_{xx}, \ 0 < x < 1, \ t > 0 \\[2mm]
\text{B.C} & v(0, t) & = & 0, \ v(1, t) = 0 \\[2mm]
\text{I.C} & v(x, 0) & = & 0, \ v_t(x, 0) = \pi \sin(\pi x)
\end{array}
\tag{192}
$$

Assuming that

$$
v(x, t) = F(x)T(t),
\tag{193}
$$

and proceeding as before we obtain

$$
F_n(x) = B_n \sin(n\pi x), n = 1, 2, \cdots
\tag{194}
$$

and

$$
T_n(t) = C_n \cos(n\pi t) + D_n \sin(n\pi t),
\tag{195}
$$

so that

$$
v(x, t) = \sum_{n=1}^{\infty} \sin(n\pi x) \left(C_n \cos(n\pi t) + D_n \sin(n\pi t) \right),
\tag{196}
$$

where C_n and D_n are as yet undetermined constants. Using the initial condition $v(x, 0) = 0$ in (196) gives

$$
C_n = 0, n \geq 0.
\tag{197}
$$

Using the initial condition $v_t(x, 0) = \pi \sin(\pi x)$ into the derivative of (196) we obtain

$$
D_1 = 1, D_k = 0, \ k \neq 1.
\tag{198}
$$

This gives the solution for $v(x,t)$ by

$$v(x,t) = \sin(\pi x)\sin(\pi t), \tag{199}$$

so that the particular solution $u(x,t)$ of (190) is given by

$$u(x,t) = 1 + x + \sin(\pi x)\sin(\pi t). \tag{200}$$

At this point, it seems reasonable to use the Adomian decomposition method to solve the initial-boundary value problem of this example. This will enable us to compare the performance of the decomposition method and the classical method of separation of variables.

Applying the inverse operator L_t^{-1} to the operator form of (190) and using the initial conditions we obtain

$$u(x,t) = 1 + x + (\pi t)\sin(\pi x) + L_t^{-1}\left(L_x u(x,t)\right). \tag{201}$$

Using the decomposition series of $u(x,t)$ into both sides of equation (201) yields

$$\sum_{n=0}^{\infty} u_n(x,t) = 1 + x + (\pi t)\sin(\pi x) + L_t^{-1}\left(L_x\left(\sum_{n=0}^{\infty} u_n(x,t)\right)\right). \tag{202}$$

Using the recursive algorithm we obtain

$$
\begin{aligned}
u_0 &= 1 + x + (\pi t)\sin(\pi x), \\
u_1 &= L_t^{-1}(L_x u_0) = -\frac{(\pi t)^3}{3!}\sin(\pi x), \\
u_2 &= L_t^{-1}(L_x u_1) = \frac{(\pi t)^5}{5!}\sin(\pi x),
\end{aligned}
\tag{203}
$$

and so on. Consequently, the solution in a series form is given by

$$u(x,t) = 1 + x + \sin(\pi x)\left((\pi t) - \frac{(\pi t)^3}{3!} + \frac{(\pi t)^5}{5!} - \cdots\right), \tag{204}$$

and in a closed form

$$u(x,t) = 1 + x + \sin(\pi x)\sin(\pi t). \tag{205}$$

It is obvious that we obtained the solution (205) by employing less computational work if compared with the method of separation of variables. The power of the decomposition method for solving differential equations is thus emphasized.

Example 7. Solve the following initial-boundary value problem

$$
\begin{aligned}
\text{PDE} \qquad u_{tt} &= u_{xx}, \ 0 < x < 1, \ t > 0 \\
\text{B.C} \qquad u(0,t) &= 2, \ u(1,t) = 3, \ t \geq 0 \\
\text{I.C} \qquad u(x,0) &= 2 + x + \sin(\pi x), \ u_t(x,0) = 0
\end{aligned}
\tag{206}
$$

Solution.

Using the transformation formula (188) where $\alpha = 2$ and $\beta = 3$, we obtain

$$u(x,t) = (2 + x) + v(x,t). \tag{207}$$

In view of (207), we find that

$$
\begin{array}{lll}
\text{PDE} & v_{tt} = v_{xx}, & 0 < x < 1,\ t > 0 \\[2mm]
\text{B.C} & v(0,t) = 0,\ v(1,t) = 0 & \tag{208} \\[2mm]
\text{I.C} & v(x,0) = \sin(\pi x),\ v_t(x,0) = 0 &
\end{array}
$$

Following the analysis presented before gives

$$F_n(x) = B_n \sin(n\pi x), n = 1, 2, \cdots \tag{209}$$

and

$$T_n(t) = C_n \cos(n\pi t) + D_n \sin(n\pi t), \tag{210}$$

so that

$$v(x,t) = \sum_{n=1}^{\infty} \sin(n\pi x)\left(C_n \cos(n\pi t) + D_n \sin(n\pi t)\right), \tag{211}$$

where C_n and D_n are as yet undetermined constants Using the initial condition $v(x,0) = \sin(\pi x)$ in (211) gives

$$C_1 = 1,\ C_k = 0, k \neq 1. \tag{212}$$

Differentiating (211) and using the initial condition $v_t(x,0) = 0$ we obtain

$$D_n = 0,\ n \geq 1. \tag{213}$$

This gives the solution for $v(x,t)$ by

$$v(x,t) = \sin(\pi x)\cos(\pi t), \tag{214}$$

so that the particular solution $u(x,t)$ of (206) is given by

$$u(x,t) = 2 + x + \sin(\pi x)\cos(\pi t). \tag{215}$$

Neumann Boundary Conditions

We next consider the second kind of boundary conditions, where $u_x(0,t) = \alpha$ and $u_x(L,t) = \beta$ are given. We point out that the transformation formula (188) works effectively for the first kind of boundary conditions, but cannot be used for the second kind of boundary conditions. To study the proper formula in this case, we consider the initial-boundary value problem

$$\begin{array}{lrl}
\text{PDE} & u_{tt} &= c^2 u_{xx}, \ 0 < x < L, \ t > 0 \\[2mm]
\text{B.C} & u_x(0,t) &= \alpha, \ u_x(L,t) = \beta, \ t \geq 0 \\[2mm]
\text{I.C} & u(x,0) &= f(x), \ u_t(x,0) = g(x)
\end{array} \qquad (216)$$

It is interesting to note that an alternative formula should be used to convert the inhomogeneous boundary conditions of (216) to homogeneous boundary conditions. We can easily prove that the transformation formula

$$u(x,t) = \alpha x + (\frac{\beta - \alpha}{2L})x^2 + c^2(\frac{\beta - \alpha}{2L})t^2 + v(x,t), \qquad (217)$$

is an appropriate formula that can be used to achieve our goal of conversion. Differentiating (217) twice with respect to t and to x we obtain

$$\begin{array}{rl}
u_{tt} &= c^2(\frac{\beta-\alpha}{L}) + v_{tt}, \\[2mm]
u_{xx} &= (\frac{\beta-\alpha}{L}) + v_{xx}.
\end{array} \qquad (218)$$

Using (217) and (218), it can be easily shown that $v(x,t)$ is governed by the initial-boundary value problem

$$\begin{array}{lrl}
\text{PDE} & v_{tt} &= c^2 v_{xx}, \ 0 < x < L, \ t > 0 \\[2mm]
\text{B.C} & v_x(0,t) &= 0, \ v_x(L,t) = 0 \\[2mm]
\text{I.C} & v(x,0) &= f(x) - \left(\alpha x + \frac{\beta-\alpha}{2L}x^2\right) \\[2mm]
& v_t(L,t) &= g(x)
\end{array} \qquad (219)$$

It is clear that an initial-boundary value problem (219) with homogeneous boundary conditions is obtained. The use of the transformation formula (217) will be explained by the following illustrative examples.

Example 8. Solve the following initial- boundary value problem

$$\begin{array}{lrl}
\text{PDE} & u_{tt} &= u_{xx}, \ 0 < x < 1, \ t > 0 \\[2mm]
\text{B.C} & u_x(0,t) &= 1, \ u_x(1,t) = 3, \ t \geq 0 \\[2mm]
\text{I.C} & u(x,0) &= x^2 + x, \ u_t(x,0) = \pi \cos(\pi x)
\end{array} \qquad (220)$$

Solution.

Using the transformation formula (217), where $\alpha = 1, \beta = 3$, we find

$$u(x,t) = x + x^2 + t^2 + v(x,t) \qquad (221)$$

In view of (221), the initial-boundary value problem for $v(x,t)$ is given by

$$
\begin{array}{llll}
\text{PDE} & v_{tt} & = & v_{xx}, \ 0 < x < L, \ t > 0 \\[2mm]
\text{B.C} & v_x(0,t) & = & 0, \ v_x(1,t) = 0, \ t \geq 0 \\[2mm]
\text{I.C} & v(x,0) & = & 0, \ v_t(x,0) = \pi \cos(\pi x)
\end{array}
\qquad (222)
$$

Proceeding as discussed before we find

$$v(x,t) = \sum_{n=0}^{\infty} \cos(n\pi x)\left(C_n \cos(n\pi t) + D_n \sin(n\pi t)\right). \qquad (223)$$

Using the initial conditions gives

$$C_n = 0, \ n \geq 0. \qquad (224)$$

and

$$D_1 = 1, D_k = 0, \ k \neq 1. \qquad (225)$$

This leads to

$$v(x,t) = \cos(\pi x)\sin(\pi t), \qquad (226)$$

which gives

$$u(x,t) = x + x^2 + t^2 + \cos(\pi x)\sin(\pi t). \qquad (227)$$

Exercises 5.3.2

Use the method of separation of variables to solve Exercises 1 – 6, where the first kind of boundary conditions are given:

1. $u_{tt} = u_{xx}, \ 0 < x < 1, \ t > 0$
   $u(0,t) = 1, \ u(1,t) = 1$
   $u(x,0) = 1, \ u_t(x,0) = \pi \sin(\pi x)$

2. $u_{tt} = u_{xx}, \ 0 < x < 1, \ t > 0$
   $u(0,t) = 2, \ u(1,t) = 3$
   $u(x,0) = 2 + x + 2\sin(\pi x), \ u_t(x,0) = 0$

3. $u_{tt} = u_{xx}, \ 0 < x < 1, \ t > 0$
   $u(0,t) = 0, \ u(1,t) = 3$
   $u(x,0) = 3x, \ u_t(x,0) = 4\pi \sin(\pi x)$

4. $u_{tt} = u_{xx}, \ 0 < x < 1, \ t > 0$
   $u(0,t) = 4, \ u(1,t) = 1$
   $u(x,0) = 4 - 3x, \ u_t(x,0) = \pi \sin(\pi x)$

5. $u_{tt} = 4u_{xx}, \ 0 < x < 1, \ t > 0$
   $u(0,t) = 3, \ u(1,t) = 7$
   $u(x,0) = 3 + 4x, \ u_t(x,0) = 2\pi \sin(\pi x)$

6. $u_{tt} = 4u_{xx}, \ 0 < x < 1, \ t > 0$
   $u(0,t) = 1, \ u(1,t) = 2$
   $u(x,0) = 1 + x, \ u_t(x,0) = 4\pi \sin(2\pi x)$

In Exercises 7   12, where the boundary conditions $u_x(0,t)$ and $u_x(1,t)$ are given, use the method of separation of variables to solve the initial-boundary value problems:

7. $u_{tt} = u_{xx}, \ 0 < x < 1, \ t > 0$
   $u_x(0,t) = 3, \ u_x(1,t) = 5$
   $u(x,0) = 3x + x^2 + \cos(\pi x), \ u_t(x,0) = 0$

8. $u_{tt} = u_{xx}, \ 0 < x < 1, \ t > 0$
   $u_x(0,t) = 4, \ u_x(1,t) = 4$
   $u(x,0) = 4x, \ u_t(x,0) = \pi \cos(\pi x)$

9. $u_{tt} = u_{xx}, \ 0 < x < 1, \ t > 0$
   $u_x(0,t) = 2, \ u_x(1,t) = 6$
   $u(x,0) = 2x + 2x^2 + \cos(\pi x), \ u_t(x,0) = 0$

10. $u_{tt} = 4u_{xx}, \ 0 < x < 1, \ t > 0$
    $u_x(0,t) = 1, \ u_x(1,t) = 1$
    $u(x,0) = x, \ u_t(x,0) = 2\pi \cos(\pi x)$

11. $u_{tt} = 4u_{xx}, \ 0 < x < 1, \ t > 0$
    $u_x(0,t) = 2, \ u_x(1,t) = 2$
    $u(x,0) = 2x + \cos(\pi x), \ u_t(x,0) = 0$

12. $u_{tt} = 9u_{xx}, \ 0 < x < 1, \ t > 0$
    $u_x(0,t) = 1, \ u_x(1,t) = 3$
    $u(x,0) = x + x^2, \ u_t(x,0) = 3\pi \cos(\pi x)$

5.3.3 Wave Equation In An Infinite Domain: D'Alembert Solution

In Section 5.2.3, the motion of a very long string, that is considered not to have boundaries, has been handled by using the decomposition method. The physical model that controls the wave motion of a very long string is governed by a PDE and initial conditions only. As mentioned before, the method of separation of variables is not applicable in this case.

However, a standard method, known as D'Alembert solution, allows us to solve the initial value problem on an infinite domain.

To derive D'Alembert formula, we consider a typical wave equation in an infinite domain given by

$$
\begin{array}{lll}
\text{PDE} & u_{tt} = c^2 u_{xx}, \; -\infty < x < \infty, \; t > 0 \\
\text{I.C} & u(x,0) = f(x), \; u_t(x,0) = g(x)
\end{array}
\tag{228}
$$

As stated in a previous section, the attention will be focused upon the disturbance occurred at the center of the very long string. The initial displacement $u(x,0)$ and the initial velocity $u_t(x,0)$ are prescribed by $f(x)$ and $g(x)$ respectively.

To derive D'Alembert solution, we consider two new variables ξ and η defined by

$$
\begin{array}{ll}
\xi = x + ct, \\
\eta = x - ct.
\end{array}
\tag{229}
$$

Using the chain rule we obtain

$$
\begin{array}{ll}
u_{xx} = u_{\xi\xi} + 2u_{\xi\eta} + u_{\eta\eta}, \\
u_{tt} = c^2 \left(u_{\xi\xi} - 2u_{\xi\eta} + u_{\eta\eta} \right).
\end{array}
\tag{230}
$$

Substituting (230) into (228) gives

$$
u_{\xi\eta} = 0.
\tag{231}
$$

Integrating (231) first with respect ξ then with respect η we obtain the general solution given by

$$
u(\xi,\eta) = F(\xi) + G(\eta),
\tag{232}
$$

where F and G are arbitrary functions. Using (229), equation (232) can be rewritten as

$$
u(x,t) = F(x + ct) + G(x - ct),
\tag{233}
$$

Using the initial condition $u(x,0) = f(x)$ into (233) yields

$$
F(x) + G(x) = f(x).
\tag{234}
$$

Substituting the initial condition $u_t(x,0) = g(x)$ into (233) gives

$$
cF'(x) - cG'(x) = g(x).
\tag{235}
$$

Integrating both sides of (235) from 0 to x gives

$$
F(x) - G(x) = \frac{1}{c} \int_0^x g(r)\, dr + K.
\tag{236}
$$

where K is the constant of integration. Solving (234) and (236) we find

$$
\begin{aligned}
F(x) &= \tfrac{1}{2}f(x) + \tfrac{1}{2c}\int_0^x g(r)\,dr + \tfrac{1}{2}K, \\
G(x) &= \tfrac{1}{2}f(x) - \tfrac{1}{2c}\int_0^x g(r)\,dr - \tfrac{1}{2}K.
\end{aligned}
\tag{237}
$$

This means that

$$
\begin{aligned}
F(x+ct) &= \tfrac{1}{2}f(x+ct) + \tfrac{1}{2c}\int_0^{x+ct} g(r)\,dr + \tfrac{1}{2}K, \\
G(x-ct) &= \tfrac{1}{2}f(x-ct) - \tfrac{1}{2c}\int_0^{x-ct} g(r)\,dr - \tfrac{1}{2}K.
\end{aligned}
\tag{238}
$$

so that by using (233) we obtain the D'Alembert formula given by

$$
u(x,t) = \frac{f(x+ct) + f(x-ct)}{2} + \frac{1}{2c}\int_{x-ct}^{x+ct} g(r)\,dr.
\tag{239}
$$

This completes the formal derivation of D'Alembert solution. It is clearly seen that the formula defines the solution $u(x,t)$ of (228) in terms of the initial conditions as shown by (239).

To explain D'Alembert's formula, we consider the following examples.

Example 9. Use the D'Alembert formula to solve the initial value problem

$$
\begin{aligned}
\text{PDE} \qquad & u_{tt} = u_{xx}, \quad -\infty < x < \infty,\ t > 0 \\
\text{I.C} \qquad & u(x,0) = \sin x,\ u_t(x,0) = 0
\end{aligned}
\tag{240}
$$

Solution.

Substituting $c = 1$, $f(x) = \sin x$ and $g(x) = 0$ into (239) gives the particular solution by

$$
u(x,t) = \frac{\sin(x+t) + \sin(x-t)}{2},
\tag{241}
$$

which gives the solution

$$
u(x,t) = \sin x \cos t,
\tag{242}
$$

obtained by converting the sum of two trigonometric functions into the product of two trigonometric functions.

Example 10. Use the D'Alembert formula to solve the initial value problem

$$
\begin{aligned}
\text{PDE} \qquad & u_{tt} = 4u_{xx}, \quad -\infty < x < \infty,\ t > 0 \\
\text{I.C} \qquad & u(x,0) = \sin x,\ u_t(x,0) = 4
\end{aligned}
\tag{243}
$$

Solution.

Substituting $c = 2$, $f(x) = \sin x$ and $g(x) = 4$ into (239) gives the particular solution

$$u(x,t) = \frac{\sin(x+2t) + \sin(x-2t)}{2} + \frac{1}{4}\int_{x-2t}^{x+2t} 4\,dr, \tag{244}$$

which gives

$$u(x,t) = \sin x \cos(2t) + 4t. \tag{245}$$

Example 11. Use the D'Alembert formula to solve the initial value problem

$$
\begin{aligned}
\text{PDE} && u_{tt} &= 9u_{xx}, && -\infty < x < \infty,\ t > 0 \\
\text{I.C} && u(x,0) &= \sin x,\ u_t(x,0) = 3\cos x
\end{aligned} \tag{246}
$$

Solution.

Note that $c = 3$, $f(x) = \sin x$ and $g(x) = 3\cos x$. Substituting into (239) gives

$$
\begin{aligned}
u(x,t) &= \frac{\sin(x+3t) + \sin(x-3t)}{2} + \frac{1}{6}\int_{x-3t}^{x+3t} 3\cos r\,dr, \\
&= \frac{\sin(x+3t) + \sin(x-3t)}{2} + \frac{\sin(x+3t) - \sin(x-3t)}{2}
\end{aligned} \tag{247}
$$

which gives

$$u(x,t) = \sin(x+3t). \tag{248}$$

Example 12. Use the D'Alembert formula to solve the initial value problem

$$
\begin{aligned}
\text{PDE} && u_{tt} &= u_{xx}, && -\infty < x < \infty,\ t > 0 \\
\text{I.C} && u(x,0) &= e^{-x},\ u_t(x,0) = \frac{2}{1+x^2}
\end{aligned} \tag{249}
$$

Solution.

Substituting $c = 1$, $f(x) = e^{-x}$ and $g(x) = \frac{1}{1+x^2}$ into (239) gives

$$
\begin{aligned}
u(x,t) &= \frac{e^{-(x+t)} + e^{-(x-t)}}{2} + \frac{1}{2}\int_{x-t}^{x+t} \frac{2}{1+r^2}\,dr, \\
&= \frac{e^{-(x+t)} + e^{-(x-t)}}{2} + [\arctan r]_{x-t}^{x+t} \\
&= \frac{e^{-x}(e^t + e^{-t})}{2} + \arctan(x+t) - \arctan(x-t),
\end{aligned} \tag{250}
$$

which gives the particular solution

$$u(x,t) = e^{-x}\cosh t + \arctan(x+t) - \arctan(x-t). \tag{251}$$

Exercises 5.3.3

Use the D'Alembert formula to solve the following initial value problems:

1. $u_{tt} = u_{xx}$, $-\infty < x < \infty$, $t > 0$
   $u(x,0) = 0$, $u_t(x,0) = 2 + \sin x$

2. $u_{tt} = u_{xx}$, $-\infty < x < \infty$, $t > 0$
   $u(x,0) = \sin x$, $u_t(x,0) = \cos x$

3. $u_{tt} = u_{xx}$, $-\infty < x < \infty$, $t > 0$
   $u(x,0) = \cos x$, $u_t(x,0) = -\sin x$

4. $u_{tt} = u_{xx}$, $-\infty < x < \infty$, $t > 0$
   $u(x,0) = \sin x$, $u_t(x,0) = 4\cos x$

5. $u_{tt} = u_{xx}$, $-\infty < x < \infty$, $t > 0$
   $u(x,0) = \sin x$, $u_t(x,0) = 2$

6. $u_{tt} = 4u_{xx}$, $-\infty < x < \infty$, $t > 0$
   $u(x,0) = \cos x$, $u_t(x,0) = -2\sin x$

7. $u_{tt} = u_{xx}$, $-\infty < x < \infty$, $t > 0$
   $u(x,0) = \sinh x$, $u_t(x,0) = \cosh x$

8. $u_{tt} = u_{xx}$, $-\infty < x < \infty$, $t > 0$
   $u(x,0) = x$, $u_t(x,0) = e^{-x}$

9. $u_{tt} = u_{xx}$, $-\infty < x < \infty$, $t > 0$
   $u(x,0) = \cosh x$, $u_t(x,0) = 0$

10. $u_{tt} = 4u_{xx}$, $-\infty < x < \infty$, $t > 0$
    $u(x,0) = 0$, $u_t(x,0) = 2\sinh x$

11. $u_{tt} = u_{xx}$, $-\infty < x < \infty$, $t > 0$
    $u(x,0) = \cos x$, $u_t(x,0) = 1 + 2x$

12. $u_{tt} = u_{xx}$, $-\infty < x < \infty$, $t > 0$
    $u(x,0) = \sin x$, $u_t(x,0) = 4 + 4x$

Chapter 6

Higher Dimensional Wave Equation

6.1 Introduction

In this chapter we will discuss the initial-boundary value problems that control the wave propagation in two and three dimensional spaces. The methods that will be applied are the Adomian decomposition method and the method of separation of variables. The two methods have been outlined before and were applied to the one dimensional wave equation in Chapter 5.

The decomposition method expands the solution u of any equation into an infinite series of components $u_0, u_1, u_2, \cdots$ where these components are elegantly computed. The determination of these components can be achieved in an easy way through a recursive relation that involves simple integrals.

The method of separation of variables provides the solution of a partial differential equation through converting the partial differential equation into several easily solvable ordinary differential equations. In addition, the method requires that the problem and the boundary conditions be linear and homogeneous, hence transformation formulas are usually used to justify this need. The method of separation of variables cannot handle initial value problems because boundary conditions are not prescribed. However, the decomposition method can easily handle these problems.

6.2 Adomian Decomposition Method

In previous chapters we have discussed Adomian decomposition method and have applied it to partial differential equations of any order, homogeneous and inhomogeneous. The decomposition method consists of decomposing the unknown function u into an infinite sum of components $u_0, u_1, u_2, \cdots$, and concerns itself with determin-

167

ing these components recurrently. The zeroth component u_0 is usually identified by the terms arising from integrating inhomogeneous terms and from initial/boundary conditions. The successive components $u_1, u_2, \cdots$ are determined in a recursive manner. It was found that few components can give an insight into the character and behavior of the solution. For numerical purposes, accuracy can be easily enhanced by determining as many components as we like.

Throughout this section, the decomposition method will be applied to two dimensional and three dimensional wave equations.

6.2.1   Two Dimensional Wave Equation

The propagation of waves in a two dimensional vibrating membrane of length a and width b is governed by the following initial-boundary value problem

$$
\begin{aligned}
\text{PDE} \qquad u_{tt} &= c^2(u_{xx} + u_{yy}),\ 0 < x < a, 0 < y < b, t > 0 \\[4pt]
\text{B.C} \qquad u(0, y, t) &= u(a, y, t) = 0 \\[4pt]
u(x, 0, t) &= u(x, b, t) = 0 \\[4pt]
\text{I.C} \qquad u(x, y, 0) &= f(x, y),\ u_t(x, y, 0) = g(x, y)
\end{aligned}
\tag{1}
$$

where $u = u(x, y, t)$ is the displacement function of any point located at the position (x, y) of a vibrating membrane at any time t, and c is related to the elasticity of the material of the rectangular plate.

As discussed before, the solution in the t direction, in the x space, or in the y space will lead to identical results. However, the solution in the t direction reduces the size of calculations compared with the other space solutions because it uses the initial conditions only. For this reason the solution in the t direction will be followed in this chapter.

We first rewrite (1) in an operator form by

$$
L_t u(x, y, t) = c^2 \left(L_x u(x, y, t) + L_y u(x, y, t) \right),
\tag{2}
$$

where the differential operators L_t, L_x, and L_y are defined by

$$
L_t = \frac{\partial^2}{\partial t^2},\ L_x = \frac{\partial^2}{\partial x^2},\ L_y = \frac{\partial^2}{\partial y^2},
\tag{3}
$$

so that the integral operator L_t^{-1} exists and given by

$$
L_t^{-1}(.) = \int_0^t \int_0^t (.) dt\, dt.
\tag{4}
$$

This means that

$$
L_t^{-1} L_t\, u(x, y, t) = u(x, y, t) - u(x, y, 0) - t u_t(x, y, 0).
\tag{5}
$$

Applying L_t^{-1} to both sides of (2) and using the initial conditions leads to

$$u(x, y, t) = f(x, y) + tg(x, y) + c^2 L_t^{-1} (L_x u + L_y u). \tag{6}$$

The decomposition method defines the solution $u(x, y, t)$ as an infinite series given by

$$u(x, y, t) = \sum_{n=0}^{\infty} u_n(x, y, t), \tag{7}$$

where the components $u_n(x, y, t), n \geq 0$ will be easily computed by using a recursive relation. Substituting (7) into both sides of (6) yields

$$\sum_{n=0}^{\infty} u_n = f(x, y) + tg(x, y) + c^2 L_t^{-1} \left(L_x \left(\sum_{n=0}^{\infty} u_n \right) + L_y \left(\sum_{n=0}^{\infty} u_n \right) \right). \tag{8}$$

To construct the recursive scheme, the decomposition method suggests that the zeroth component $u_0(x, y, t)$ is identified as the terms arising from the initial/boundary conditions and from integrating inhomogeneous terms if exist. The components $u_n(x, y, t), n \geq 0$ can be completely determined by using the recursive relation

$$\begin{aligned} u_0(x, y, t) &= f(x, y) + tg(x, y), \\ u_{k+1}(x, y, t) &= c^2 L_t^{-1} (L_x u_k + L_y u_k), k \geq 0. \end{aligned} \tag{9}$$

Consequently, the successive components can be completely computed, hence the solution in a series form follows immediately.

To give a clear overview of the implementation of the decomposition method, we have chosen several examples, homogeneous and inhomogeneous, to illustrate the discussion given above.

Homogeneous Wave Equations

The Adomian decomposition method will be used to solve the following homogeneous wave equations in two dimensional vibrating membrane with homogeneous or inhomogeneous boundary conditions.

Example 1. Use the Adomian decomposition method to solve the initial-boundary value problem.

$$\begin{aligned} \text{PDE} \qquad u_{tt} &= 2(u_{xx} + u_{yy}), 0 < x, y < \pi, t > 0 \\ \text{B.C} \qquad u(0, y, t) &= u(\pi, y, t) = 0 \\ u(x, 0, t) &= u(x, \pi, t) = 0 \\ \text{I.C} \qquad u(x, y, 0) &= \sin x \sin y, \ u_t(x, y, 0) = 0 \end{aligned} \tag{10}$$

Solution.

In an operator form, Eq. (10) becomes

$$L_t u(x, y, t) = 2(L_x u(x, y, t) + L_y u(x, y, t)). \tag{11}$$

Applying the inverse operator L_t^{-1} to (11) gives

$$u(x, y, t) = \sin x \sin y + 2L_t^{-1}\left(L_x u + L_y u\right). \tag{12}$$

The decomposition method decomposes the solution $u(x, y, t)$ by the decomposition series

$$u(x, y, t) = \sum_{n=0}^{\infty} u_n(x, y, t). \tag{13}$$

Substituting (13) into both sides of (12) yields

$$\sum_{n=0}^{\infty} u_n = \sin x \sin y + 2\, L_t^{-1}\left(L_x\left(\sum_{n=0}^{\infty} u_n\right) + L_y\left(\sum_{n=0}^{\infty} u_n\right)\right). \tag{14}$$

The zeroth component $u_0(x, y, t)$ is usually identified by all terms that are not included under the inverse operator L_t^{-1} in (14). Consequently, we set the recursive relation

$$
\begin{aligned}
u_0(x, y, t) &= \sin x \sin y, \\
u_{k+1}(x, y, t) &= 2L_t^{-1}\left(L_x u_k + L_y u_k\right), k \geq 0.
\end{aligned}
\tag{15}
$$

This in turn gives

$$
\begin{aligned}
u_0(x, y, t) &= \sin x \sin y, \\
u_1(x, y, t) &= 2L_t^{-1}\left(L_x u_0 + L_y u_0\right) = -\frac{(2t)^2}{2!}\sin x \sin y, \\
u_2(x, y, t) &= 2L_t^{-1}\left(L_x u_1 + L_y u_1\right) = \frac{(2t)^4}{4!}\sin x \sin y,
\end{aligned}
\tag{16}
$$

and so on. The solution in a series form is given by

$$u(x, y, t) = \sin x \sin y \left(1 - \frac{(2t)^2}{2!} + \frac{(2t)^4}{4!} - \cdots\right), \tag{17}$$

and in a closed form by

$$u(x, y, t) = \sin x \sin y \cos(2t), \tag{18}$$

obtained upon using the Taylor expansion of $\cos(2t)$.

Example 2. Use the Adomian decomposition method to solve the initial-boundary value problem.

$$
\begin{aligned}
\text{PDE} \qquad u_{tt} &= 8(u_{xx} + u_{yy}),\, 0 < x, y < \pi,\, t > 0 \\
\text{B.C} \qquad u(0, y, t) &= u(\pi, y, t) = 0 \\
u(x, 0, t) &= u(x, \pi, t) = 0 \\
\text{I.C} \qquad u(x, y, 0) &= 0,\, u_t(x, y, 0) = 4 \sin x \sin y
\end{aligned}
\tag{19}
$$

Solution.

Proceeding as in Example 1 we find

$$
u(x, y, t) = 4t \sin x \sin y + 8L_t^{-1} \left(L_x u + L_y u \right).
\tag{20}
$$

We next define $u(x, y, t)$ by an infinite series

$$
u(x, y, t) = \sum_{n=0}^{\infty} u_n(x, y, t),
\tag{21}
$$

that carries (20) into

$$
\sum_{n=0}^{\infty} u_n = 4t \sin x \sin y + 8L_t^{-1} \left(L_x \left(\sum_{n=0}^{\infty} u_n \right) + L_y \left(\sum_{n=0}^{\infty} u_n \right) \right).
\tag{22}
$$

Following Adomian's assumptions we find

$$
\begin{aligned}
u_0(x, y, t) &= 4t \sin x \sin y, \\
u_{k+1}(x, y, t) &= 8L_t^{-1} \left(L_x u_k + L_y u_k \right),\, k \geq 0.
\end{aligned}
\tag{23}
$$

It follows that

$$
\begin{aligned}
u_0(x, y, t) &= 4t \sin x \sin y, \\
u_1(x, y, t) &= 8L_t^{-1} \left(L_x u_0 + L_y u_0 \right) = -\frac{(4t)^3}{3!} \sin x \sin y, \\
u_2(x, y, t) &= 8L_t^{-1} \left(L_x u_1 + L_y u_1 \right) = \frac{(4t)^5}{5!} \sin x \sin y,
\end{aligned}
\tag{24}
$$

and so on. Combining (24) and (21), the solution in a series form is given by

$$
u(x, y, t) = \sin x \sin y \left((4t) - \frac{(4t)^3}{3!} + \frac{(4t)^5}{5!} - \cdots \right),
\tag{25}
$$

and the exact solution

$$u(x, y, t) = \sin x \sin y \sin(4t), \tag{26}$$

follows immediately.

Example 3. Use the Adomian decomposition method to solve the initial-boundary value problem.

$$
\begin{aligned}
\text{PDE} \qquad u_{tt} &= \tfrac{1}{2}(u_{xx} + u_{yy}), \ 0 < x, y < \pi, \ t > 0 \\[4pt]
\text{B.C} \quad u(0, y, t) &= u(\pi, y, t) = 1 \\[4pt]
u(x, 0, t) &= 1 + \sin x \sin t, \ u(x, \pi, t) = 1 - \sin x \sin t \\[4pt]
\text{I.C} \quad u(x, y, 0) &= 1, \ u_t(x, y, 0) = \sin x \cos y
\end{aligned}
\tag{27}
$$

Solution.

We note that the wave equation is homogeneous and the boundary conditions are inhomogeneous. The decomposition method will be applied in a direct way as used before.

Applying the inverse operator L_t^{-1} to the operator form of (27) leads to

$$u(x, y, t) = 1 + t \sin x \cos y + \frac{1}{2}L_t^{-1}\left(L_x u + L_y u\right), \tag{28}$$

where by using the decomposition series for $u(x, y, t)$ we obtain

$$\sum_{n=0}^{\infty} u_n = 1 + t \sin x \cos y + \frac{1}{2}L_t^{-1}\left(L_x\left(\sum_{n=0}^{\infty} u_n\right) + L_y\left(\sum_{n=0}^{\infty} u_n\right)\right). \tag{29}$$

The components of $u(x, y, t)$ can be easily determined in a recursive manner by

$$
\begin{aligned}
u_0(x, y, t) &= 1 + t \sin x \cos y, \\[4pt]
u_{k+1}(x, y, t) &= \tfrac{1}{2}L_t^{-1}\left(L_x u_k + L_y u_k\right), k \geq 0.
\end{aligned}
\tag{30}
$$

The first few components of the solution $u(x, y, t)$ are given by

$$
\begin{aligned}
u_0(x, y, t) &= 1 + t \sin x \cos y, \\[4pt]
u_1(x, y, t) &= \tfrac{1}{2}L_t^{-1}\left(L_x u_0 + L_y u_0\right) = -\tfrac{1}{3!}t^3 \sin x \cos y, \\[4pt]
u_2(x, y, t) &= \tfrac{1}{2}L_t^{-1}\left(L_x u_1 + L_y u_1\right) = \tfrac{1}{5!}t^5 \sin x \cos y, \\[4pt]
u_3(x, y, t) &= \tfrac{1}{2}L_t^{-1}\left(L_x u_2 + L_y u_2\right) = -\tfrac{1}{7!}t^7 \sin x \cos y.
\end{aligned}
\tag{31}
$$

The solution in a series form is given by

$$u(x, y, t) = 1 + \sin x \cos y \left(t - \frac{1}{3!} t^3 + \frac{1}{5!} t^5 - \cdots \right), \tag{32}$$

and in a closed form by

$$u(x, y, t) = 1 + \sin x \cos y \sin t. \tag{33}$$

Example 4. Use the Adomian decomposition method to solve the initial-boundary value problem.

$$
\begin{array}{lrcl}
\text{PDE} & u_{tt} & = & u_{xx} + u_{yy} - 2u, \ 0 < x, y < \pi, \ t > 0 \\[2mm]
\text{B.C} & u(0, y, t) & = & -u(\pi, y, t) = \cos y \sin(2t) \\[2mm]
& u(x, 0, t) & = & -u(x, \pi, t) = \cos x \sin(2t) \\[2mm]
\text{I.C} & u(x, y, 0) & = & 0, \ u_t(x, y, 0) = 2 \cos x \cos y
\end{array}
\tag{34}
$$

Solution.

We note that an additional term $-2u$ is included in the standard wave equation. This arises when each element of the membrane is subjected to an additional force which is proportional to its displacement $u(x, y, t)$.

Applying the inverse operator L_t^{-1} to the operator form of (34) and using the initial conditions we obtain

$$u(x, y, t) = 2t \cos x \cos y + L_t^{-1} \left(L_x u + L_y u - 2u \right). \tag{35}$$

Using the decomposition series of $u(x, y, t)$ into both sides of (35) gives

$$\sum_{n=0}^{\infty} u_n = 2t \cos x \cos y$$

$$+ L_t^{-1} \left(L_x \left(\sum_{n=0}^{\infty} u_n \right) + L_y \left(\sum_{n=0}^{\infty} u_n \right) - 2 \left(\sum_{n=0}^{\infty} u_n \right) \right). \tag{36}$$

The components of $u(x, y, t)$ can be recursively determined by

$$
\begin{array}{rcl}
u_0(x, y, t) & = & 2t \cos x \cos y, \\[2mm]
u_{k+1}(x, y, t) & = & L_t^{-1} \left(L_x u_k + L_y u_k - 2u_k \right), k \geq 0,
\end{array}
\tag{37}
$$

so that

$$
\begin{aligned}
u_0(x, y, t) &= 2t \cos x \cos y, \\
u_1(x, y, t) &= -\tfrac{1}{3!}(2t)^3 \cos x \cos y, \\
u_2(x, y, t) &= \tfrac{1}{5!}(2t)^5 \cos x \cos y.
\end{aligned}
\tag{38}
$$

In view of (38), the solution in a series form is given by

$$
u(x, y, t) = \cos x \cos y \left((2t) - \frac{1}{3!}(2t)^3 + \frac{1}{5!}(2t)^5 - \cdots \right),
\tag{39}
$$

and in a closed form by

$$
u(x, y, t) = \cos x \cos y \sin(2t).
\tag{40}
$$

Example 5. Use the Adomian decomposition method to solve the initial-boundary value problem.

$$
\begin{aligned}
\text{PDE} \qquad u_{tt} &= 2(u_{xx} + u_{yy}),\ 0 < x, y < \pi,\ t > 0 \\
\text{B.C} \qquad u(0, y, t) &= y,\ u(\pi, y, t) = \pi + y \\
u(x, 0, t) &= x,\ u(x, \pi, t) = \pi + x \\
\text{I.C} \qquad u(x, y, 0) &= x + y + \sin x \sin y,\ u_t(x, y, 0) = 0
\end{aligned}
\tag{41}
$$

Solution.

Note that the boundary conditions are inhomogeneous and given by functions and not constants. The decomposition method will attack the problem directly without any need to convert the inhomogeneous conditions to homogeneous conditions.

Operating with L_t^{-1} on (41) gives

$$
u(x, y, t) = x + y + \sin x \sin y + 2L_t^{-1}\left(L_x u + L_y u\right).
\tag{42}
$$

It then follows

$$
\sum_{n=0}^{\infty} u_n = x + y + \sin x \sin y + 2L_t^{-1}\left(L_x\left(\sum_{n=0}^{\infty} u_n\right) + L_y\left(\sum_{n=0}^{\infty} u_n\right)\right).
\tag{43}
$$

Consequently, we set the relation

$$
\begin{aligned}
u_0(x, y, t) &= x + y + \sin x \sin y, \\
u_{k+1}(x, y, t) &= 2L_t^{-1}\left(L_x u_k + L_y u_k\right),\ k \geq 0.
\end{aligned}
\tag{44}
$$

This gives the first few components of the solution $u(x, y, t)$ by

$$
\begin{aligned}
u_0(x, y, t) &= x + y + \sin x \sin y, \\
u_1(x, y, t) &= 2L_t^{-1}\left(L_x u_0 + L_y u_0\right) = -\tfrac{1}{2!}(2t)^2 \sin x \sin y, \\
u_2(x, y, t) &= 2L_t^{-1}\left(L_x u_1 + L_y u_1\right) = \tfrac{1}{4!}(2t)^4 \sin x \sin y.
\end{aligned}
\tag{45}
$$

The series solution is given by

$$
u(x, y, t) = x + y + \sin x \sin y \left(1 - \frac{1}{2!}(2t)^2 + \frac{1}{4!}(2t)^4 - \cdots\right),
\tag{46}
$$

and in a closed form by

$$
u(x, y, t) = x + y + \sin x \sin y \cos(2t).
\tag{47}
$$

Example 6. Use the Adomian decomposition method to solve the initial-boundary value problem.

$$
\begin{array}{llll}
\text{PDE} & u_{tt} &=& \frac{x^2}{4}u_{xx} + \frac{y^2}{4}u_{yy}, \ 0 < x, y < 1, \ t > 0 \\[2mm]
\text{B.C} & u(0, y, t) &=& 0, u(1, y, t) = y^2 \cosh t \\[2mm]
& u(x, 0, t) &=& 0, u(x, 1, t) = x^2 \cosh t \\[2mm]
\text{I.C} & u(x, y, 0) &=& x^2 y^2, \ u_t(x, y, 0) = 0
\end{array}
\tag{48}
$$

Solution.

It is important to note that the coefficients of u_{xx} and u_{yy} are functions and not constants. Applying the inverse operator L_t^{-1} to the operator form of the PDE of (48) yields

$$
u(x, y, t) = x^2 y^2 + L_t^{-1}\left(\frac{x^2}{4}L_x u + \frac{y^2}{4}L_y u\right).
\tag{49}
$$

Using the decomposition series of $u(x, y, t)$ into both sides of (49) gives

$$
\sum_{n=0}^{\infty} u_n = x^2 y^2 + L_t^{-1}\left(\frac{x^2}{4}L_x\left(\sum_{n=0}^{\infty} u_n\right) + \frac{y^2}{4}L_y\left(\sum_{n=0}^{\infty} u_n\right)\right).
\tag{50}
$$

The recursive relation

$$
\begin{aligned}
u_0(x, y, t) &= x^2 y^2, \\
u_{k+1}(x, y, t) &= L_t^{-1}\left(\frac{x^2}{4}L_x u_k + \frac{y^2}{4}L_y u_k\right), k \geq 0,
\end{aligned}
\tag{51}
$$

follows immediately. It then follows that

$$
\begin{aligned}
u_0(x,y,t) &= x^2 y^2, \\
u_1(x,y,t) &= L_t^{-1}\left(\tfrac{x^2}{4}L_x u_0 + \tfrac{y^2}{4}L_y u_0\right) = \tfrac{1}{2!}t^2 x^2 y^2, \\
u_2(x,y,t) &= L_t^{-1}\left(\tfrac{x^2}{4}L_x u_1 + \tfrac{y^2}{4}L_y u_1\right) = \tfrac{1}{4!}t^4 x^2 y^2, \\
u_3(x,y,t) &= L_t^{-1}\left(\tfrac{x^2}{4}L_x u_2 + \tfrac{y^2}{4}L_y u_2\right) = \tfrac{1}{6!}t^6 x^2 y^2,
\end{aligned}
\tag{52}
$$

$$\vdots$$

Consequently, the series solution

$$
u(x,y,t) = x^2 y^2 \left(1 + \frac{1}{2!}t^2 + \frac{1}{4!}t^4 + \frac{1}{6!}t^6 + \cdots\right),
\tag{53}
$$

and the exact solution

$$
u(x,y,t) = x^2 y^2 \cosh t,
\tag{54}
$$

is readily obtained.

Inhomogeneous Wave Equations

We now consider the inhomogeneous wave equation of the form

$$
u_{tt} = c^2(u_{xx} + u_{yy}) + h(x,y,t),
\tag{55}
$$

where $h(x,y,t)$ is the inhomogeneous term. One significant advantage of Adomian decomposition method is that it handles the inhomogeneous partial differential equations in an identical manner to that used before in handling homogeneous differential equations. However, it is interesting to note that the zeroth component u_0 is identified by all terms that arise from using initial conditions and from integrating inhomogeneous terms as well.

In the following, the decomposition method will be illustrated by discussing the inhomogeneous equations.

Example 7. Use the Adomian decomposition method to solve the initial-boundary value problem.

$$
\begin{aligned}
\text{PDE} \qquad u_{tt} &= \tfrac{1}{2}(u_{xx} + u_{yy}) - 2,\ 0 < x,y < \pi,\ t > 0 \\[4pt]
\text{B.C} \qquad u(0,y,t) &= y^2,\ u(\pi,y,t) = \pi^2 + y^2 \\[4pt]
u(x,0,t) &= x^2,\ u(x,\pi,t) = \pi^2 + x^2 \\[4pt]
\text{I.C} \qquad u(x,y,0) &= x^2 + y^2,\ u_t(x,y,0) = \sin x \sin y
\end{aligned}
\tag{56}
$$

Solution.

Note that the PDE (56) contains the term -2; hence it is inhomogeneous equation. In addition, the boundary conditions are inhomogeneous.

Applying the inverse operator L_t^{-1} to the operator form of (56) gives

$$u(x, y, t) = x^2 + y^2 + t \sin x \sin y - t^2 + \frac{1}{2} L_t^{-1} (L_x u + L_y u), \tag{57}$$

obtained by using the initial conditions. Using the series representation of $u(x, y, t)$ in both sides of (57) we obtain

$$\sum_{n=0}^{\infty} u_n = x^2 + y^2 + t \sin x \sin y - t^2 + \frac{1}{2} L_t^{-1} \left(L_x \left(\sum_{n=0}^{\infty} u_n \right) + L_y \left(\sum_{n=0}^{\infty} u_n \right) \right). \tag{58}$$

To determine the components of u, we use the recursive relation

$$
\begin{aligned}
u_0(x, y, t) &= x^2 + y^2 + t \sin x \sin y - t^2, \\
u_{k+1}(x, y, t) &= \tfrac{1}{2} L_t^{-1} (L_x u_k + L_y u_k), \, k \geq 0,
\end{aligned}
\tag{59}
$$

that gives

$$
\begin{aligned}
u_0(x, y, t) &= x^2 + y^2 + t \sin x \sin y - t^2, \\
u_1(x, y, t) &= \tfrac{1}{2} L_t^{-1} (L_x u_0 + L_y u_0) = t^2 - \tfrac{1}{3!} t^3 \sin x \sin y, \\
u_2(x, y, t) &= \tfrac{1}{2} L_t^{-1} (L_x u_1 + L_y u_1) = \tfrac{1}{5!} t^5 \sin x \sin y.
\end{aligned}
\tag{60}
$$

It follows that the solution in a series form is

$$u(x, y, t) = x^2 + y^2 + \sin x \sin y \left(t - \tfrac{1}{3!} t^3 + \tfrac{1}{5!} t^5 - \cdots \right), \tag{61}$$

and in a closed form is

$$u(x, y, t) = x^2 + y^2 + \sin x \sin y \sin t. \tag{62}$$

Unlike Example 7 where the inhomogeneous term is a constant, the inhomogeneous in the following example is a function of y.

Example 8. Use the Adomian decomposition method to solve the initial-boundary value problem.

$$
\begin{array}{lll}
\text{PDE} & u_{tt} = (u_{xx} + u_{yy}) + \cos y, \ 0 < x, y < \pi, \ t > 0 \\[4pt]
\text{B.C} & u(0, y, t) = u(\pi, y, t) = \cos y \\[4pt]
& u(x, 0, t) = 1 + \sin x \sin t, \ u(x, \pi, t) = -1 - \sin x \sin t \\[4pt]
\text{I.C} & u(x, y, 0) = \cos y, \ u_t(x, y, 0) = \sin x
\end{array}
\tag{63}
$$

Solution.

Proceeding as before we obtain

$$u(x, y, t) = \cos y + t \sin x + \frac{1}{2!} t^2 \cos y + L_t^{-1} \left(L_x u + L_y u \right). \tag{64}$$

It then follows that

$$\sum_{n=0}^{\infty} u_n = \cos y + t \sin x + \frac{1}{2!} t^2 \cos y + L_t^{-1} \left(L_x \left(\sum_{n=0}^{\infty} u_n \right) + L_y \left(\sum_{n=0}^{\infty} u_n \right) \right). \tag{65}$$

Identifying the zeroth component $u_0(x, y, t)$ by all terms that arise from initial conditions and from applying L_t^{-1} to the inhomogeneous term $\cos y$, the recursive relationship is therefore defined by

$$
\begin{aligned}
u_0(x, y, t) &= \cos y + t \sin x + \tfrac{1}{2!} t^2 \cos y, \\
u_{k+1}(x, y, t) &= L_t^{-1} \left(L_x u_k + L_y u_k \right), k \geq 0,
\end{aligned}
\tag{66}
$$

hence we find

$$
\begin{aligned}
u_0(x, y, t) &= \cos y + t \sin x + \tfrac{1}{2!} t^2 \cos y, \\
u_1(x, y, t) &= L_t^{-1} \left(L_x u_0 + L_y u_0 \right) = -\tfrac{1}{2!} t^2 \cos y - \tfrac{1}{3!} t^3 \sin x - \tfrac{1}{4!} t^4 \cos y, \quad (67) \\
u_2(x, y, t) &= L_t^{-1} \left(L_x u_1 + L_y u_1 \right) = \tfrac{1}{4!} t^4 \cos y + \tfrac{1}{5!} t^5 \sin x + \tfrac{1}{6!} t^6 \cos y.
\end{aligned}
$$

This gives the solution in a series form by

$$u(x, y, t) = \cos y + \sin x \left(t - \tfrac{1}{3!} t^3 + \tfrac{1}{5!} t^5 - \cdots \right) \tag{68}$$

and in a closed form by

$$u(x, y, t) = \cos y + \sin x \sin t. \tag{69}$$

Example 9. Use the Adomian decomposition method to solve the initial-boundary value problem.

$$
\begin{aligned}
\text{PDE} \qquad u_{tt} &= 2(u_{xx} + u_{yy}) + 6t + 2x + 4y, \ 0 < x, y < \pi, \ t > 0 \\
\\
\text{B.C} \qquad u(0, y, t) &= t^3 + 2t^2 y, \ u(\pi, y, t) = t^3 + \pi t^2 + 2t^2 y \\
\\
u(x, 0, t) &= t^3 + t^2 x, \ u(x, \pi, t) = t^3 + t^2 x + 2\pi t^2 \\
\\
\text{I.C} \qquad u(x, y, 0) &= 0, \ u_t(x, y, 0) = 2 \sin x \sin y
\end{aligned}
\tag{70}
$$

Solution.

We follow the discussion introduced before to obtain

$$u(x, y, t) = 2t \sin x \sin y + t^3 + t^2 x + 2t^2 y + 2L_t^{-1} \left(L_x u + L_y u \right), \qquad (71)$$

where we find

$$\sum_{n=0}^{\infty} u_n = 2t \sin x \sin y + t^3 + t^2 x + 2t^2 y$$

$$+ L_t^{-1} \left(L_x \left(\sum_{n=0}^{\infty} u_n \right) + L_y \left(\sum_{n=0}^{\infty} u_n \right) \right). \qquad (72)$$

We next set the recursive relation

$$\begin{aligned} u_0(x, y, t) &= 2t \sin x \sin y + t^3 + t^2 x + 2t^2 y, \\ u_{k+1}(x, y, t) &= 2L_t^{-1} \left(L_x u_k + L_y u_k \right)), \, k \geq 0, \end{aligned} \qquad (73)$$

that gives the first few components

$$\begin{aligned} u_0(x, y, t) &= 2t \sin x \sin y + t^3 + t^2 x + 2t^2 y, \\ u_1(x, y, t) &= -\frac{(2t)^3}{3!} \sin x \sin y, \\ u_2(x, y, t) &= \frac{(2t)^5}{5!} \sin x \sin y. \end{aligned} \qquad (74)$$

In view of of (74), the solution in a series form is

$$u(x, y, t) = t^3 + t^2 x + 2t^2 y + \sin x \sin y \left((2t) - \frac{(2t)^3}{3!} + \frac{(2t)^5}{5!} - \cdots \right), \qquad (75)$$

and in a closed form is

$$u(x, y, t) = t^3 + t^2 x + 2t^2 y + \sin x \sin y \sin(2t). \qquad (76)$$

Exercises 6.2.1

In Exercises 1 – 8, use the decomposition method to solve the homogeneous initial-boundary value problems:

1. $u_{tt} = 2(u_{xx} + u_{yy})$, $0 < x, y < \pi$, $t > 0$
   $u(0, y, t) = u(\pi, y, t) = 0$
   $u(x, 0, t) = u(x, \pi, t) = 0$
   $u(x, y, 0) = 0$, $u_t(x, y, 0) = 2 \sin x \sin y$

2. $u_{tt} = 2(u_{xx} + u_{yy}),\ 0 < x, y < \pi,\ t > 0$
   $u(0, y, t) = u(\pi, y, t) = 0$
   $u(x, 0, t) = u(x, \pi, t) = 0$
   $u(x, y, 0) = 0,\ u_t(x, y, 0) = 4\sin(2x)\sin(2y)$

3. $u_{tt} = 2(u_{xx} + u_{yy}),\ 0 < x, y < \pi,\ t > 0$
   $u(0, y, t) = u(\pi, y, t) = 0$
   $u(x, 0, t) = u(x, \pi, t) = 0$
   $u(x, y, 0) = \sin(2x)\sin(2y),\ u_t(x, y, 0) = 0$

4. $u_{tt} = \frac{1}{2}(u_{xx} + u_{yy}),\ 0 < x, y < \pi,\ t > 0$
   $u(0, y, t) = u(\pi, y, t) = 2$
   $u(x, 0, t) = u(x, \pi, t) = 2$
   $u(x, y, 0) = 2,\ u_t(x, y, 0) = \sin x \sin y$

5. $u_{tt} = 2(u_{xx} + u_{yy}),\ 0 < x, y < \pi,\ t > 0$
   $u(0, y, t) = u(\pi, y, t) = 1 + y$
   $u(x, 0, t) = 1,\ u(x, \pi, t) = 1 + \pi$
   $u(x, y, 0) = 1 + y,\ u_t(x, y, 0) = 2\sin x \sin y$

6. $u_{tt} = u_{xx} + u_{yy},\ 0 < x, y < \pi,\ t > 0$
   $u(0, y, t) = 1 + \sin y \sin t,\ u(\pi, y, t) = 1 + \pi + \sin y \sin t$
   $u(x, 0, t) = u(x, \pi, t) = 1 + x$
   $u(x, y, 0) = 1 + x,\ u_t(x, y, 0) = \sin y$

7. $u_{tt} = u_{xx} + u_{yy} - 2u,\ 0 < x, y < \pi,\ t > 0$
   $u(0, y, t) = u(\pi, y, t) = 0$
   $u(x, 0, t) = u(x, \pi, t) = 0$
   $u(x, y, 0) = 0,\ u_t(x, y, 0) = 2\sin x \sin y$

8. $u_{tt} = u_{xx} + u_{yy} - 7u,\ 0 < x, y < \pi,\ t > 0$
   $u(0, y, t) = u(\pi, y, t) = 0$
   $u(x, 0, t) = u(x, \pi, t) = 0$
   $u(x, y, 0) = \sin x \sin y,\ u_t(x, y, 0) = 0$

In Exercises 9 – 14, use the decomposition method to solve the inhomogeneous initial-boundary value problems:

9. $u_{tt} = \frac{1}{2}(u_{xx} + u_{yy}) + \frac{1}{2}\sin x,\ 0 < x, y < \pi,\ t > 0$
   $u(0, y, t) = u(\pi, y, t) = 0$
   $u(x, 0, t) = u(x, \pi, t) = \sin x$
   $u(x, y, 0) = \sin x,\ u_t(x, y, 0) = \sin x \sin y$

10. $u_{tt} = u_{xx} + u_{yy} + \cos x,\ 0 < x, y < \pi,\ t > 0$
    $u(0, y, t) = 1 + \sin y \sin t,\ u(\pi, y, t) = -1 - \sin y \sin t$
    $u(x, 0, t) = u(x, \pi, t) = \cos x$
    $u(x, y, 0) = \cos x,\ u_t(x, y, 0) = \sin y$

11. $u_{tt} = u_{xx} + u_{yy} - 4,\ 0 < x, y < \pi,\ t > 0$
$u(0, y, t) = y^2, u(\pi, y, t) = \pi^2 + y^2$
$u(x, 0, t) = x^2 + \sin x \sin t, u(x, \pi, t) = \pi^2 + x^2 + \sin x \sin t$
$u(x, y, 0) = x^2 + y^2, u_t(x, y, 0) = \sin x$

12. $u_{tt} = u_{xx} + u_{yy} - 8,\ 0 < x, y < \pi,\ t > 0$
$u(0, y, t) = 2y^2, u(\pi, y, t) = 2\pi^2 + 2y^2$
$u(x, 0, t) = 2x^2, u(x, \pi, t) = 2\pi^2 + 2x^2$
$u(x, y, 0) = 2x^2 + 2y^2 + 2\sin x \sin y, u_t(x, y, 0) = 0$

13. $u_{tt} = \frac{1}{2}(u_{xx} + u_{yy}) + 2,\ 0 < x, y < \pi,\ t > 0$
$u(0, y, t) = t^2 + ty, u(\pi, y, t) = t^2 + \pi t + ty$
$u(x, 0, t) = t^2 + tx, u(x, \pi, t) = t^2 + \pi t + tx$
$u(x, y, 0) = 0, u_t(x, y, 0) = x + y + \sin x \sin y$

14. $u_{tt} = 2(u_{xx} + u_{yy}) + 6t + 2x,\ 0 < x, y < \pi,\ t > 0$
$u(0, y, t) = t^3 + ty, u(\pi, y, t) = t^3 + \pi t^2 + ty$
$u(x, 0, t) = t^3 + t^2 x, u(x, \pi, t) = t^3 + t^2 x + \pi t$
$u(x, y, 0) = \sin x \sin y, u_t(x, y, 0) = y$

In Exercises 15 – 20, use the decomposition method to solve the initial-boundary value problems:

15. $u_{tt} = u_{xx} + u_{yy} - 2,\ 0 < x, y < \pi,\ t > 0$
$u(0, y, t) = u(\pi, y, t) = y^2$
$u(x, 0, t) = \sin x \cos t, u(x, \pi, t) = \pi^2 + \sin x \cos t$
$u(x, y, 0) = y^2 + \sin x, u_t(x, y, 0) = 0$

16. $u_{tt} = u_{xx} + u_{yy} - 2,\ 0 < x, y < \pi,\ t > 0$
$u(0, y, t) = \sin y \sin t, u(\pi, y, t) = \pi^2 + \sin y \sin t$
$u(x, 0, t) = u(x, \pi, t) = x^2$
$u(x, y, 0) = x^2, u_t(x, y, 0) = \sin y$

17. $u_{tt} = u_{xx} + u_{yy} + \sin x,\ 0 < x, y < \pi,\ t > 0$
$u(0, y, t) = u(\pi, y, t) = \sin y \sin t$
$u(x, 0, t) = u(x, \pi, t) = \sin x$
$u(x, y, 0) = \sin x, u_t(x, y, 0) = \sin y$

18. $u_{tt} = 2(u_{xx} + u_{yy}),\ 0 < x, y < \pi,\ t > 0$
$u(0, y, t) = -u(\pi, y, t) = \cos y \sin(2t)$
$u(x, 0, t) = -u(x, \pi, t) = \cos x \sin(2t)$
$u(x, y, 0) = 0, u_t(x, y, 0) = 2\cos x \cos y$

19. $u_{tt} = \frac{1}{2}(u_{xx} + u_{yy}) + 12t^2 + 2y,\ 0 < x, y < \pi,\ t > 0$
$u(0, y, t) = u(\pi, y, t) = t^4 + t^2 y$
$u(x, 0, t) = t^4, u(x, \pi, t) = t^4 + \pi t^2$
$u(x, y, 0) = 0, u_t(x, y, 0) = \sin x \sin y$

20. $u_{tt} = \frac{1}{2}(u_{xx} + u_{yy})$, $0 < x, y < \pi$, $t > 0$
    $u(0, y, t) = t^2 + y^2$, $u(\pi, y, t) = \pi^2 + t^2 + y^2$
    $u(x, 0, t) = t^2 + x^2$, $u(x, \pi, t) = \pi^2 + t^2 + x^2$
    $u(x, y, 0) = x^2 + y^2$, $u_t(x, y, 0) = \sin x \sin y$

In Exercises 21 – 24, solve the partial differential equations where coefficients of u_{xx} and u_{yy} are functions and constants:

21. $u_{tt} = \frac{x^2}{4} u_{xx} + \frac{y^2}{4} u_{yy}$, $0 < x, y < 1$, $t > 0$
    $u(0, y, t) = 0$, $u(1, y, t) = y^2 \sinh t$
    $u(x, 0, t) = 0$, $u(x, 1, t) = x^2 \sinh t$
    $u(x, y, 0) = 0$, $u_t(x, y, 0) = x^2 y^2$

22. $u_{tt} = \frac{x^2}{4} u_{xx} + \frac{y^2}{4} u_{yy}$, $0 < x, y < 1$, $t > 0$
    $u(0, y, t) = 0$, $u(1, y, t) = y^2 e^t$
    $u(x, 0, t) = 0$, $u(x, 1, t) = x^2 e^t$
    $u(x, y, 0) = x^2 y^2$, $u_t(x, y, 0) = x^2 y^2$

23. $u_{tt} = \frac{x^2}{2} u_{xx} + \frac{y^2}{2} u_{yy}$, $0 < x, y < 1$, $t > 0$
    $u(0, y, t) = y^2 \cosh t$, $u(1, y, t) = \sinh t + y^2 \cosh t$
    $u(x, 0, t) = x^2 \sinh t$, $u(x, 1, t) = x^2 \sinh t + \cosh t$
    $u(x, y, 0) = y^2$, $u_t(x, y, 0) = x^2$

24. $u_{tt} = \frac{x^2}{2} u_{xx} + \frac{y^2}{2} u_{yy}$, $0 < x, y < 1$, $t > 0$
    $u(0, y, t) = y^2 e^t$, $u(1, y, t) = e^{-t} + y^2 e^t$
    $u(x, 0, t) = x^2 e^{-t}$, $u(x, 1, t) = e^t + x^2 e^{-t}$
    $u(x, y, 0) = x^2 + y^2$, $u_t(x, y, 0) = y^2 - x^2$

6.2.2   Three Dimensional Wave Equation

The propagation of waves in a three dimensional volume of length a, width b, and height d is governed by the following initial boundary value problem

$$
\begin{array}{rrcl}
\text{PDE} & u_{tt} & = & c^2(u_{xx} + u_{yy} + u_{zz}), t > 0 \\[2mm]
\text{B.C} & u(0, y, z, t) & = & u(a, y, z, t) = 0 \\[2mm]
& u(x, 0, z, t) & = & u(x, b, z, t) = 0 \\[2mm]
& u(x, y, 0, t) & = & u(x, y, d, t) = 0 \\[2mm]
\text{I.C} & u(x, y, z, 0) & = & f(x, y, z), \, u_t(x, y, z, 0) = g(x, y, z)
\end{array}
\tag{77}
$$

where $0 < x < a$, $0 < y < b$, $0 < z < d$, and $u = u(x, y, z, t)$ is the displacement of any point located at the position (x, y, z) of a rectangular volume at any time t, and c is the velocity of a propagation wave.

As discussed before, the solution in the t space minimizes the volume of calculations. Accordingly, the operator L_t^{-1} will be applied here. We first rewrite (77) in an operator form by

$$L_t u = c^2 (L_x u + L_y u + L_z u), \tag{78}$$

where the differential operators L_x, L_y, and L_z are defined by

$$L_t = \frac{\partial^2}{\partial t^2}, L_x = \frac{\partial^2}{\partial x^2}, L_y = \frac{\partial^2}{\partial y^2}, L_z = \frac{\partial^2}{\partial z^2}, \tag{79}$$

so that the integral operator L_t^{-1} represents a two-fold integration from 0 to t given by

$$L_t^{-1}(.) = \int_0^t \int_0^t (.) dt. \tag{80}$$

This means that

$$L_t^{-1} L_t u(x, y, z, t) = u(x, y, z, t) - u(x, y, z, 0) - t u_t(x, y, z, 0). \tag{81}$$

Applying L_t^{-1} to both sides of (78), noting (81) and using the initial conditions we find

$$u(x, y, z, t) = f(x, y, z) + t g(x, y, z) + c^2 L_t^{-1} (L_x u + L_y u + L_z u). \tag{82}$$

The decomposition method defines the solution $u(x, y, z, t)$ as a series given by

$$u(x, y, z, t) = \sum_{n=0}^{\infty} u_n(x, y, z, t). \tag{83}$$

Substituting (83) into both sides of (82) yields

$$\sum_{n=0}^{\infty} u_n = f(x, y, z) + t g(x, y, z)$$

$$+ c^2 L_t^{-1} \left(L_x \left(\sum_{n=0}^{\infty} u_n \right) + L_y \left(\sum_{n=0}^{\infty} u_n \right) + L_z \left(\sum_{n=0}^{\infty} u_n \right) \right). \tag{84}$$

The components $u_n(x, y, z, t), n \geq 0$ can be completely determined by using the recursive relation

$$u_0(x, y, z, t) = f(x, y, z) + t g(x, y, z),$$
$$u_{k+1}(x, y, z, t) = c^2 L_t^{-1} (L_x u_k + L_y u_k + L_z u_k), k \geq 0. \tag{85}$$

Having determined the components $u_n, n \geq 0$ by applying the scheme (85), the solution in a series form follows immediately.

Homogeneous Wave Equations

The decomposition method will be used to discuss the following homogeneous wave equations in three dimensional space with homogeneous or inhomogeneous boundary conditions.

Example 10. Use the Adomian decomposition method to solve the initial boundary value problem

$$
\begin{array}{rlrl}
\text{PDE} & & u_{tt} & = 3(u_{xx} + u_{yy} + u_{zz}),\ 0 < x, y, z < \pi,\ t > 0 \\[2mm]
\text{B.C} & u(0, y, z, t) & = u(\pi, y, z, t) = 0 \\[2mm]
& u(x, 0, z, t) & = u(x, \pi, z, t) = 0 & \qquad (86) \\[2mm]
& u(x, y, 0, t) & = u(x, y, \pi, t) = 0 \\[2mm]
\text{I.C} & u(x, y, z, 0) & = 0,\ u_t(x, y, z, 0) = 3 \sin x \sin y \sin z
\end{array}
$$

Solution.

The PDE of (86) can be rewritten by

$$
L_t u = 3 \left(L_x u + L_y u + L_z u \right). \tag{87}
$$

Applying the inverse operator L_t^{-1} to (87), using (81) and substituting the initial conditions we obtain

$$
u(x, y, z, t) = 3t \sin x \sin y \sin z + 3 L_t^{-1} \left(L_x u + L_y u + L_z u \right). \tag{88}
$$

Using the decomposition series

$$
u(x, y, z, t) = \sum_{n=0}^{\infty} u_n(x, y, z, t), \tag{89}
$$

into both sides of (88) yields

$$
\sum_{n=0}^{\infty} u_n = 3t \sin x \sin y \sin z
$$

$$
+ 3 L_t^{-1} \left(L_x \left(\sum_{n=0}^{\infty} u_n \right) + L_y \left(\sum_{n=0}^{\infty} u_n \right) + L_z \left(\sum_{n=0}^{\infty} u_n \right) \right). \tag{90}
$$

Identifying the zeroth component as discussed before we then set the relation

$$
\begin{array}{rll}
u_0(x, y, z, t) & = & 3t \sin x \sin y \sin z, \\[2mm]
u_{k+1}(x, y, z, t) & = & 3 L_t^{-1} \left(L_x u_k + L_y u_k + L_z u_k \right),\ k \geq 0.
\end{array} \tag{91}
$$

The first few components of the decomposition of u are given by

$$
\begin{aligned}
u_0(x, y, z, t) &= 3t \sin x \sin y \sin z, \\
u_1(x, y, z, t) &= 3L_t^{-1}(L_x u_0 + L_y u_0 + L_z u_0) = -\frac{(3t)^3}{3!} \sin x \sin y \sin z, \qquad (92) \\
u_2(x, y, z, t) &= 3L_t^{-1}(L_x u_1 + L_y u_1 + L_z u_1) = \frac{(3t)^5}{5!} \sin x \sin y \sin z.
\end{aligned}
$$

For numerical purposes, further components can be computed to improve the accuracy level of the approximation.

Combining (89) and (92), the solution in a series form is given by

$$
u(x, y, z, t) = \sin x \sin y \sin z \left((3t) - \frac{(3t)^3}{3!} + \frac{(3t)^5}{5!} - \cdots \right), \qquad (93)
$$

and in a closed form by

$$
u(x, y, z, t) = \sin x \sin y \sin z \sin(3t). \qquad (94)
$$

Example 11. Use the Adomian decomposition method to solve the initial boundary value problem

$$
\begin{aligned}
\text{PDE} \qquad u_{tt} &= u_{xx} + u_{yy} + u_{zz} - u, \ 0 < x, y, z < \pi, \ t > 0 \\
\text{B.C} \quad u(0, y, z, t) &= u(\pi, y, z, t) = 0 \\
u(x, 0, z, t) &= u(x, \pi, z, t) = 0 \qquad (95) \\
u(x, y, 0, t) &= u(x, y, \pi, t) = 0 \\
\text{I.C} \quad u(x, y, z, 0) &= 0, \ u_t(x, y, z, 0) = 2 \sin x \sin y \sin z
\end{aligned}
$$

Solution.

We note that an additional term $-u$ is contained in the standard wave equation. This term usually arises when each element of the rectangular volume is subjected to an additional force.

Applying L_t^{-1} to the operator form of the PDE of (95), using (81) and substituting the initial conditions we obtain

$$
u = 2t \sin x \sin y \sin z + L_t^{-1}(L_x u + L_y u + L_z u - u). \qquad (96)
$$

Following the analysis made above we find

$$
\sum_{n=0}^{\infty} u_n = 2t \sin x \sin y \sin z
$$

$$
- L_t^{-1} \left(L_x \left(\sum_{n=0}^{\infty} u_n \right) + L_y \left(\sum_{n=0}^{\infty} u_n \right) + L_z \left(\sum_{n=0}^{\infty} u_n \right) - \left(\sum_{n=0}^{\infty} u_n \right) \right). \qquad (97)
$$

Therefore we set the relation

$$
\begin{aligned}
u_0(x, y, z, t) &= 2t \sin x \sin y \sin z, \\
u_{k+1}(x, y, z, t) &= L_t^{-1} (L_x u_k + L_y u_k + L_z u_k - u_k), \, k \geq 0,
\end{aligned}
\tag{98}
$$

that gives

$$
\begin{aligned}
u_0(x, y, z, t) &= 2t \sin x \sin y \sin z, \\
u_1(x, y, z, t) &= L_t^{-1} (L_x u_0 + L_y u_0 + L_z u_0 - u_0) = -\frac{(2t)^3}{3!} \sin x \sin y \sin z, \\
u_2(x, y, z, t) &= L_t^{-1} (L_x u_1 + L_y u_1 + L_z u_1 - u_1) = \frac{(2t)^5}{5!} \sin x \sin y \sin z, \\
u_3(x, y, z, t) &= L_t^{-1} (L_x u_2 + L_y u_2 + L_z u_2 - u_2) = -\frac{(2t)^7}{7!} \sin x \sin y \sin z.
\end{aligned}
\tag{99}
$$

The series solution is given by

$$
u(x, y, z, t) = \sin x \sin y \sin z \left((2t) - \frac{(2t)^3}{3!} + \frac{(2t)^5}{5!} - \cdots \right),
\tag{100}
$$

and the exact solution is

$$
u(x, y, z, t) = \sin x \sin y \sin z \sin(2t).
\tag{101}
$$

Example 12. Use the Adomian decomposition method to solve the initial boundary value problem

$$
\begin{aligned}
\text{PDE} \qquad u_{tt} &= u_{xx} + u_{yy} + u_{zz} - u, \, 0 < x, y, z < \pi, \, t > 0 \\
\text{B.C} \quad u(0, y, z, t) &= -u(\pi, y, z, t) = \sin y \sin(z + 2t) \\
u(x, 0, z, t) &= -u(x, \pi, z, t) = \sin x \sin(z + 2t) \\
u(x, y, 0, t) &= -u(x, y, \pi, t) = \sin(x + y) \sin(2t) \\
\text{I.C} \quad u(x, y, z, 0) &= \sin(x + y) \sin z, \, u_t(x, y, z, 0) = 2 \sin(x + y) \cos z
\end{aligned}
\tag{102}
$$

Solution.

It is interesting to note that the boundary conditions are inhomogeneous and given by functions and not constants.

Applying L_t^{-1} to (102), using (81) gives

$$
\begin{aligned}
u(x, y, z, t) &= \sin(x + y) \sin z + 2t \sin(x + y) \cos z \\
&\quad + L_t^{-1} (L_x u + L_y u + L_z u - u),
\end{aligned}
\tag{103}
$$

therefore we find

$$\sum_{n=0}^{\infty} u_n = \sin(x+y)\sin z + 2t\sin(x+y)\cos z$$

$$+L_t^{-1}\left(L_x\left(\sum_{n=0}^{\infty} u_n\right) + L_y\left(\sum_{n=0}^{\infty} u_n\right) + L_z\left(\sum_{n=0}^{\infty} u_n\right) - \left(\sum_{n=0}^{\infty} u_n\right)\right). \quad (104)$$

This means that

$$\begin{aligned}
u_0(x,y,z,t) &= \sin(x+y)\sin z + 2t\sin(x+y)\cos z, \\
u_{k+1}(x,y,z,t) &= L_t^{-1}\left(L_x u_k + L_y u_k + L_z u_k - u_k\right), k \geq 0,
\end{aligned} \quad (105)$$

and therefore we obtain

$$\begin{aligned}
u_0(x,y,z,t) &= \sin(x+y)\sin z + 2t\sin(x+y)\cos z, \\
u_1(x,y,z,t) &= L_t^{-1}\left(L_x u_0 + L_y u_0 + L_z u_0 - u_0\right) \\
&= -\frac{(2t)^2}{2!}\sin(x+y)\sin z - \frac{(2t)^3}{3!}\sin(x+y)\cos z, \\
u_2(x,y,z,t) &= L_t^{-1}\left(L_x u_1 + L_y u_1 + L_z u_1 - u_1\right), \\
&= \frac{(2t)^4}{4!}\sin(x+y)\sin z + \frac{(2t)^5}{5!}\sin(x+y)\cos z, \\
u_3(x,y,z,t) &= L_t^{-1}\left(L_x u_2 + L_y u_2 + L_z u_2 - u_2\right), \\
&= -\frac{(2t)^6}{6!}\sin(x+y)\sin z - \frac{(2t)^7}{7!}\sin(x+y)\cos z, \\
&\vdots
\end{aligned} \quad (106)$$

In view of (106), the solution in a series form is given by

$$\begin{aligned}
u(x,y,z,t) &= \sin(x+y)\sin z \left(1 - \frac{(2t)^2}{2!} + \frac{(2t)^4}{4!} - \cdots\right) \\
&\quad + \sin(x+y)\cos z \left((2t) - \frac{(2t)^3}{3!} + \frac{(2t)^5}{5!} - \cdots\right),
\end{aligned} \quad (107)$$

and the exact solution

$$\begin{aligned}
u(x,y,z,t) &= \sin(x+y)\left(\sin z \cos(2t) + \cos z \sin(2t)\right), \\
&= \sin(x+y)\sin(z+2t),
\end{aligned} \quad (108)$$

is readily obtained.

Example 13. Use the Adomian decomposition method to solve the initial boundary value problem

$$\text{PDE} \qquad u_{tt} = \frac{x^2}{18}u_{xx} + \frac{y^2}{18}u_{yy} + \frac{z^2}{18}u_{zz} - u, \ 0 < x, y, z < 1, t > 0$$

$$\text{B.C} \quad u(0, y, z, t) = 0, u(1, y, z, t) = y^4 z^4 \sinh t$$

$$u(x, 0, z, t) = 0, u(x, 1, z, t) = x^4 z^4 \sinh t$$

$$u(x, y, 0, t) = 0, u(x, y, 1, t) = x^4 y^4 \sinh t$$

$$\text{I.C} \quad u(x, y, z, 0) = 0, u_t(x, y, z, 0) = x^4 y^4 z^4$$

$$\tag{109}$$

Solution.

The coefficients of the derivatives of u are functions and not constants. Following the discussions presented above we find

$$u(x, y, z, t) = tx^4 y^4 z^4 + L_t^{-1}\left(\frac{x^2}{18}L_x u + \frac{y^2}{18}L_y u + \frac{z^2}{18}L_z u - u\right), \tag{110}$$

that leads to gives

$$\sum_{n=0}^{\infty} u_n = x^4 y^4 z^4 - L_t^{-1}\left(\sum_{n=0}^{\infty} u_n\right)$$

$$+ L_t^{-1}\left(\frac{x^2}{18}L_x\left(\sum_{n=0}^{\infty} u_n\right) + \frac{y^2}{18}L_y\left(\sum_{n=0}^{\infty} u_n\right) + \frac{z^2}{18}L_z\left(\sum_{n=0}^{\infty} u_n\right)\right). \tag{111}$$

The recursive relation

$$u_0(x, y, z, t) = tx^4 y^4 z^4,$$

$$u_{k+1}(x, y, z, t) = L_t^{-1}\left(\frac{x^2}{18}L_x u_k + \frac{y^2}{18}L_y u_k + \frac{z^2}{18}L_z u_k - u_k\right), k \geq 0. \tag{112}$$

gives the first few components by

$$u_0(x, y, z, t) = tx^4 y^4 z^4,$$

$$u_1(x, y, z, t) = L_t^{-1}\left(\frac{x^2}{18}L_x u_0 + \frac{y^2}{18}L_y u_0 + \frac{z^2}{18}L_z u_0 - u_0\right) = \frac{t^3}{3!}x^4 y^4 z^4, \tag{113}$$

$$u_2(x, y, z, t) = L_t^{-1}\left(\frac{x^2}{18}L_x u_1 + \frac{y^2}{18}L_y u_1 + \frac{z^2}{18}L_z u_1 - u_1\right) = \frac{t^5}{5!}x^4 y^4 z^4.$$

In view of (113), the solution in a series form is given by

$$u(x, y, z, t) = x^4 y^4 z^4\left(t + \frac{t^3}{3!} + \frac{t^5}{5!} + \cdots\right), \tag{114}$$

and in a closed form by

$$u(x, y, z, t) = x^4 y^4 z^4 \sinh t. \tag{115}$$

Inhomogeneous Wave Equations

We now consider the inhomogeneous wave equation in a three dimensional space of the form

$$u_{tt} = c^2 (u_{xx} + u_{yy} + u_{zz}) + h(x, y, z), \tag{116}$$

where $h(x, y, z)$ is an inhomogeneous term. The decomposition method can be applied without any need to transform this equation to a homogeneous equation. The following examples will be used to explain the implementation of the method.

Example 14. Use the Adomian decomposition method to solve the initial boundary value problem

$$
\begin{aligned}
\text{PDE} \quad u_{tt} &= u_{xx} + u_{yy} + u_{zz} + \sin x + \sin y, \ 0 < x, y, z < \pi \\
\text{B.C} \ u(0, y, z, t) &= u(\pi, y, z, t) = \sin y + \sin z \sin t \\
u(x, 0, z, t) &= u(x, \pi, z, t) = \sin x + \sin z \sin t \\
u(x, y, 0, t) &= u(x, y, \pi, t) = \sin x + \sin y \\
\text{I.C} \ u(x, y, z, 0) &= \sin x + \sin y, \ u_t(x, y, z, 0) = \sin z
\end{aligned}
\tag{117}
$$

Solution.

Operating with L_t^{-1} on (117) gives

$$u = \sin x + \sin y + t \sin z + \frac{t^2}{2!} \sin x + \frac{t^2}{2!} \sin y + L_t^{-1} (L_x u + L_y u + L_z u). \tag{118}$$

Following the analysis made above, we substitute the decomposition series

$$u(x, y, z, t) = \sum_{n=0}^{\infty} u_n(x, y, z, t), \tag{119}$$

into both sides of (118) to obtain

$$
\sum_{n=0}^{\infty} u_n = \sin x + \sin y + t \sin z + \frac{t^2}{2!} \sin x + \frac{t^2}{2!} \sin y
$$
$$
+ L_t^{-1} \left(L_x \left(\sum_{n=0}^{\infty} u_n \right) + L_y \left(\sum_{n=0}^{\infty} u_n \right) + L_z \left(\sum_{n=0}^{\infty} u_n \right) \right). \tag{120}
$$

This leads to

$$
\begin{aligned}
u_0(x, y, z, t) &= \sin x + \sin y + t \sin z + \tfrac{t^2}{2!} \sin x + \tfrac{t^2}{2!} \sin y, \\
u_1(x, y, z, t) &= L_t^{-1} \left(L_x u_0 + L_y u_0 + L_z u_0 \right) \\
&= -\tfrac{t^2}{2!} \sin x - \tfrac{t^2}{2!} \sin y - \tfrac{t^3}{3!} \sin z - \tfrac{t^4}{4!} \sin x - \tfrac{t^4}{4!} \sin y, \quad (121) \\
u_2(x, y, z, t) &= L_t^{-1} \left(L_x u_1 + L_y u_1 + L_z u_1 \right), \\
&= \tfrac{t^4}{4!} \sin x + \tfrac{t^4}{4!} \sin y + \tfrac{t^5}{5!} \sin z + \tfrac{t^6}{6!} \sin x + \tfrac{t^6}{6!} \sin y.
\end{aligned}
$$

The series solution and the exact solution are given by

$$
u(x, y, z, t) = \sin x + \sin y + \sin z \left(t - \frac{t^3}{3!} + \frac{t^5}{5!} - \cdots \right), \quad (122)
$$

and

$$
u(x, y, z, t) = \sin x + \sin y + \sin z \sin t, \quad (123)
$$

respectively.

Exercises 6.2.2

In Exercises 1 – 8, use the decomposition method to solve the homogeneous initial-boundary value problems:

1. $u_{tt} = 3(u_{xx} + u_{yy} + u_{zz}), \ 0 < x, y, z < \pi, \ t > 0$
   $u(0, y, z, t) = u(\pi, y, z, t) = 0$
   $u(x, 0, z, t) = u(x, \pi, z, t) = 0$
   $u(x, y, 0, t) = u(x, y, \pi, t) = 0$
   $u(x, y, z, 0) = 0, u_t(x, y, z, 0) = 6 \sin(2x) \sin(2y) \sin(2z)$

2. $u_{tt} = \tfrac{1}{3}(u_{xx} + u_{yy} + u_{zz}), \ 0 < x, y, z < \pi, \ t > 0$
   $u(0, y, z, t) = u(\pi, y, z, t) = 1$
   $u(x, 0, z, t) = u(x, \pi, z, t) = 1$
   $u(x, y, 0, t) = u(x, y, \pi, t) = 1$
   $u(x, y, z, 0) = 1, u_t(x, y, z, 0) = \sin x \sin y \sin z$

3. $u_{tt} = 3(u_{xx} + u_{yy} + u_{zz}), \ 0 < x, y, z < \pi, \ t > 0$
   $u(0, y, z, t) = u(\pi, y, z, t) = 3$
   $u(x, 0, z, t) = u(x, \pi, z, t) = 3$
   $u(x, y, 0, t) = u(x, y, \pi, t) = 3$
   $u(x, y, z, 0) = 3 + \sin x \sin y \sin z, u_t(x, y, z, 0) = 0$

4. $u_{tt} = u_{xx} + u_{yy} + u_{zz} - 6u, \ 0 < x, y, z < \pi, \ t > 0$
   $u(0, y, z, t) = u(\pi, y, z, t) = 0$
   $u(x, 0, z, t) = u(x, \pi, z, t) = 0$

$$u(x, y, 0, t) = u(x, y, \pi, t) = 0$$
$$u(x, y, z, 0) = 0, u_t(x, y, z, 0) = 3 \sin x \sin y \sin z$$

5. $u_{tt} = \frac{1}{2}(u_{xx} + u_{yy} + u_{zz}) - u,\ 0 < x, y, z < \pi,\ t > 0$
    $u(0, y, z, t) = -u(\pi, y, z, t) = \sin(2y) \sin(z + 2t)$
    $u(x, 0, z, t) = u(x, \pi, z, t) = \sin x \sin(z + 2t)$
    $u(x, y, 0, t) = -u(x, y, \pi, t) = \sin(x + 2y) \sin(2t)$
    $u(x, y, z, 0) = \sin(x + 2y) \sin z, u_t(x, y, z, 0) = 2 \sin(x + 2y) \cos z$

6. $u_{tt} = u_{xx} + u_{yy} + u_{zz} - 2u,\ 0 < x, y, z < \pi,\ t > 0$
    $u(0, y, z, t) = u(\pi, y, z, t) = 0$
    $u(x, 0, z, t) = u(x, \pi, z, t) = 0$
    $u(x, y, 0, t) = u(x, y, \pi, t) = 0$
    $u(x, y, z, 0) = \sin x \sin(2y) \sin(3z), u_t(x, y, z, 0) = 0$

7. $u_{tt} = \frac{1}{3}(u_{xx} + u_{yy} + u_{zz}),\ 0 < x, y, z < \pi,\ t > 0$
    $u(0, y, z, t) = -u(\pi, y, z, t) = \cos y \sin(z + t)$
    $u(x, 0, z, t) = -u(x, \pi, z, t) = \cos x \sin(z + t)$
    $u(x, y, 0, t) = -u(x, y, \pi, t) = \cos(x + y) \sin t$
    $u(x, y, z, 0) = \cos(x + y) \sin z, u_t(x, y, z, 0) = \cos(x + y) \cos z$

8. $u_{tt} = \frac{1}{3}(u_{xx} + u_{yy} + u_{zz}),\ 0 < x, y, z < \pi,\ t > 0$
    $u(0, y, z, t) = u(\pi, y, z, t) = 1 + z$
    $u(x, 0, z, t) = u(x, \pi, z, t) = 1 + z$
    $u(x, y, 0, t) = 1, u(x, y, \pi, t) = 1 + \pi$
    $u(x, y, z, 0) = 1 + z, u_t(x, y, z, 0) = \sin x \sin y \sin z$

In Exercises 9 – 14, solve the inhomogeneous initial-boundary value problems:

9. $u_{tt} = u_{xx} + u_{yy} + u_{zz} + 2 \sin x \sin y,\ 0 < x, y, z < \pi,\ t > 0$
    $u(0, y, z, t) = u(\pi, y, z, t) = \sin z \sin t$
    $u(x, 0, z, t) = u(x, \pi, z, t) = \sin z \sin t$
    $u(x, y, 0, t) = u(x, y, \pi, t) = \sin x \sin y$
    $u(x, y, z, 0) = \sin x \sin y, u_t(x, y, z, 0) = \sin z$

10. $u_{tt} = u_{xx} + u_{yy} + u_{zz} + \cos x + \cos y,\ 0 < x, y, z < \pi,\ t > 0$
    $u_x(0, y, z, t) = u_x(\pi, y, z, t) = 0$
    $u_y(x, 0, z, t) = u_y(x, \pi, z, t) = 0$
    $u_z(x, y, 0, t) = -u_z(x, y, \pi, t) = \sin t$
    $u(x, y, z, 0) = \cos x + \cos y, u_t(x, y, z, 0) = \sin z$

11. $u_{tt} = u_{xx} + u_{yy} + u_{zz} - 6 - \sin t,\ 0 < x, y, z < \pi,\ t > 0$
    $u_x(0, y, z, t) = 0, u_x(\pi, y, z, t) = 2\pi$
    $u_y(x, 0, z, t) = 0, u_y(x, \pi, z, t) = 2\pi$
    $u_z(x, y, 0, t) = 0, u_z(x, y, \pi, t) = 2\pi$
    $u(x, y, z, 0) = x^2 + y^2 + z^2, u_t(x, y, z, 0) = 1$

12. $u_{tt} = u_{xx} + u_{yy} + u_{zz} + 2 - \sin t,\ 0 < x, y, z < \pi,\ t > 0$
$u_x(0, y, z, t) = u_x(\pi, y, z, t) = t$
$u_y(x, 0, z, t) = u_y(x, \pi, z, t) = t$
$u_z(x, y, 0, t) = u_z(x, y, \pi, t) = t$
$u(x, y, z, 0) = 0, u_t(x, y, z, 0) = 1 + x + y + z$

13. $u_{tt} = u_{xx} + u_{yy} + u_{zz} + 2(x + y + z),\ 0 < x, y, z < \pi,\ t > 0$
$u_x(0, y, z, t) = u_x(\pi, y, z, t) = t^2$
$u_y(x, 0, z, t) = -u_y(x, \pi, z, t) = t^2 - \sin t$
$u_z(x, y, 0, t) = u_z(x, y, \pi, t) = t^2$
$u(x, y, z, 0) = 0, u_t(x, y, z, 0) = \sin y$

14. $u_{tt} = u_{xx} + u_{yy} + u_{zz} - 6,\ 0 < x, y, z < \pi,\ t > 0$
$u_x(0, y, z, t) = 0, u_x(\pi, y, z, t) = 2\pi$
$u_y(x, 0, z, t) = 0, u_y(x, \pi, z, t) = 2\pi$
$u_z(x, y, 0, t) = 0, u_z(x, y, \pi, t) = 2\pi$
$u(x, y, z, 0) = x^2 + y^2 + z^2 + \cos y, u_t(x, y, z, 0) = 0$

15. $u_{tt} = \frac{1}{2}(u_{xx} + u_{yy} + u_{zz}) - 1,\ 0 < x, y, z < \pi,\ t > 0$
$u_x(0, y, z, t) = 0, u_x(\pi, y, z, t) = 2\pi$
$u_y(x, 0, z, t) = -u_y(x, \pi, z, t) = \sin z \cos t$
$u_z(x, y, 0, t) = -u_z(x, y, \pi, t) = \sin y \cos t$
$u(x, y, z, 0) = x^2 + \sin x \sin z, u_t(x, y, z, 0) = 0$

In Exercises 15 – 20, use the decomposition method to solve the initial-boundary value problems:

16. $u_{tt} = u_{xx} + u_{yy} + u_{zz} - u,\ 0 < x, y, z < \pi,\ t > 0$
$u_x(0, y, z, t) = u_x(\pi, y, z, t) = 0$
$u_y(x, 0, z, t) = u_y(x, \pi, z, t) = 0$
$u_z(x, y, 0, t) = u_z(x, y, \pi, t) = 0$
$u(x, y, z, 0) = 1, u_t(x, y, z, 0) = 2 \cos x \cos y \cos z$

17. $u_{tt} = u_{xx} + u_{yy} + u_{zz} - u,\ 0 < x, y, z < \pi,\ t > 0$
$u(0, y, z, t) = u(\pi, y, z, t) = 1$
$u(x, 0, z, t) = u_y(x, \pi, z, t) = 1$
$u(x, y, 0, t) = u(x, y, \pi, t) = 1$
$u(x, y, z, 0) = 1 + \sin x \sin y \sin z, u_t(x, y, z, 0) = 0$

18. $u_{tt} = u_{xx} + u_{yy} + u_{zz} + 2 \sin x \sin y,\ 0 < x, y, z < \pi,\ t > 0$
$u(0, y, z, t) = u(\pi, y, z, t) = 1 + \sin z \sin t$
$u(x, 0, z, t) = u(x, \pi, z, t) = 1 + \sin z \sin t$
$u(x, y, 0, t) = u(x, y, \pi, t) = 1 + \sin x \sin y$
$u(x, y, z, 0) = 1 + \sin x \sin y, u_t(x, y, z, 0) = \sin z$

19. $u_{tt} = \frac{1}{3}(u_{xx} + u_{yy} + u_{zz}),\ 0 < x, y, z < \pi,\ t > 0$
$u_x(0, y, z, t) = 0, u_x(\pi, y, z, t) = 2\pi$

$$u_y(x, 0, z, t) = 0, u_y(x, \pi, z, t) = 2\pi$$
$$u_z(x, y, 0, t) = 0, u_z(x, y, \pi, t) = 2\pi$$
$$u(x, y, z, 0) = x^2 + y^2 + z^2 + \cos x \cos y \cos z, u_t(x, y, z, 0) = 0$$

20. $u_{tt} = u_{xx} + u_{yy} + u_{zz} - 12x^2 - 12y^2, \ 0 < x, y, z < \pi, \ t > 0$
$$u_x(0, y, z, t) = 0, u_x(\pi, y, z, t) = 4\pi^3$$
$$u_y(x, 0, z, t) = 0, u_y(x, \pi, z, t) = 4\pi^3$$
$$u_z(x, y, 0, t) = u_z(x, y, \pi, t) = 0$$
$$u(x, y, z, 0) = x^4 + y^4 + \cos z, u_t(x, y, z, 0) = 0$$

In Exercises 21 – 24, solve the inhomogeneous initial-boundary value problems where the coefficients of the derivatives are functions and not constants:

21. $u_{tt} = \frac{x^2}{6} u_{xx} + \frac{y^2}{6} u_{yy} + \frac{z^2}{6} u_{zz}, \ 0 < x, y, z < 1, \ t > 0$
$$u(0, y, z, t) = 0, u(1, y, z, t) = y^2 z^2 \cosh t$$
$$u(x, 0, z, t) = 0, u(x, 1, z, t) = x^2 z^2 \cosh t$$
$$u(x, y, 0, t) = 0, u(x, y, 1, t) = x^2 y^2 \cosh t$$
$$u(x, y, z, 0) = x^2 y^2 z^2, u_t(x, y, z, 0) = 0$$

22. $u_{tt} = \frac{x^2}{6} u_{xx} + \frac{y^2}{6} u_{yy} + \frac{z^2}{6} u_{zz}, \ 0 < x, y, z < 1, \ t > 0$
$$u_x(0, y, z, t) = 0, u_x(1, y, z, t) = 3 \sinh t$$
$$u_y(x, 0, z, t) = 0, u_y(x, 1, z, t) = 3 \cosh t$$
$$u_z(x, y, 0, t) = 0, u_z(x, y, 1, t) = 3 \cosh t$$
$$u(x, y, z, 0) = \cosh t (y^3 + z^3), u_t(x, y, z, 0) = x^3$$

23. $u_{tt} = \frac{x^2}{2} u_{xx} + \frac{y^2}{2} u_{yy} + \frac{z^2}{2} u_{zz}, \ 0 < x, y, z < 1, \ t > 0$
$$u_x(0, y, z, t) = 0, u_x(1, y, z, t) = 2e^t$$
$$u_y(x, 0, z, t) = 0, u_y(x, 1, z, t) = 2e^{-t}$$
$$u_z(x, y, 0, t) = 0, u_z(x, y, 1, t) = 2e^t$$
$$u(x, y, z, 0) = x^2 + y^2 + z^2, u_t(x, y, z, 0) = x^2 - y^2 + z^2$$

24. $u_{tt} = \frac{x^2}{6} u_{xx} + \frac{y^2}{6} u_{yy} + \frac{z^2}{6} u_{zz}, \ 0 < x, y, z < 1, \ t > 0$
$$u_x(0, y, z, t) = 0, u_x(1, y, z, t) = 3 \sinh t$$
$$u_y(x, 0, z, t) = 0, u_y(x, 1, z, t) = 3 \sinh t$$
$$u_z(x, y, 0, t) = 0, u_z(x, y, 1, t) = 3 \sinh t$$
$$u(x, y, z, 0) = 0, u_t(x, y, z, 0) = x^3 + y^3 + z^3$$

6.3 Method of Separation of Variables

In this section, the homogeneous partial differential equations that describe the wave propagation in a two dimensional space and in a three dimensional space will be discussed by using the classical method of *separation of variables*. The most significant feature of this method is that it reduces the partial differential equation into a system of ordinary differential equations, where each ODE depends on one variable only, and can be solved independently. The boundary conditions and the initial conditions are then used to determine the constants of integration.

The complete details of the method can be found in the preceding chapters, hence emphasis will be focused on applying the method.

6.3.1   Two Dimensional Wave Equation

The propagation of waves in a two dimensional vibrating membrane of length a and width b is governed by the following initial boundary value problem

$$
\begin{aligned}
\text{PDE} \quad u_{tt} &= c^2(u_{xx} + u_{yy}),\ 0 < x < a,\ 0 < y < b, t > 0 \\[2mm]
\text{B.C}\ u(0, y, t) &= u(a, y, t) = 0 \\[2mm]
u(x, 0, t) &= u(x, b, t) = 0 \\[2mm]
\text{I.C}\ u(x, y, 0) &= f(x, y),\ u_t(x, 0, t) = g(x, y)
\end{aligned}
\tag{124}
$$

where $u = u(x, y, t)$ defines the displacement function of any point at the position (x, y) of a vibrating membrane at any time t, and c is related to the elasticity of the material of the membrane.

The method of separation of variables is based on an assumption that the solution $u(x, y, t)$ can be expressed as the product of distinct functions $F(x), G(y)$, and $T(t)$, such that each function depends on one variable only. Based on this assumption, we first set

$$
u(x, y, t) = F(x)G(y)T(t).
\tag{125}
$$

Differentiating both sides of (125) twice with respect to t, x and y respectively, we obtain

$$
\begin{aligned}
u_{tt} &= F(x)G(y)T^{''}(t), \\[2mm]
u_{xx} &= F^{''}(x)G(y)T(t), \\[2mm]
u_{yy} &= F(x)G^{''}(y)T(t).
\end{aligned}
\tag{126}
$$

Substituting (126) into the PDE of (124) gives

$$
F(x)G(y)T^{''}(t) = c^2\left(F^{''}(x)G(y)T(t) + F(x)G^{''}(y)T(t)\right).
\tag{127}
$$

Dividing both sides of (127) by $c^2 F(x)G(y)T(t)$ yields

$$
\frac{T^{''}(t)}{c^2 T(t)} = \frac{F^{''}(x)}{F(x)} + \frac{G^{''}(y)}{G(y)}.
\tag{128}
$$

It is easily observed from (128) that the left hand side depends only on t and the right hand side depends only on x and y. This means that the equality holds only if both sides are equal to the same constant. Assuming that the right hand side is a constant, it is valid to assume that it is the sum of two constants. This admits the use of

$$\frac{F''(x)}{F(x)} = -\nu^2, \tag{129}$$

and

$$\frac{G''(y)}{G(y)} = -\mu^2. \tag{130}$$

Consequently, we find

$$F''(x) + \nu^2 F(x) = 0, \tag{131}$$

and

$$G''(y) + \mu^2 G(y) = 0. \tag{132}$$

The left hand side of (128) is thus equal to the constant $-(\nu^2 + \mu^2)$, hence we set

$$\frac{T''(t)}{c^2 T(t)} = -(\nu^2 + \mu^2), \tag{133}$$

or equivalently

$$T''(t) + c^2(\nu^2 + \mu^2)T(t) = 0. \tag{134}$$

The selection of $-(\nu^2 + \mu^2)$ is the only selection that will provide nontrivial solutions.

It is interesting to note that the partial differential equation of (124) has been transformed to three second order ordinary differential equations given by (131), (132), and (134).

The second order differential equations (131) and (132) give the solutions

$$F(x) = A\cos(\nu x) + B\sin(\nu x), \tag{135}$$

and

$$G(y) = \alpha\cos(\mu y) + \beta\sin(\mu y), \tag{136}$$

where A, B, α, and β are constants.

To determine the constants A and B, we use the boundary conditions at $x = 0$ and at $x = a$ to find that

$$F(0) \;=\; 0, \; F(a) \;=\; 0. \tag{137}$$

Substituting (137) into (135) gives

$$A = 0, \tag{138}$$

and

$$\nu_n = \frac{n\pi}{a}, \; n = 1, 2, 3, \cdots. \tag{139}$$

It is important to note here that we exclude $n = 0$ and $B = 0$ because each will lead to the trivial solution $u(x, y, t) = 0$. Using the results obtained for the constants A and ν_n, we therefore find

$$F_n(x) = B_n \sin(\frac{n\pi}{a}x), \ n = 1, 2, 3, \cdots. \tag{140}$$

In a parallel manner, we use the second boundary condition at $y = 0$ and at $y = b$ into (136) to find that

$$\alpha = 0, \tag{141}$$

and

$$\mu_m = \frac{m\pi}{b}, m = 1, 2, 3, \cdots. \tag{142}$$

We exclude $m = 0$ and $\beta = 0$, because each will lead to the trivial solution as indicated before. Consequently, we obtain

$$G_m(y) = \beta_m \sin\frac{m\pi}{b}, \ m = 1, 2, 3, \cdots. \tag{143}$$

The solution of (134) is therefore given by

$$T_{nm}(t) = C_{nm} \cos(c\lambda_{nm}t) + D_{nm} \sin(c\lambda_{nm}t), \tag{144}$$

where

$$\lambda_{mn}^2 = (\frac{n\pi}{a})^2 + (\frac{m\pi}{b})^2, \tag{145}$$

and C_{nm} and D_{nm} are constants.

Combining the results obtained for $F_n(x)$, $G_m(y)$, and $T_{nm}(t)$ we obtain the infinite sequence of product functions

$$
\begin{aligned}
u_{nm} &= F_n(x)G_m(y)T_{nm}(t) \\
&= \sin(\tfrac{n\pi}{a}x) \sin(\tfrac{m\pi}{b}y) \left(C_{nm} \cos(c\lambda_{nm}t) + D_{nm} \sin(c\lambda_{nm}t) \right),
\end{aligned}
\tag{146}
$$

that satisfies the PDE of (124) and the boundary conditions. Using the superposition principle gives the general solution
$u(x, y, t) =$

$$\sum_{n=1}^{\infty} \sum_{m=1}^{\infty} \sin(\frac{n\pi}{a}x) \sin(\frac{m\pi}{b}y)(C_{nm} \cos(c\lambda_{nm}t) + D_{nm} \sin(c\lambda_{nm}t)), \tag{147}$$

where the arbitrary constants C_{nm} and D_{nm} are as yet undetermined.

To determine the constants C_{nm} and D_{nm}, we use the given initial conditions to find

$$\sum_{n=1}^{\infty} \sum_{m=1}^{\infty} C_{nm} \sin(\frac{n\pi}{a}x) \sin(\frac{m\pi}{b}y) = f(x, y), \tag{148}$$

and

$$\sum_{n=1}^{\infty}\sum_{m=1}^{\infty} c\lambda_{nm} D_{nm} \sin(\frac{n\pi}{a}x) \sin(\frac{m\pi}{b}y) = g(x,y). \tag{149}$$

Consequently, the arbitrary constants C_{nm} and D_{nm} are completely determined by using double Fourier coefficients where we find

$$C_{nm} = \frac{4}{ab} \int_0^b \int_0^a f(x,y) \sin(\frac{n\pi}{a}x) \sin(\frac{m\pi}{b}y) dx\, dy, \tag{150}$$

and

$$D_{nm} = \frac{4}{\lambda_{nm}abc} \int_0^b \int_0^a g(x,y) \sin(\frac{n\pi}{a}x) \sin(\frac{m\pi}{b}y) dx\, dy. \tag{151}$$

Having determined the constants C_{nm} and D_{nm}, the particular solution $u(x,y,t)$ that satisfies the initial boundary value problem (124) is readily obtained.

It is interesting to point out that the constants C_{nm} and D_{nm} can also be determined by expanding the double Fourier series of (147), applying the initial conditions, and then by equating the coefficients of like terms on both sides. Clearly this works if the initial conditions are defined explicitly in terms of trigonometric functions of sines and cosines.

For illustration, several examples will be discussed to emphasize the use of the method.

Example 1. Use the method of separation of variables to solve the initial-boundary value problem:

$$\begin{aligned}
\text{PDE} \quad u_{tt} &= 2(u_{xx} + u_{yy}), \ 0 < x, y < \pi, \ t > 0 \\
\text{B.C} \quad u(0,y,t) &= u(\pi,y,t) = 0 \\
u(x,0,t) &= u(x,\pi,t) = 0 \\
\text{I.C} \quad u(x,y,0) &= \sin x \sin y, \ u_t(x,y,0) = 0
\end{aligned} \tag{152}$$

Solution.

The method of separation of variables assumes that

$$u(x,y,t) = F(x)G(y)T(t). \tag{153}$$

Proceeding as before, we obtain

$$F''(x) + \nu^2 F(x) = 0, \tag{154}$$

$$G''(y) + \mu^2 G(y) = 0, \tag{155}$$

and
$$T''(t) + 2\lambda^2 T(t) = 0, \tag{156}$$

where
$$\lambda^2 = \nu^2 + \mu^2, \tag{157}$$

and ν, μ and λ are constants.

From (154) and (155), we find
$$F(x) = A\cos(\nu x) + B\sin(\nu x), \tag{158}$$

and
$$G(y) = \alpha\cos(\mu y) + \beta\sin(\mu y), \tag{159}$$

respectively, where A, B, α and β are constants. Inserting the boundary conditions at $x = 0$ and at $x = \pi$ into (158) gives
$$A = 0, \tag{160}$$

and
$$\nu_n = n, \ n = 1, 2, 3, \cdots, \tag{161}$$

so that
$$F_n(x) = B_n\sin(nx), \ n = 1, 2, 3, \cdots. \tag{162}$$

Likewise, using the boundary conditions at $y = 0$ and at $y = \pi$ into (159) gives
$$\alpha = 0, \tag{163}$$

and
$$\mu_m = m, \ m = 1, 2, 3, \cdots, \tag{164}$$

so that
$$G_m(y) = \beta_m\sin(my), \ m = 1, 2, 3, \cdots. \tag{165}$$

The solution of (156) is therefore given by
$$T_{nm}(t) = C_{nm}\cos(\sqrt{2}\lambda_{nm}t) + D_{nm}\sin(\sqrt{2}\lambda_{nm}t), \tag{166}$$

where
$$\lambda_{nm} = \sqrt{n^2 + m^2}. \tag{167}$$

Combining the results obtained above, we obtain
$$u_{nm}(x, y, t) = \sin(nx)\sin(my)(C_{nm}\cos(\sqrt{2}\lambda_{nm}t) + D_{nm}\sin(\sqrt{2}\lambda_{nm}t)). \tag{168}$$

Using the superposition principle, the general solution of the problem is given by the double series
$$u = \sum_{n=1}^{\infty}\sum_{m=1}^{\infty}\sin(nx)\sin(my)(C_{nm}\cos(\sqrt{2}\lambda_{nm}t) + D_{nm}\sin(\sqrt{2}\lambda_{nm}t)). \tag{169}$$

To determine the constants C_{nm}, we use $u(x, y, 0) = \sin x \sin y$ and expand the double series (169) to find

$$C_{11} \sin(x) \sin(y) + C_{12} \sin(x) \sin(2y) + \cdots = \sin(x) \sin(y). \tag{170}$$

Equating the coefficients of like terms on both sides yields

$$C_{11} = 1, \; C_{ij} = 0, i \neq 1, j \neq 1, n = m = 1. \tag{171}$$

Likewise, inserting the second initial condition $u_t(x, y, 0) = 0$ into the derivative of (169) gives

$$D_{nm} = 0, \; n \geq 1, \; m \geq 1. \tag{172}$$

The particular solution is therefore given by

$$u(x, y, t) = \sin x \sin y \cos(2t). \tag{173}$$

Example 2. Use the method of separation of variables to solve the initial-boundary value problem:

$$
\begin{aligned}
\text{PDE} \quad u_{tt} &= 2(u_{xx} + u_{yy}), \; 0 < x, y < \pi, \; t > 0 \\
\text{B.C} \quad u_x(0, y, t) &= u_x(\pi, y, t) = 0 \\
u_y(x, 0, t) &= u_y(x, \pi, t) = 0 \\
\text{I.C} \quad u(x, y, 0) &= 1 + \cos x \cos y, \; u_t(x, y, 0) = 0
\end{aligned}
\tag{174}
$$

Solution.

Note first that the boundary conditions are of the second kind where the derivatives of $u(x, y, t)$ are given instead of the initial displacements. As discussed before, we set

$$u(x, y, t) = F(x)G(y)T(t). \tag{175}$$

Substituting (175) into (174) and proceeding as before we obtain

$$F''(x) + \nu^2 F(x) = 0, \tag{176}$$

$$G''(y) + \mu^2 G(y) = 0, \tag{177}$$

and

$$T''(t) + 2(\nu^2 + \mu^2)T(t) = 0. \tag{178}$$

Solving (176) we find

$$F(x) = A \cos(\nu x) + B \sin(\nu x). \tag{179}$$

It is important to note that the boundary conditions

$$u_x(0, y, t) = u_x(\pi, y, t) = 0, \tag{180}$$

imply that

$$F'(0) = 0, F'(\pi) = 0. \tag{181}$$

Substituting (181) into (179) gives

$$B = 0, \nu = n, \, n = 0, 1, 2, \cdots, \tag{182}$$

so that $\nu = 0$ is included because it does not provide the trivial solution. Consequently, we find

$$F_n(x) = A_n \cos(nx), \, n = 0, 1, 2, \cdots. \tag{183}$$

Solving (177) and using the proper boundary conditions we obtain

$$G_m(y) = \alpha_m \cos(my), \, m = 0, 1, 2, 3, \cdots. \tag{184}$$

The solution of the second order differential equation (178) is therefore given by

$$T_{nm}(t) = C_{nm} \cos(\sqrt{2}\lambda_{nm}t) + D_{nm} \sin(\sqrt{2}\lambda_{nm}t), n \geq 0, m \geq 0, \tag{185}$$

where

$$\lambda_{nm} = \sqrt{n^2 + m^2}. \tag{186}$$

The general solution is given by

$$u = \sum_{n=0}^{\infty} \sum_{m=0}^{\infty} \cos(nx) \cos(my) \left(C_{nm} \cos(\sqrt{2}\lambda_{nm}t) + D_{nm} \sin(\sqrt{2}\lambda_{nm}t) \right). \tag{187}$$

To determine the constants C_{nm}, we first expand the series (187) and use the first initial condition to find

$$C_{00} + C_{11} \cos x \cos y + \cdots = 1 + \cos x \cos y. \tag{188}$$

Equating the coefficients of like terms on both sides gives

$$\begin{aligned} C_{00} &= 1, n = 0, m = 0, \\ C_{11} &= 1, n = 1, m = 1, \end{aligned} \tag{189}$$

where other coefficients are zeros.

To determine the constants D_{nm}, we use the second boundary condition in the derivative of (187) to find that

$$D_{nm} = 0, \, n \geq 0, m \geq 0. \tag{190}$$

Accordingly, the particular solution is given by

$$u(x, y, t) = 1 + \cos x \cos y \cos(2t). \tag{191}$$

Example 3. Use the method of separation of variables to solve the initial-boundary value problem:

$$
\begin{array}{rlrl}
\text{PDE} & u_{tt} & = & u_{xx} + u_{yy}, \ 0 < x, y < \pi, \ t > 0 \\[4pt]
\text{B.C} & u(0, y, t) & = & u(\pi, y, t) = 0 \\[4pt]
& u(x, 0, t) & = & u(x, \pi, t) = 0 \\[4pt]
\text{I.C} & u(x, y, 0) & = & 1, \ u_t(x, y, 0) = 0
\end{array}
\tag{192}
$$

Solution.

We first set

$$
u(x, y, t) = F(x)G(y)T(t),
\tag{193}
$$

Proceeding as before we obtain

$$
\begin{array}{rll}
F_n(x) & = & B_n \sin(nx), \ \lambda_n = n, \ n = 1, 2, \cdots, \tag{194} \\[4pt]
G_m(y) & = & \beta_m \sin(my), \ \mu_m = m, \ m = 1, 2, \cdots, \tag{195} \\[4pt]
T_{nm}(t) & = & C_{nm} \cos(\lambda_{nm}t) + D_{nm} \sin(\lambda_{nm}t), \tag{196}
\end{array}
$$

where

$$
\lambda_{nm} = \sqrt{n^2 + m^2}.
\tag{197}
$$

Combining (194), (195), and (196) and using the superposition principle, the general solution of the problem is given by

$$
u = \sum_{n=1}^{\infty} \sum_{m=1}^{\infty} \sin(nx) \sin(my)(C_{nm} \cos(\lambda_{nm}t) + D_{nm} \sin(\lambda_{nm}t)).
\tag{198}
$$

To determine the constants C_{nm}, we use the double series coefficients method to obtain

$$
C_{nm} = \frac{4}{\pi^2} \int_0^{\pi} \int_0^{\pi} \sin(nx) \sin(my) \, dx \, dy,
\tag{199}
$$

which gives

$$
C_{nm} = \begin{cases} 0 & \text{for } n \text{ is even, } m \text{ is even,} \\[6pt] \frac{16}{\pi^2 nm} & \text{for } n \text{ is odd, } m \text{ is odd.} \end{cases}
\tag{200}
$$

To determine D_{nm}, we use the second boundary condition to find

$$
D_{nm} = 0, \ n \geq 1, \ m \geq 1.
\tag{201}
$$

Accordingly, the particular solution in a series form is given by

$$
u = \frac{16}{\pi^2} \sum_{n=0}^{\infty} \sum_{m=0}^{\infty} \frac{1}{(2n+1)(2m+1)} \sin(2n+1)x \, \sin(2m+1)y \cos(\lambda_{nm}t),
\tag{202}
$$

where
$$\lambda_{nm} = \sqrt{(2n+1)^2 + (2m+1)^2}, \; n \geq 0, \; m \geq 0. \tag{203}$$

Example 4. Use the method of separation of variables to solve the initial-boundary value problem:

$$
\begin{array}{rrl}
\text{PDE} & u_{tt} = & u_{xx} + u_{yy}, \; 0 < x, y < \pi, \; t > 0 \\[6pt]
\text{B.C} \quad u(0,y,t) = & u(\pi,y,t) = 0 \\[6pt]
u(x,0,t) = & u(x,\pi,t) = 0 \\[6pt]
\text{I.C} \quad u(x,y,0) = & 0, \; u_t(x,y,0) = 1
\end{array}
\tag{204}
$$

Solution.

We first set
$$u(x,y,t) = F(x)G(y)T(t), \tag{205}$$

Following Example 3, we obtain

$$F_n(x) = B_n \sin(nx), \; \nu_n = n, \; n = 1, 2, \cdots, \tag{206}$$

$$G_m(y) = B_m \sin(my), \; \mu_m = m, \; m = 1, 2, \cdots, \tag{207}$$

and

$$T_{nm}(t) = C_{nm} \cos(\lambda_{nm}t) + D_{nm} \sin(\lambda_{nm}t), \tag{208}$$

where

$$\lambda_{nm} = \sqrt{n^2 + m^2}. \tag{209}$$

Combining (206), (207), and (208) and using the superposition principle, the general solution of the problem is given by

$$u = \sum_{n=1}^{\infty} \sum_{m=1}^{\infty} \sin(nx) \sin(my) \left(C_{nm} \cos(\lambda_{nm}t) + D_{nm} \sin(\lambda_{nm}t) \right). \tag{210}$$

To determine the constants C_{nm}, we use the double Fourier coefficients method to obtain

$$C_{nm} = 0, \; n \geq 1, \; m \geq 1. \tag{211}$$

Using the second boundary condition we find

$$\lambda_{nm} D_{nm} = \frac{4}{\pi^2} \int_0^\pi \int_0^\pi \sin(nx) \sin(my) \, dx \, dy, \tag{212}$$

which gives

$$
D_{nm} =
\begin{cases}
0 & \text{for } n \text{ is even, } m \text{ is even,} \\[10pt]
\frac{16}{\lambda_{nm} \pi^2 nm} & \text{for } n \text{ is odd, } m \text{ is odd.}
\end{cases}
\tag{213}
$$

Accordingly, the particular solution of the initial-boundary value problem (204) is given by the double series form

$$u(x, y, t) = \frac{16}{\pi^2} \sum_{n=0}^{\infty} \sum_{m=0}^{\infty} \frac{1}{\lambda_{nm}(2n+1)(2m+1)}$$

$$\times \sin(2n+1)x \, \sin(2m+1)y \, \sin(\lambda_{nm}t), \tag{214}$$

where

$$\lambda_{nm} = \sqrt{(2n+1)^2 + (2m+1)^2}. \tag{215}$$

Exercises 6.3.1

Use the method of separation of variables in the following initial-boundary value problems:

1. $u_{tt} = 2(u_{xx} + u_{yy}), \, 0 < x, y < \pi$
   $u(0, y, t) = u(\pi, y, t) = 0$
   $u(x, 0, t) = u(x, \pi, t) = 0$
   $u(x, y, 0) = \sin(2x)\sin(2y), \, u_t(x, y, 0) = 0$

2. $u_{tt} = 5(u_{xx} + u_{yy}), \, 0 < x, y < \pi$
   $u(0, y, t) = u(\pi, y, t) = 0$
   $u(x, 0, t) = u(x, \pi, t) = 0$
   $u(x, y, 0) = \sin x \sin(2y), \, u_t(x, y, 0) = 0$

3. $u_{tt} = 2(u_{xx} + u_{yy}), \, 0 < x, y < \pi$
   $u(0, y, t) = u(\pi, y, t) = 0$
   $u(x, 0, t) = u(x, \pi, t) = 0$
   $u(x, y, 0) = 0, \, u_t(x, y, 0) = 2\sin x \sin y$

4. $u_{tt} = 5(u_{xx} + u_{yy}), \, 0 < x, y < \pi$
   $u(0, y, t) = u(\pi, y, t) = 0$
   $u(x, 0, t) = u(x, \pi, t) = 0$
   $u(x, y, 0) = 0, \, u_t(x, y, 0) = 5\sin x \sin(2y)$

5. $u_{tt} = 2(u_{xx} + u_{yy}), \, 0 < x, y < \pi$
   $u_x(0, y, t) = u_x(\pi, y, t) = 0$
   $u_y(x, 0, t) = u_y(x, \pi, t) = 0$
   $u(x, y, 0) = 2, \, u_t(x, y, 0) = 2\cos x \cos y$

6. $u_{tt} = 8(u_{xx} + u_{yy}), \, 0 < x, y < \pi$
   $u_x(0, y, t) = u_x(\pi, y, t) = 0$
   $u_y(x, 0, t) = u_y(x, \pi, t) = 0$
   $u(x, y, 0) = 1 + \cos x \cos y, \, u_t(x, y, 0) = 0$

7. $u_{tt} = 2(u_{xx} + u_{yy}), 0 < x, y < \pi$
   $u(0, y, t) = u(\pi, y, t) = 0$
   $u_y(x, 0, t) = u_y(x, \pi, t) = 0$
   $u(x, y, 0) = 0, u_t(x, y, 0) = 2 \sin x \cos y$

8. $u_{tt} = 2(u_{xx} + u_{yy}), 0 < x, y < \pi$
   $u_x(0, y, t) = u_x(\pi, y, t) = 0$
   $u(x, 0, t) = u(x, \pi, t) = 0$
   $u(x, y, 0) = \cos x \sin y, u_t(x, y, 0) = 0$

9. $u_{tt} = 2(u_{xx} + u_{yy}), 0 < x, y < \pi$
   $u_x(0, y, t) = u_x(\pi, y, t) = 0$
   $u(x, 0, t) = u(x, \pi, t) = 0$
   $u(x, y, 0) = 0, u_t(x, y, 0) = 2 \cos x \sin y$

10. $u_{tt} = 5(u_{xx} + u_{yy}), 0 < x, y < \pi$
    $u_x(0, y, t) = u_x(\pi, y, t) = 0$
    $u_y(x, 0, t) = u_y(x, \pi, t) = 0$
    $u(x, y, 0) = 3, u_t(x, y, 0) = 5 \cos x \cos(2y)$

11. $u_{tt} = 2(u_{xx} + u_{yy}), 0 < x, y < \pi$
    $u(0, y, t) = u(\pi, y, t) = 0$
    $u(x, 0, t) = u(x, \pi, t) = 0$
    $u(x, y, 0) = 2, u_t(x, y, 0) = 0$

12. $u_{tt} = 2(u_{xx} + u_{yy}), 0 < x, y < \pi$
    $u(0, y, t) = u(\pi, y, t) = 0$
    $u(x, 0, t) = u(x, \pi, t) = 0$
    $u(x, y, 0) = 0, u_t(x, y, 0) = 3$

6.3.2   Three Dimensional Wave Equation

The propagation of waves in a three dimensional space of length a, width b and of height d is governed by the initial boundary value problem

$$
\begin{aligned}
\text{PDE} \qquad u_{tt} &= c^2(u_{xx} + u_{yy} + u_{zz}), t > 0 \\
\text{B.C} \qquad u(0, y, z, t) &= u(a, y, z, t) = 0 \\
u(x, 0, z, t) &= u(x, b, z, t) = 0 \qquad (216) \\
u(x, y, 0, t) &= u(x, y, d, t) = 0 \\
\text{I.C} \qquad u(x, y, z, 0) &= f(x, y, z), u_t(x, y, z, 0) = g(x, y, z)
\end{aligned}
$$

where $0 < x < a$, $0 < y < b$, $0 < z < d$ and $u = u(x, y, z, t)$ defines the displacement of any point at the position (x, y, z) of a rectangular volume at any time t, c is the velocity of a propagation wave.

The method of separation of variables assumes that $u(x, y, z, t)$ consists of the product of four distinct functions each depends on one variable only. This means that we can set

$$u(x, y, z, t) = F(x)G(y)H(z)T(t). \tag{217}$$

Substituting (217) into (216), and dividing both sides of the resulting equation by $c^2 F(x)G(y)H(z)T(t)$ we obtain

$$\frac{T''(t)}{c^2 T(t)} = \frac{F''(x)}{F(x)} + \frac{G''(y)}{G(y)} + \frac{H''(z)}{H(z)}. \tag{218}$$

The equality in (218) holds only if both sides are equal to the same constant. This allows us to set

$$F''(x) + \nu^2 F(x) = 0, \tag{219}$$

$$G''(y) + \mu^2 G(y) = 0, \tag{220}$$

$$H''(z) + \eta^2 H(z) = 0, \tag{221}$$

$$T''(t) + c^2 \lambda^2 T(t) = 0, \tag{222}$$

where ν, μ, η, and λ are constants, and

$$\lambda^2 = (\nu^2 + \mu^2 + \eta^2). \tag{223}$$

Solving the second order normal forms (219)   (221), we obtain the following solutions

$$F(x) = A\cos(\nu x) + B\sin(\nu x), \tag{224}$$

$$G(y) = \alpha\cos(\mu y) + \beta\sin(\mu y), \tag{225}$$

$$H(z) = \gamma\cos(\eta z) + \delta\sin(\eta z), \tag{226}$$

respectively, where $A, B, \alpha, \beta, \gamma$, and δ are constants. Using the proper boundary conditions into (224) – (226) as applied before we find

$$A = 0, \quad \nu_n = \frac{n\pi}{a}, \; n = 1, 2, 3, \cdots, \tag{227}$$

$$\alpha = 0 \quad \mu_m = \frac{m\pi}{b}, \; m = 1, 2, 3, \cdots, \tag{228}$$

$$\gamma = 0, \quad \eta_r = \frac{r\pi}{d}, \; r = 1, 2, 3, \cdots, \tag{229}$$

so that

$$F_n(x) = B_n \sin(\frac{n\pi}{a} x), \; n = 1, 2, 3, \cdots, \tag{230}$$

$$G_m(y) = \beta_m \sin(\frac{m\pi}{b} y), \; m = 1, 2, 3, \cdots, \tag{231}$$

$$H_r(z) = \delta_r \sin(\frac{r\pi}{d} z), \; r = 1, 2, 3, \cdots. \tag{232}$$

The solution of (222) is therefore given by

$$T_{nmr}(t) = C_{nmr}\cos(c\lambda_{nmr}t) + D_{nmr}\sin(c\lambda_{nmr}t), \tag{233}$$

where

$$\lambda_{nmr} = (\frac{n\pi}{a})^2 + (\frac{m\pi}{b})^2 + (\frac{r\pi}{d})^2. \tag{234}$$

Combining the results obtained above for $F_n(x), G_m(y), H_r(z)$, and $T_{nmr}(t)$ and using the superposition principle we can formulate the general solution of (216) in the form

$$u(x,y,z,t) = \sum_{r=1}^{\infty}\sum_{m=1}^{\infty}\sum_{n=1}^{\infty} \sin(\frac{n\pi}{a}x)\sin(\frac{m\pi}{b}y)\sin(\frac{r\pi}{d}z)$$

$$\times (C_{nmr}\cos(c\lambda_{nmr}t) + D_{nmr}\sin(c\lambda_{nmr}t)). \tag{235}$$

It remains now to determine the constants C_{nmr} and D_{nmr}. Using the initial condition $u(x,y,z,0) = f(x,y,z)$ into (235), the coefficients C_{nmr} are given by

$$C_{nmr} =$$

$$\frac{8}{abd}\int_0^d\int_0^b\int_0^a f(x,y,z)\sin(\frac{n\pi}{a}x)\sin(\frac{m\pi}{b}y)\sin(\frac{r\pi}{d}z)\,dx\,dy\,dz. \tag{236}$$

Using the initial condition $u_t(x,y,z,0) = g(x,y,z)$ into the derivative of (235), the coefficients D_{nmr} can be determined in the following form

$$D_{nmr} =$$

$$\frac{8}{\lambda_{nmr}cabd}\int_0^d\int_0^b\int_0^a g(x,y,z)\sin(\frac{n\pi}{a}x)\sin(\frac{m\pi}{b}y)\sin(\frac{r\pi}{d}z)\,dx\,dy\,dz. \tag{237}$$

Having determined the coefficients C_{nmr} and D_{nmr}, the particular solution of the initial-boundary value problem follows immediately upon substituting (236) and (237) into (235).

To explain the use of the method of separation of variables, several illustrative examples will be discussed as follows:

Example 5. Solve the initial-boundary value problem

$$\begin{aligned}
\text{PDE} \qquad u_{tt} &= 3(u_{xx} + u_{yy} + u_{zz}),\ 0 < x,y,z < \pi,\ t > 0 \\[4pt]
\text{B.C} \quad u(0,y,z,t) &= u(\pi,y,z,t) = 0 \\[2pt]
u(x,0,z,t) &= u(x,\pi,z,t) = 0 \\[2pt]
u(x,y,0,t) &= u(x,y,\pi,t) = 0 \\[2pt]
\text{I.C} \quad u(x,y,z,0) &= \sin x\sin y\sin z,\ u_t(x,y,z,0) = 0
\end{aligned} \tag{238}$$

Solution.

Proceeding as before, we set

$$u(x, y, z, t) = F(x)G(y)H(z)T(t). \tag{239}$$

Following the discussions presented above we find

$$
\begin{align}
F_n(x) &= B_n \sin(nx), \ \nu_n = n, \ n = 1, 2, 3, \cdots, \tag{240} \\
G_m(y) &= \beta_m \sin(my), \ \mu_m = m, \ m = 1, 2, 3, \cdots, \tag{241} \\
H_r(z) &= \delta_r \sin(rz), \ \eta_r = r, \ r = 1, 2, 3, \cdots, \tag{242} \\
T_{nmr}(t) &= C_{nmr} \cos(\sqrt{3}\lambda_{nmr}t) + D_{nmr} \sin(\sqrt{3}\lambda_{nmr}t), \tag{243}
\end{align}
$$

where

$$\lambda_{nmr} = \sqrt{n^2 + m^2 + r^2}. \tag{244}$$

Recall that we exclude $n = 0, m = 0$, and $r = 0$.

Proceeding as before and using the superposition principle, we can formulate the general solution in the form
$u(x, y, z, t) =$

$$
\sum_{r=1}^{\infty} \sum_{m=1}^{\infty} \sum_{n=1}^{\infty} \sin(nx) \sin(my) \sin(rz) \tag{245}
$$
$$
\times \left(C_{nmr} \cos(\sqrt{3}\lambda_{nmr}t) + D_{nmr} \sin(\sqrt{3}\lambda_{nmr}t) \right).
$$

To determine the constants C_{nmr}, we use the first initial condition and expand (245) to find

$$C_{111} \sin x \sin y \sin z + \cdots = \sin x \sin y \sin z \tag{246}$$

Equation the coefficients of like terms in both sides we obtain

$$
\begin{align}
C_{111} &= 1, \text{ for } n = 1, m = 1, r = 1, \\
C_{ijk} &= 0, \text{ for } i \neq 1, j \neq 1, k \neq 1. \tag{247}
\end{align}
$$

To determine the constants D_{nmr}, we use the second initial condition into the derivative of (245) to find

$$D_{nmr} = 0, \text{ for } n \geq 1, m \geq 1, r \geq 1. \tag{248}$$

Substituting the results (247) and (248) into (245) gives the particular solution

$$u(x, y, z, t) = \sin x \sin y \sin z \cos(3t). \tag{249}$$

Example 6. Solve the initial-boundary value problem

$$\text{PDE} \qquad u_{tt} \;=\; 3(u_{xx} + u_{yy} + u_{zz}),\ 0 < x, y, z < \pi,\ t > 0$$

$$
\begin{aligned}
\text{B.C} \quad u(0, y, z, t) &= u(\pi, y, z, t) = 0 \\
u(x, 0, z, t) &= u(x, \pi, z, t) = 0 \\
u(x, y, 0, t) &= u(x, y, \pi, t) = 0
\end{aligned}
\tag{250}
$$

$$\text{I.C} \quad u(x, y, z, 0) \;=\; 1,\ u_t(x, y, z, 0) = 0$$

Solution.

Proceeding as before, we set

$$u(x, y, z, t) = F(x)G(y)H(z)T(t). \tag{251}$$

Substituting (251) into (250) and proceeding as before we find

$$
\begin{aligned}
F_n(x) &= B_n \sin(nx),\ \nu_n = n,\ n = 1, 2, 3, \cdots, \tag{252} \\
G_m(y) &= \beta_m \sin(my),\ \mu_m = m,\ m = 1, 2, 3, \cdots, \tag{253} \\
H_r(z) &= \delta_r \sin(rz),\ \eta_r = r,\ r = 1, 2, 3, \cdots, \tag{254} \\
T_{nmr}(t) &= C_{nmr} \cos(\lambda_{nmr}t) + D_{nmr} \sin(\lambda_{nmr}t), \tag{255}
\end{aligned}
$$

where

$$\lambda_{nmr} = \sqrt{n^2 + m^2 + r^2}. \tag{256}$$

Using the superposition principle, the general solution is expressed in the form

$$
\begin{aligned}
u(x, y, z, t) \;=\; & \sum_{r=1}^{\infty} \sum_{m=1}^{\infty} \sum_{n=1}^{\infty} \sin(nx) \sin(my) \sin(rz) \\
& \times \left(C_{nmr} \cos(\lambda_{nmr}t) + D_{nmr} \sin(\lambda_{nmr}t) \right).
\end{aligned}
\tag{257}
$$

We next use the initial condition $u(x, y, z, 0) = 1$ into (257) to determine the coefficients C_{nmr}, hence we find

$$C_{nmr} = \frac{8}{\pi^3} \int_0^{\pi} \int_0^{\pi} \int_0^{\pi} f(x, y, z) \sin(nx) \sin(my) \sin(nz)\, dx\, dy\, dz, \tag{258}$$

which gives

$$
C_{nm} =
\begin{cases}
0 & \text{for } n \text{ is even, } m \text{ is even, } r \text{ is even,} \\[2mm]
\dfrac{64}{\pi^3 nmr} & \text{for } n \text{ is odd, } m \text{ is odd, } r \text{ is odd.}
\end{cases}
\tag{259}
$$

It is clear that the coefficients D_{nmr} are given by

$$D_{nmr} = 0, \text{ for } n \geq 1, m \geq 1, r \geq 1. \tag{260}$$

To obtain a nontrivial solution, we note that n, m, and r should be odd as shown in (259). Consequently, the particular solution is given by

$u(x, y, z, t) =$

$$\frac{64}{\pi^3} \sum_{r=0}^{\infty} \sum_{m=0}^{\infty} \sum_{n=0}^{\infty} \frac{1}{(2n+1)(2m+1)(2r+1)} \sin(2n+1)x \tag{261}$$

$$\times \sin(2m+1)y \sin(2r+1)z \cos(\lambda_{nmr}t),$$

where
$$\lambda_{nmr} = \sqrt{(2n+1)^2 + (2m+1)^2 + (2r+1)^2}, \tag{262}$$

obtained upon combining (259) and (256).

Exercises 6.3.2

Use the method of separation of variables in the following initial-boundary value problems:

1. $u_{tt} = 12(u_{xx} + u_{yy} + u_{zz}), \ 0 < x, y, z < \pi$
   $u(0, y, z, t) = u(\pi, y, z, t) = 0$
   $u(x, 0, z, t) = u(x, \pi, z, t) = 0$
   $u(x, y, 0, t) = u(x, y, \pi, 0) = 0$
   $u(x, y, z, 0) = 0, u_t(x, y, z, 0) = 12 \sin(2x) \sin(2y) \sin(2z)$

2. $u_{tt} = 14(u_{xx} + u_{yy} + u_{zz}), \ 0 < x, y, z < \pi$
   $u(0, y, z, t) = u(\pi, y, z, t) = 0$
   $u(x, 0, z, t) = u(x, \pi, z, t) = 0$
   $u(x, y, 0, t) = u(x, y, \pi, 0) = 0$
   $u(x, y, z, 0) = \sin x \sin(2y) \sin(3z), u_t(x, y, z, 0) = 0$

3. $u_{tt} = 6(u_{xx} + u_{yy} + u_{zz}), \ 0 < x, y, z < \pi$
   $u(0, y, z, t) = u(\pi, y, z, t) = 0$
   $u(x, 0, z, t) = u(x, \pi, z, t) = 0$
   $u(x, y, 0, t) = u(x, y, \pi, 0) = 0$
   $u(x, y, z, 0) = 0, u_t(x, y, z, 0) = 6 \sin x \sin y \sin(2z)$

4. $u_{tt} = 4(u_{xx} + u_{yy} + u_{zz}), \ 0 < x, y, z < \pi$
   $u_x(0, y, z, t) = u_x(\pi, y, z, t) = 0$
   $u_y(x, 0, z, t) = u_y(x, \pi, z, t) = 0$
   $u_z(x, y, 0, t) = u_z(x, y, \pi, 0) = 0$
   $u(x, y, z, 0) = 0, u_t(x, y, z, 0) = 6 \cos x \cos(2y) \cos(2z)$

5. $u_{tt} = 12(u_{xx} + u_{yy} + u_{zz}), 0 < x, y, z < \pi$
   $u_x(0, y, z, t) = u_x(\pi, y, z, t) = 0$
   $u_y(x, 0, z, t) = u_y(x, \pi, z, t) = 0$
   $u_z(x, y, 0, t) = u_z(x, y, \pi, 0) = 0$
   $u(x, y, z, 0) = 3, u_t(x, y, z, 0) = 6 \cos x \cos y \cos z$

6. $u_{tt} = 12(u_{xx} + u_{yy} + u_{zz}), 0 < x, y, z < \pi$
   $u_x(0, y, z, t) = u_x(\pi, y, z, t) = 0$
   $u_y(x, 0, z, t) = u_y(x, \pi, z, t) = 0$
   $u_z(x, y, 0, t) = u_z(x, y, \pi, 0) = 0$
   $u(x, y, z, 0) = 4 + \cos x \cos y \cos z, u_t(x, y, z, 0) = 0$

7. $u_{tt} = 3(u_{xx} + u_{yy} + u_{zz}), 0 < x, y, z < \pi$
   $u(0, y, z, t) = u(\pi, y, z, t) = 0$
   $u_y(x, 0, z, t) = u_y(x, \pi, z, t) = 0$
   $u_z(x, y, 0, t) = u_z(x, y, \pi, 0) = 0$
   $u(x, y, z, 0) = 0, u_t(x, y, z, 0) = 3 \sin x \cos y \cos z$

8. $u_{tt} = 3(u_{xx} + u_{yy} + u_{zz}), 0 < x, y, z < \pi$
   $u(0, y, z, t) = u(\pi, y, z, t) = 0$
   $u(x, 0, z, t) = u(x, \pi, z, t) = 0$
   $u_z(x, y, 0, t) = u_z(x, y, \pi, 0) = 0$
   $u(x, y, z, 0) = 0, u_t(x, y, z, 0) = 3 \sin x \sin y \cos z$

9. $u_{tt} = 12(u_{xx} + u_{yy} + u_{zz}), 0 < x, y, z < \pi$
   $u_x(0, y, z, t) = u_x(\pi, y, z, t) = 0$
   $u(x, 0, z, t) = u(x, \pi, z, t) = 0$
   $u(x, y, 0, t) = u(x, y, \pi, 0) = 0$
   $u(x, y, z, 0) = \cos x \sin y \sin z, u_t(x, y, z, 0) = 0$

10. $u_{tt} = 6(u_{xx} + u_{yy} + u_{zz}), 0 < x, y, z < \pi$
    $u(0, y, z, t) = u(\pi, y, z, t) = 0$
    $u(x, 0, z, t) = u(x, \pi, z, t) = 0$
    $u_z(x, y, 0, t) = u_z(x, y, \pi, 0) = 0$
    $u(x, y, z, 0) = \sin x \sin y \cos(2z), u_t(x, y, z, 0) = 0$

11. $u_{tt} = u_{xx} + u_{yy} + u_{zz}, 0 < x, y, z < \pi$
    $u(0, y, z, t) = u(\pi, y, z, t) = 0$
    $u(x, 0, z, t) = u(x, \pi, z, t) = 0$
    $u_z(x, y, 0, t) = u_z(x, y, \pi, 0) = 0$
    $u(x, y, z, 0) = 0, u_t(x, y, z, 0) = \sqrt{3} \sin x \sin y \cos z$

12. $u_{tt} = 6(u_{xx} + u_{yy} + u_{zz}), 0 < x, y, z < \pi$
    $u_x(0, y, z, t) = u_x(\pi, y, z, t) = 0$
    $u_y(x, 0, z, t) = u_y(x, \pi, z, t) = 0$
    $u_z(x, y, 0, t) = u_z(x, y, \pi, 0) = 0$
    $u(x, y, z, 0) = 0, u_t(x, y, z, 0) = 6 \cos x \cos y \cos 2z$

Chapter 7

Laplace's Equation

7.1 Introduction

In Chapter 4 we have discussed the PDEs that control the heat flow in two and three dimensional spaces given by

$$
\begin{aligned}
u_t &= \overline{k}(u_{xx} + u_{yy}), \\
u_t &= \overline{k}(u_{xx} + u_{yy} + u_{zz}),
\end{aligned}
\tag{1}
$$

respectively, where $\overline{k}$ is the thermal diffusivity. If the temperature u reaches a steady state, that is, when u does not depend on time t and depends only on the space variables, then the time derivative u_t vanishes as $t \to \infty$. In view of this, we substitute $u_t = 0$ into (1), hence we obtain the Laplace's equations in two and three dimensions given by

$$
\begin{aligned}
u_{xx} + u_{yy} &= 0, \\
u_{xx} + u_{yy} + u_{zz} &= 0.
\end{aligned}
\tag{2}
$$

The Laplace's equation is used to describe gravitational potential in absence of mass, to define electrostatic potential in absence of charges, and to describe temperature in a steady-state heat flow. The Laplace's equation is often called the potential equation because u defines the potential function.

Recall that the heat and the wave equations investigate the evolution of temperature and displacement respectively. However, it is worth noting that Laplace's equation describes physical phenomena at equilibrium. Moreover, since the solution of the Laplace's equation does not depend on time t, initial conditions are not specified and boundary conditions at the edges of a rectangle or at the faces of a rectangular volume are specified. For this reason, Laplace's equation is best described as a Boundary Value Problem (BVP).

In this chapter we will discuss the Laplace's equation in two or three dimensional spaces and in polar coordinates. The methods that will be used are the Adomian decomposition method and the method of separation of variables. The two methods have been outlined in previous chapters and have been applied in heat flow and wave equations.

7.2   Adomian Decomposition Method

The Adomian decomposition method is now well known and it has been used in details in the previous five chapters. The method, as discussed before, provides the solution in terms of a rapidly convergent series. In a manner parallel to that used in the preceding chapters, we will apply the Adomian decomposition method to Laplace's equation with specified boundary conditions.

7.2.1   Two Dimensional Laplace's Equation

The two dimensional Laplace's equation will be discussed using all types of boundary conditions. As stated above, initial conditions are irrelevant. The boundary conditions associated with Laplace's equation can be identified into three types of boundary conditions, namely:

1. Dirichlet boundary conditions:
In this type, the solution $u(x, y)$ of Laplace's equation is specified on the boundary. The Laplace's equation in this case is known as Dirichlet problem for a rectangle.

2. Neumann boundary conditions:
In this type, the normal derivative u_n is specified on the boundary. The Laplace's equation in this case is known as Neumann problem.

3. Robin boundary conditions:
In this type, the function u is specified on parts of the boundary and the directional derivative u_n is specified on other parts of the boundary.

Without loss of generality, we consider the two dimensional Laplace's equation is given by the boundary value problem

$$\text{PDE} \qquad u_{xx} + u_{yy} \;=\; 0, 0 < x < a, 0 < y < b$$

$$\text{B.C} \qquad u(0, y) \;=\; 0,\; u(a, y) = f(y) \tag{3}$$

$$u(x, 0) \;=\; 0,\; u(x, b) = 0$$

where $u = u(x, y)$ is the solution of the Laplace's equation at any point located at the position (x, y) of a rectangular plate.

We first write (3) in an operator form by

$$L_y u(x, y) = -L_x u(x, y), \tag{4}$$

where the differential operators L_x and L_y are defined by

$$L_x = \frac{\partial^2}{\partial x^2}, L_y = \frac{\partial^2}{\partial y^2}, \tag{5}$$

so that the inverse operators L_x^{-1} and L_y^{-1} are two-fold integral operators defined by

$$
\begin{aligned}
L_x^{-1}(.) &= \int_0^x \int_0^x (.) dx\, dx, \\
L_y^{-1}(.) &= \int_0^y \int_0^y (.) dy\, dy.
\end{aligned}
\tag{6}
$$

Applying the inverse operator L_y^{-1} to both sides of (4) and using the boundary conditions we obtain

$$u(x, y) = yg(x) - L_y^{-1} L_x u(x, y), \tag{7}$$

where

$$g(x) = u_y(x, 0), \tag{8}$$

a boundary condition that is not given but will be determined.

Using the decomposition series

$$u(x, y) = \sum_{n=0}^{\infty} u_n(x, y), \tag{9}$$

into both sides of (7) gives

$$\sum_{n=0}^{\infty} u_n(x, y) = yg(x) - L_y^{-1} L_x \left(\sum_{n=0}^{\infty} u_n(x, y) \right). \tag{10}$$

Adomian's analysis admits the use of the recursive relation

$$
\begin{aligned}
u_0(x, y) &= yg(x), \\
u_{k+1}(x, y) &= -L_y^{-1} L_x(u_k), k \geq 0.
\end{aligned}
\tag{11}
$$

This leads to

$$
\begin{aligned}
u_0(x, y) &= yg(x), \\
u_1(x, y) &= -L_y^{-1} L_x(u_0) = -\tfrac{1}{3!} y^3 g''(x) \\
u_2(x, y) &= -L_y^{-1} L_x(u_1) = \tfrac{1}{5!} y^5 g^{(iv)}(x),
\end{aligned}
\tag{12}
$$

and so on. We can determine as many components as we like to enhance the accuracy level.

In view of (12), we can write

$$u(x,y) = yg(x) - \frac{1}{3!}y^3 g''(x) + \frac{1}{5!}y^5 g^{(iv)}(x) - \cdots . \tag{13}$$

To complete the determination of the series solution of $u(x,y)$, we should determine $g(x)$. This can be easily done by using the inhomogeneous boundary condition $y(\pi,y) = f(y)$. Substituting $x = \pi$ into (13), using the Taylor expansion for $f(y)$, and equating the coefficients of like terms in both sides leads to the complete determination of $g(x)$.

Having determined the function $g(x)$, the series solution (9) of $u(x,y)$ is thus established.

To give a clear overview of the use of the decomposition method in Laplace's equation, we discuss below the following illustrative boundary value problems.

Example 1. Use the Adomian decomposition method to solve the boundary value problem

$$\text{PDE} \qquad u_{xx} + u_{yy} = 0, 0 < x, y < \pi$$

$$\text{B.C} \qquad u(0,y) = 0, \ u(\pi,y) = \sinh \pi \sin y \tag{14}$$

$$u(x,0) = 0, \ u(x,\pi) = 0$$

Solution.

Applying the inverse operator L_y^{-1} to the operator form of (14), and using the proper boundary conditions we find

$$u(x,y) = yg(x) - L_y^{-1} L_x u(x,y), \tag{15}$$

where

$$g(x) = u_y(x,0). \tag{16}$$

The decomposition method defines the solution $u(x,y)$ by an infinite series given by

$$u(x,y) = \sum_{n=0}^{\infty} u_n(x,y). \tag{17}$$

Substituting (17) into both sides of (15) gives

$$\sum_{n=0}^{\infty} u_n(x,y) = yg(x) - L_y^{-1}\left(L_x\left(\sum_{n=0}^{\infty} u_n(x,y)\right)\right). \tag{18}$$

This gives the recursive relation

$$u_0(x,y) = yg(x),$$

$$u_{k+1}(x,y) = -L_y^{-1} L_x(u_k(x,y)), k \geq 0, \tag{19}$$

that gives the first few components

$$
\begin{aligned}
u_0(x, y) &= yg(x), \\
u_1(x, y) &= -L_y^{-1} L_x(u_0(x, y)) = -\tfrac{1}{3!} y^3 g''(x), \\
u_2(x, y) &= -L_y^{-1} L_x(u_1(x, y)) = \tfrac{1}{5!} y^5 g^{(iv)}(x), \\
u_3(x, y) &= -L_y^{-1} L_x(u_2(x, y)) = -\tfrac{1}{7!} y^7 g^{(vi)}(x).
\end{aligned}
\tag{20}
$$

Combining the above results obtained for the components yields

$$
u(x, y) = yg(x) - \frac{1}{3!} y^3 g''(x) + \frac{1}{5!} y^5 g^{(iv)}(x) - \frac{1}{7!} y^7 g^{(vi)}(x) + \cdots.
\tag{21}
$$

To determine the function $g(x)$, we use the inhomogeneous boundary condition $u(\pi, y) = \sinh \pi \sin y$, and by using the Taylor expansion of $\sin y$ we obtain

$$
\begin{aligned}
&yg(\pi) - \frac{1}{3!} y^3 g''(\pi) + \frac{1}{5!} y^5 g^{(iv)}(\pi) - \frac{1}{7!} y^7 g^{(vi)}(\pi) + \cdots \\
&= \sinh \pi (y - \frac{1}{3!} y^3 + \frac{1}{5!} y^5 - \frac{1}{7!} y^7 + \cdots).
\end{aligned}
\tag{22}
$$

Equating the coefficients of like terms on both sides gives

$$
g(\pi) = g''(\pi) = g^{(iv)}(\pi) = \cdots = \sinh \pi.
\tag{23}
$$

This means that

$$
g(x) = \sinh x,
\tag{24}
$$

the only function that when substituted in (21) will justify the remaining boundary conditions. Consequently, the solution in a series form is given by

$$
u(x, y) = \sinh x \left(y - \frac{1}{3!} y^3 + \frac{1}{5!} y^5 - \frac{1}{7!} y^7 + \cdots \right),
\tag{25}
$$

and in a closed form by

$$
u(x, y) = \sinh x \sin y,
\tag{26}
$$

obtained upon using the Taylor expansion for $\sin y$.

Example 2. Use the Adomian decomposition method to solve the boundary value problem

$$
\begin{aligned}
\text{PDE} \qquad & u_{xx} + u_{yy} = 0, 0 < x, y < \pi \\
\text{B.C} \qquad & u(0, y) = 0, \, u(\pi, y) = 0 \\
& u(x, 0) = 0, \, u(x, \pi) = \sin x \sinh \pi
\end{aligned}
\tag{27}
$$

Solution.

We first rewrite (27) in an operator form by

$$L_x u(x, y) = -L_y u(x, y). \tag{28}$$

Applying the inverse operator L_x^{-1} to both sides of (28), and using the proper boundary conditions we find

$$u(x, y) = xh(y) - L_x^{-1} L_y u(x, y), \tag{29}$$

where

$$h(y) = u_x(0, y). \tag{30}$$

Using the series representation of $u(x, y)$ gives

$$\sum_{n=0}^{\infty} u_n(x, y) = xh(y) - L_x^{-1} \left(L_y \left(\sum_{n=0}^{\infty} u_n(x, y) \right) \right), \tag{31}$$

that admits the use of the recursive relation

$$\begin{aligned} u_0(x, y) &= xh(y), \\ u_{k+1}(x, y) &= -L_x^{-1} L_y(u_k(x, y)), k \geq 0, \end{aligned} \tag{32}$$

that in turn gives

$$\begin{aligned} u_0(x, y) &= xh(y), \\ u_1(x, y) &= -L_x^{-1} L_y(u_0(x, y)) = -\tfrac{1}{3!}x^3 h''(y), \\ u_2(x, y) &= -L_x^{-1} L_y(u_1(x, y)) = \tfrac{1}{5!}x^5 h^{(iv)}(y). \end{aligned} \tag{33}$$

Using the above results obtained for the components gives

$$u(x, y) = xh(y) - \frac{1}{3!}x^3 h''(y) + \frac{1}{5!}x^5 h^{(iv)}(y) - \frac{1}{7!}x^7 h^{(vi)}(y) + \cdots. \tag{34}$$

Using the boundary condition $u(x, \pi) = \sinh \pi \sin x$, and using the Taylor expansion of $\sin x$ we obtain

$$\begin{aligned} & xh(\pi) - \frac{1}{3!}x^3 h''(\pi) + \frac{1}{5!}x^5 h^{(iv)}(\pi) - \frac{1}{7!}x^7 h^{(vi)}(\pi) + \cdots \\ & = \sinh \pi (x - \frac{1}{3!}x^3 + \frac{1}{5!}x^5 - \frac{1}{7!}x^7 + \cdots). \end{aligned} \tag{35}$$

Equating the coefficients of like terms on both sides gives

$$h(\pi) = h''(\pi) = h^{(iv)}(\pi) = \cdots = \sinh \pi. \tag{36}$$

This means, considering the remaining boundary condition, that

$$h(y) = \sinh y. \tag{37}$$

Combining the results obtained above in (34) and (36), the solution in a series form is given by

$$u(x, y) = \sinh y \left(x - \frac{1}{3!}x^3 + \frac{1}{5!}x^5 - \frac{1}{7!}x^7 + \cdots \right), \tag{38}$$

and in a closed form by

$$u(x, y) = \sin x \sinh y. \tag{39}$$

Example 3. Use the Adomian decomposition method to solve the boundary value problem

$$
\begin{array}{llll}
\text{PDE} & u_{xx} + u_{yy} & = & 0, 0 < x, y < \pi \\[2mm]
\text{B.C} & u_x(0, y) & = & 0, \; u_x(\pi, y) = 0 \\[2mm]
& u_y(x, 0) & = & \cos x, \; u_y(x, \pi) = \cosh \pi \cos x
\end{array} \tag{40}
$$

Solution.

We point out that the boundary conditions are the Neumann boundary conditions where the directional derivatives are specified. The decomposition method can be applied in a direct way. Applying the inverse operator L_y^{-1} to both sides of the operator form of (40) we find

$$u(x, y) = g(x) + y \cos x - L_y^{-1} L_x u(x, y), \tag{41}$$

where

$$g(x) = u(x, 0). \tag{42}$$

This in turn gives

$$\sum_{n=0}^{\infty} u_n(x, y) = g(x) + y \cos x - L_y^{-1} \left(L_x \left(\sum_{n=0}^{\infty} u_n(x, y) \right) \right), \tag{43}$$

so that the recursive relation is given by

$$
\begin{array}{lll}
u_0(x, y) & = & g(x) + y \cos x, \\[2mm]
u_{k+1}(x, y) & = & -L_y^{-1} L_x(u_k(x, y)), \; k \geq 0.
\end{array} \tag{44}
$$

It then follows that

$$
\begin{array}{lll}
u_0(x, y) & = & g(x) + y \cos x, \\[2mm]
u_1(x, y) & = & -L_y^{-1} L_x(u_0(x, y)) = -\frac{1}{2!}y^2 g''(x) + \frac{1}{3!}y^3 \cos x, \\[2mm]
u_2(x, y) & = & -L_y^{-1} L_x(u_1(x, y)) = \frac{1}{4!}y^4 g^{(iv)}(x) + \frac{1}{5!}y^5 \cos x.
\end{array} \tag{45}
$$

and so on. This gives

$$
\begin{aligned}
u(x, y) \;=\;& \cos x \left(y + \frac{1}{3!} y^3 + \frac{1}{5!} y^5 + \cdots \right) \\
& + g(x) - \frac{1}{2!} y^2 g''(x) + \frac{1}{4!} y^4 g^{(iv)}(x) + \cdots,
\end{aligned}
\tag{46}
$$

or equivalently

$$
u(x, y) = \cos x \sinh y + g(x) - \frac{1}{2!} y^2 g''(x) + \frac{1}{4!} y^4 g^{(iv)}(x) - \cdots .
\tag{47}
$$

To determine $g(x)$, we use the boundary condition $u_y(x, \pi) = \cosh \pi \cos x$ to obtain

$$
\cosh \pi \cos x + \left(-\pi g''(x) + \frac{1}{3!} \pi^3 g^{(iv)}(x) + \cdots \right) = \cosh \pi \cos x.
\tag{48}
$$

Equating the coefficients of like terms on both sides and following the discussion presented in Ex. 4, we find

$$
g(x) = 0.
\tag{49}
$$

Consequently, we obtain

$$
u(x, y) = \cos x \sinh y.
\tag{50}
$$

It is important to note that Neumann problem has a property that the solution is determined up to an additive constant. An arbitrary constant C_0 cannot be determined by the decomposition method and by the classic method of separation of variables as will be seen later. Based on this, the solution should be given by

$$
u(x, y) = C_0 + \cos x \sinh y,
\tag{51}
$$

that satisfies the equation and the boundary conditions.

Example 4. Use the Adomian decomposition method to solve the boundary value problem

$$
\begin{array}{lrcl}
\text{PDE} & u_{xx} + u_{yy} & = & 0, 0 < x, y < \pi \\[4pt]
\text{B.C} & u_x(0, y) & = & 0,\; u_x(\pi, y) = 0 \\[4pt]
& u(x, 0) & = & \cos x,\; u(x, \pi) = \cosh \pi \cos x
\end{array}
\tag{52}
$$

Solution.

We point out that the boundary conditions are the mixed boundary conditions where the solution $u(x, y)$ is specified at two edges and the derivatives $u_x(x, y)$ are specified at the remaining two edges.

Applying the inverse operator L_y^{-1} to both sides of the operator form of (52), and using the proper boundary conditions we find

$$u(x,y) = \cos x + yg(x) - L_y^{-1} L_x u(x,y), \tag{53}$$

where

$$g(x) = u_y(x,0). \tag{54}$$

Using the series representation of $u(x,y)$ gives

$$\sum_{n=0}^{\infty} u_n(x,y) = \cos x + yg(x) - L_y^{-1}\left(L_x\left(\sum_{n=0}^{\infty} u_n(x,y)\right)\right). \tag{55}$$

Following the analysis presented above we find

$$
\begin{aligned}
u_0(x,y) &= \cos x + yg(x), \\
u_1(x,y) &= -L_y^{-1}L_x(u_0(x,y)) = \tfrac{1}{2!}y^2\cos x - \tfrac{1}{3!}y^3 g''(x), \\
u_2(x,y) &= -L_y^{-1}L_x(u_1(x,y)) = \tfrac{1}{4!}y^4\cos x + \tfrac{1}{5!}y^5 g^{(iv)}(x).
\end{aligned}
\tag{56}
$$

This gives

$$
\begin{aligned}
u(x,y) &= \cos x\left(1 + \frac{1}{2!}y^2 + \frac{1}{4!}y^4 + \cdots\right) \\
&\quad + yg(x) - \frac{1}{3!}y^3 g''(x) + \frac{1}{5!}y^5 g^{(iv)}(x) - \cdots,
\end{aligned}
\tag{57}
$$

or equivalently

$$u(x,y) = \cos x \cosh y + yg(x) - \frac{1}{3!}y^3 g''(x) + \frac{1}{5!}y^5 g^{(iv)}(x) - \cdots. \tag{58}$$

To determine $g(x)$, we use the boundary condition $u(x,\pi) = \cosh\pi\cos x$ to find

$$g(x) = 0. \tag{59}$$

The solution in a closed form is given by

$$u(x,y) = \cos x \cosh y. \tag{60}$$

Exercises 7.2.1

Use the decomposition method to solve the following Laplace's equations:

1. $u_{xx} + u_{yy} = 0,\ 0 < x, y < \pi$
   $u(0,y) = 0, u(\pi,y) = \sinh\pi\cos y$
   $u(x,0) = \sinh x, u(x,\pi) = -\sinh x$

2. $u_{xx} + u_{yy} = 0,\ 0 < x, y < \pi$
$u(0, y) = \sin y,\ u(\pi, y) = \cosh \pi \sin y$
$u(x, 0) = 0,\ u(x, \pi) = 0$

3. $u_{xx} + u_{yy} = 0,\ 0 < x, y < \pi$
$u(0, y) = \cos y,\ u(\pi, y) = \cosh \pi \cos y$
$u(x, 0) = \cosh x,\ u(x, \pi) = -\cosh x$

4. $u_{xx} + u_{yy} = 0,\ 0 < x, y < \pi$
$u(0, y) = 0,\ u(\pi, y) = 0$
$u(x, 0) = 0,\ u(x, \pi) = \sinh(2\pi) \sin(2x)$

5. $u_{xx} + u_{yy} = 0,\ 0 < x, y < \pi$
$u(0, y) = 0,\ u(\pi, y) = 0$
$u(x, 0) = \sin(2x),\ u(x, \pi) = \cosh(2\pi) \sin(2x)$

6. $u_{xx} + u_{yy} = 0,\ 0 < x, y < \pi$
$u(0, y) = \cosh(3y),\ u(\pi, y) = -\cosh(3y)$
$u(x, 0) = \cos(3x),\ u(x, \pi) = \cosh(3\pi) \cos(3x)$

7. $u_{xx} + u_{yy} = 0,\ 0 < x, y < \pi$
$u(0, y) = 0,\ u(\pi, y) = \sinh(2\pi) \cos(2y)$
$u(x, 0) = \sinh(2x),\ u(x, \pi) = \sinh(2x)$

8. $u_{xx} + u_{yy} = 0,\ 0 < x, y < \pi$
$u(0, y) = \cos(2y),\ u(\pi, y) = \cosh(2\pi) \cos(2y)$
$u(x, 0) = \cosh(2x),\ u(x, \pi) = \cosh(2x)$

9. $u_{xx} + u_{yy} = 0,\ 0 < x, y < \pi$
$u_x(0, y) = 0,\ u_x(\pi, y) = 0$
$u_y(x, 0) = 0,\ u_y(x, \pi) = \cos x \sinh \pi$

10. $u_{xx} + u_{yy} = 0,\ 0 < x, y < \pi$
$u_x(0, y) = \cosh y,\ u_x(\pi, y) = -\cosh y$
$u_y(x, 0) = 0,\ u_y(x, \pi) = \sin x \sinh \pi$

11. $u_{xx} + u_{yy} = 0,\ 0 < x, y < \pi$
$u_x(0, y) = \cos y,\ u_x(\pi, y) = \cosh \pi \cos y$
$u(x, 0) = \sinh x,\ u(x, \pi) = -\sinh x$

12. $u_{xx} + u_{yy} = 0,\ 0 < x, y < \pi$
$u_x(0, y) = \cosh y,\ u_x(\pi, y) = -\cosh y$
$u(x, 0) = \sin x,\ u(x, \pi) = \cosh \pi \sin x$

13. $u_{xx} + u_{yy} = 0,\ 0 < x, y < \pi$
$u(0, y) = 0,\ u(\pi, y) = \pi$
$u(x, 0) = x,\ u(x, \pi) = x + \sinh \pi \sin x$

14. $u_{xx} + u_{yy} = 0,\ 0 < x, y < \pi$
    $u(0, y) = y, u(\pi, y) = y$
    $u(x, 0) = \sin x, u(x, \pi) = \pi + \cosh \pi \sin x$

15. $u_{xx} + u_{yy} = 0,\ 0 < x, y < \pi$
    $u(0, y) = 1, u(\pi, y) = 1$
    $u(x, 0) = 1, u(x, \pi) = 1 + \sinh \pi \sin x$

16. $u_{xx} + u_{yy} = 0,\ 0 < x, y < \pi$
    $u(0, y) = 1 + \sinh y, u(\pi, y) = 1 - \sinh y$
    $u(x, 0) = 1, u(x, \pi) = 1 + \sinh \pi \cos x$

7.3 Method of Separation of Variables

In this section, the method of separation of variables will be used to solve the Laplace's equation in two and three dimensional spaces and in a circular disc. Recall that the method reduces the partial differential equation into a system of ordinary differential equations, where each ordinary differential equation depends on one variable only. We then solve each ordinary differential equation independently. The homogeneous boundary conditions are used to evaluate the constants of integration. The superposition principle will be used to establish a general solution. The remaining inhomogeneous boundary condition will be employed to determine the particular solution that will satisfy the equation and the boundary conditions.

Because the solution of the Laplace's equation does not depend on time, initial conditions are irrelevant and only boundary conditions are specified at the edges of a rectangle or faces of a rectangular volume.

The complete details of the method of separation of variables have been outlined before, therefore we will focus our discussion on the implementation of the method.

7.3.1 Laplace's Equation in Two Dimensions

The two dimensional Laplace's equation is governed by the following boundary value problem

$$\begin{aligned}
\text{PDE} \qquad u_{xx} + u_{yy} &= 0, 0 < x < a, 0 < y < b \\[6pt]
\text{B.C} \qquad u(0, y) &= 0,\ u(a, y) = 0 \qquad\qquad (61) \\[6pt]
u(x, 0) &= 0,\ u(x, b) = g(x)
\end{aligned}$$

where $u = u(x, y)$ is the solution of the Laplace's equation at any point located at the position (x, y) of a rectangle.

As indicated before, the method of separation of variables suggests that the solution $u(x, y)$ can be assumed as the product of distinct functions $F(x)$ and $G(y)$

such that each function depends on one variable only. This means that we can set

$$u(x, y) = F(x)G(y). \tag{62}$$

Differentiating both sides of (62) twice with respect to x and y respectively and substituting in the PDE of (61) we find

$$F''(x)G(y) + F(x)G''(y) = 0. \tag{63}$$

Dividing both sides by $F(x)G(y)$ gives

$$\frac{F''(x)}{F(x)} = -\frac{G''(y)}{G(y)}. \tag{64}$$

It is obvious that the left hand side depends only on the variable x and the right hand side depends only on the variable y. The equality holds only if both sides are equal to the same constant. Accordingly, we set

$$\frac{F''(x)}{F(x)} = -\frac{G''(y)}{G(y)} = -\lambda^2. \tag{65}$$

In view of (65) we obtain the two second order ordinary differential equations

$$F''(x) + \lambda^2 F(x) = 0, \tag{66}$$

and

$$G''(y) - \lambda^2 G(y) = 0. \tag{67}$$

The second order differential equations (66) and (67) give the solutions

$$F(x) = A\cos(\lambda x) + B\sin(\lambda x), \tag{68}$$

and

$$G(y) = \alpha\cosh(\lambda y) + \beta\sinh(\lambda y), \tag{69}$$

where A, B, α and β are constants. To achieve this goal, we use the boundary conditions $u(0, y) = u(a, y) = 0$ into (68) gives

$$A = 0, \tag{70}$$

and

$$\lambda_n = \frac{n\pi}{a}, n = 1, 2, 3, \cdots. \tag{71}$$

It is obvious that $n = 0$ and $B = 0$ are excluded because each will give the trivial solution. We therefore conclude that

$$F_n(x) = \sin(\frac{n\pi}{a}x), n = 1, 2, 3, \cdots. \tag{72}$$

Using the boundary condition $u(x,0) = 0$ in (69) yields

$$\alpha = 0, \tag{73}$$

and hence

$$G_n(y) = \sinh(\frac{n\pi}{a}y), n = 1, 2, 3, \cdots. \tag{74}$$

Combining (72) and (74) we obtain the fundamental solutions

$$u_n(x,y) = \sin(\frac{n\pi}{a}x)\sinh(\frac{n\pi}{a}y), n = 1, 2, 3, \cdots, \tag{75}$$

that satisfy the partial differential equation in (61) and the three homogeneous boundary conditions for each value of n.

Using the superposition principle we obtain

$$u(x,y) = \sum_{n=1}^{\infty} C_n \sin(\frac{n\pi}{a}x)\sinh(\frac{n\pi}{a}y), \tag{76}$$

where the constants C_n are as yet undetermined. To determine C_n, we use the inhomogeneous boundary condition to find

$$\sum_{n=1}^{\infty} C_n \sin(\frac{n\pi}{a}x)\sinh(\frac{n\pi}{a}b) = g(x). \tag{77}$$

The constants C_n are then determined by using Fourier series to find

$$C_n \sinh(\frac{n\pi}{a}b) = \frac{2}{a}\int_0^a \sin(\frac{n\pi}{a}x)g(x)\,dx. \tag{78}$$

Consequently, the solution of the Laplace's equation is given by (76) with C_n defined by (78).

The method of separation of variables will be illustrated by discussing the following examples.

Example 1. Use the method of separation of variables to solve the boundary value problem

$$\begin{aligned}
\text{PDE} \qquad u_{xx} + u_{yy} &= 0, 0 < x, y < \pi \\
\text{B.C} \qquad u(0,y) &= 0, \; u(\pi, y) = 0 \qquad\qquad (79)\\
u(x,0) &= 0, \; u(x,\pi) = \sinh\pi\sin x
\end{aligned}$$

Solution.

The method of separation of variables suggests

$$u(x,y) = F(x)G(y), \tag{80}$$

that gives the second order ordinary differential equations

$$F^{''}(x) + \lambda^2 F(x) = 0, \tag{81}$$

and

$$G^{''}(y) - \lambda^2 G(y) = 0. \tag{82}$$

The second order differential equations (81) and (82) give the solutions

$$F(x) = A\cos(\lambda x) + B\sin(\lambda x), \tag{83}$$

and

$$G(y) = \alpha\cosh(\lambda y) + \beta\sinh(\lambda y), \tag{84}$$

where A, B, α and β are constants. To determine these constants, we use the boundary conditions $u(0,y) = 0, u(\pi,y) = 0$ into (83) gives

$$A = 0, \tag{85}$$

and

$$\lambda_n = n, n = 1, 2, 3, \cdots. \tag{86}$$

This gives

$$F_n(x) = \sin(nx), n = 1, 2, 3, \cdots. \tag{87}$$

Using the boundary condition $u(x,0) = 0$ into (84) yields

$$\alpha = 0, \tag{88}$$

and hence

$$G_n(y) = \sinh(ny), n = 1, 2, 3, \cdots. \tag{89}$$

Combining (87) and (89) we obtain the fundamental solutions

$$u_n(x,y) = \sin(nx)\sinh(ny), n = 1, 2, 3, \cdots, \tag{90}$$

that satisfy the partial differential equation and the three homogeneous boundary conditions for each value of n.

Using the superposition principle, the solution $u(x,y)$ can be written in the form

$$u(x,y) = \sum_{n=1}^{\infty} C_n \sin(nx)\sinh(ny), \tag{91}$$

where the constants C_n are as yet undetermined. To determine C_n, we use the inhomogeneous boundary condition $u(x,\pi) = \sinh\pi\sin y$ to find

$$\sum_{n=1}^{\infty} C_n \sinh(n\pi)\sin(nx) = \sinh\pi\sin x. \tag{92}$$

The constants C_n are then determined by using Fourier series or by expanding the series and equating coefficients of like terms, hence we obtain

$$C_1 = 1, C_k = 0, k \neq 1. \tag{93}$$

Consequently, the solution of the Laplace's equation is given by

$$u(x, y) = \sin x \sinh y. \tag{94}$$

Example 2. Use the method of separation of variables to solve the boundary value problem

$$\begin{aligned}
\text{PDE} \qquad u_{xx} + u_{yy} &= 0, 0 < x, y < \pi \\
\text{B.C} \qquad u(0, y) &= 0, \ u(\pi, y) = 0 \\
u(x, 0) &= 0, \ u(x, \pi) = 1
\end{aligned} \tag{95}$$

Solution.

We first set

$$u(x, y) = F(x)G(y). \tag{96}$$

Substituting (96) into (95) and proceeding as before we find

$$F_n(x) = \sin(nx), n = 1, 2, 3, \cdots, \tag{97}$$

and

$$G_n(y) = \sinh(ny), n = 1, 2, 3, \cdots. \tag{98}$$

Accordingly, we obtain the the fundamental solutions

$$u_n(x, y) = \sin(nx) \sinh(ny), n = 1, 2, 3, \cdots. \tag{99}$$

Using the superposition principle we obtain

$$u(x, y) = \sum_{n=1}^{\infty} C_n \sin(nx) \sinh(ny), \tag{100}$$

where $C_n, n \geq 1$ are constants. To determine C_n, we use the inhomogeneous boundary condition $u(x, \pi) = 1$ in (100) to find

$$\sum_{n=1}^{\infty} C_n \sin(nx) \sinh(n\pi) = 1. \tag{101}$$

The constants C_n can be determined by using the Fourier series, hence we find

$$C_n \sinh(n\pi) = \frac{2}{\pi} \int_0^{\pi} \sin(nx) dx, \tag{102}$$

so that

$$C_n \sinh(n\pi) = \begin{cases} \frac{4}{n\pi} & \text{if } n \text{ is odd} \\ 0 & \text{if } n \text{ is even} \end{cases} \tag{103}$$

Consequently, the solution is given by

$$u(x,y) = \sum_{m=0}^{\infty} \frac{4}{(2m+1)\pi \sinh(2m+1)\pi} \sin(2m+1)x \sinh(2m+1)y. \tag{104}$$

Example 3. Use the method of separation of variables to solve the boundary value problem

$$\text{PDE} \qquad u_{xx} + u_{yy} = 0, 0 < x, y < \pi$$

$$\text{B.C} \qquad u(0,y) = 0, \ u(\pi,y) = 0 \tag{105}$$

$$u(x,0) = \sinh \pi \sin x, \ u(x,\pi) = 0$$

Solution.

We first set

$$u(x,y) = F(x)G(y). \tag{106}$$

Proceeding as before we find

$$F''(x) + \lambda^2 F(x) = 0, \tag{107}$$

and

$$G''(y) - \lambda^2 G(y) = 0. \tag{108}$$

Solving these equations and using the boundary conditions we find

$$F_n(x) = \sin(nx), n = 1, 2, 3, \cdots. \tag{109}$$

Solving (108) we obtain

$$G_n(y) = \alpha_n \cosh(ny) + \beta_n \sinh(ny). \tag{110}$$

To properly use the boundary condition $u(x,\pi) = 0$, we first rewrite (110) in the form

$$G_n(y) = C_n \sinh n(K - y). \tag{111}$$

Using the boundary condition $u(x,\pi) = 0$ in (111) yields

$$K = \pi, C_n \neq 0. \tag{112}$$

This gives

$$G_n(y) = C_n \sinh n(\pi - y), n = 1, 2, 3, \cdots. \tag{113}$$

Using the superposition principle we obtain

$$u(x, y) = \sum_{n=1}^{\infty} C_n \sin(nx) \sinh n(\pi - y), \tag{114}$$

where the constants C_n are as yet undermined. Using the inhomogeneous boundary condition $u(x, 0) = \sin x \sinh \pi$ in (114) to find

$$\sum_{n=1}^{\infty} C_n \sin(nx) \sinh(n\pi) = \sin x \sinh \pi. \tag{115}$$

The constants C_n can be determined by expanding the series and equating the coefficients of like terms on both sides where we obtain

$$C_1 = 1, C_k = 0, k \neq 1. \tag{116}$$

Consequently, the solution is given by

$$u(x, y) = \sin x \sinh(\pi - y). \tag{117}$$

Example 4. Use the method of separation of variables to solve the boundary value problem

$$
\begin{array}{rlll}
\text{PDE} & u_{xx} + u_{yy} & = & 0, 0 < x, y < \pi \\[2mm]
\text{B.C} & u_x(0, y) & = & 0, \, u_x(\pi, y) = 0 \\[2mm]
& u_y(x, 0) & = & 0, \, u_y(x, \pi) = \sinh \pi \cos x
\end{array} \tag{118}
$$

Solution.

It is important to note that this type of problems is well known as the Neumann problem where the directional derivatives of $u(x, y)$ are prescribed on the boundary. A necessary condition for this problem to be solvable is that the integral of the inhomogeneous boundary condition vanishes. This can be easily satisfied by noting that

$$\int_0^{\pi} \sinh \pi \cos x \, dx = 0. \tag{119}$$

In addition, we will show that the solution of the Neumann problem will be determined up to an additive constant.

We first set

$$u(x, y) = F(x)G(y). \tag{120}$$

Proceeding as before we find

$$F''(x) + \lambda^2 F(x) = 0, \tag{121}$$

and
$$G''(y) - \lambda^2 G(y) = 0. \tag{122}$$

Solving these equations and using the boundary conditions we find
$$F_n(x) = \cos(nx), n = 0, 1, 2, 3, \cdots, \tag{123}$$

and
$$G_n(y) = C_n \cosh(ny), n = 0, 1, 2, 3, \cdots. \tag{124}$$

Using the superposition principle we obtain
$$u(x, y) = \sum_{n=0}^{\infty} C_n \cos(nx) \cosh(ny), \tag{125}$$

or equivalently
$$u(x, y) = C_0 + \sum_{n=1}^{\infty} C_n \cos(nx) \cosh(ny), \tag{126}$$

where the constants C_n are as yet undetermined. Using the inhomogeneous boundary condition $u_y(x, \pi) = \sinh \pi \cos x$ in the derivative of (126) and expand the series to find
$$C_1 \sinh \pi \cos x + C_2 \sinh(2\pi) \cos(2x) + \cdots = \sinh \pi \cos x. \tag{127}$$

Equating the coefficients of like terms on both sides gives
$$C_1 = 1, C_k = 0, k \neq 1. \tag{128}$$

It is important to note that the constant C_0 is eliminated when the boundary condition $u_y(x, \pi)$ is used, and there is no prescribed condition that will determine C_0. Accordingly, the constant C_0 remains arbitrary and the solution is therefore given by
$$u(x, y) = C_0 + \cos x \cosh y. \tag{129}$$

This confirms the well-known property that Neumann problem is solved up to an additive arbitrary constant.

Exercises 7.3.1

Use the method of separation of variables to solve the following Laplace's equations:

1. $u_{xx} + u_{yy} = 0, 0 < x < \pi, \ 0 < y < \pi$
   $u(0, y) = 0, u(\pi, y) = 0$
   $u(x, 0) = 0, u(x, \pi) = \sinh(2\pi) \sin(2x)$

2. $u_{xx} + u_{yy} = 0, 0 < x < \pi, \ 0 < y < \pi$
   $u(0, y) = 0, u(\pi, y) = \sinh(3\pi) \sin(3y)$
   $u(x, 0) = 0, u(x, \pi) = 0$

3. $u_{xx} + u_{yy} = 0,\, 0 < x < \pi,\, 0 < y < \pi$
   $u(0,y) = 0,\, u(\pi,y) = 4\sinh 2\pi \sin 2y$
   $u(x,0) = 0,\, u(x,\pi) = 0$

4. $u_{xx} + u_{yy} = 0,\, 0 < x < \pi,\, 0 < y < \pi$
   $u(0,y) = \cosh y,\, u(\pi,y) = -\cosh y$
   $u(x,0) = \cos x,\, u(x,\pi) = \cosh \pi \cos x$

5. $u_{xx} + u_{yy} = 0,\, 0 < x < \pi,\, 0 < y < \pi$
   $u(0,y) = 0,\, u(\pi,y) = 0$
   $u(x,0) = \sin x,\, u(x,\pi) = \cosh \pi \sin x$

6. $u_{xx} + u_{yy} = 0,\, 0 < x < \pi,\, 0 < y < \pi$
   $u(0,y) = \cosh 2y,\, u(\pi,y) = \cosh 2y$
   $u(x,0) = \cos 2x,\, u(x,\pi) = \cosh 2\pi \cos 2x$

7. $u_{xx} + u_{yy} = 0,\, 0 < x < \pi,\, 0 < y < \pi$
   $u(0,y) = 0,\, u(\pi,y) = \sinh \pi \sin(\pi - y)$
   $u(x,0) = 0,\, u(x,\pi) = 0$

8. $u_{xx} + u_{yy} = 0,\, 0 < x < \pi,\, 0 < y < \pi$
   $u(0,y) = 0,\, u(\pi,y) = 0$
   $u(x,0) = \sinh(2\pi)\sin(2x),\, u(x,\pi) = 0$

9. $u_{xx} + u_{yy} = 0,\, 0 < x < \pi,\, 0 < y < \pi$
   $u_x(0,y) = 0,\, u_x(\pi,y) = 0$
   $u_y(x,0) = 0,\, u_y(x,\pi) = 2\sinh(2\pi)\cos(2x)$

10. $u_{xx} + u_{yy} = 0,\, 0 < x < \pi,\, 0 < y < \pi$
    $u_x(0,y) = 0,\, u_x(\pi,y) = 2\sinh(2\pi)\cos(2y)$
    $u_y(x,0) = 0,\, u_y(x,\pi) = 0$

7.3.2 Laplace's Equation in Three Dimensions

The Laplace's equation in three dimensional space is governed by the boundary value problem

$$\text{PDE} \qquad u_{xx} + u_{yy} + u_{zz} = 0,$$

$$0 < x < a, 0 < y < b, 0 < z < c$$

$$\text{B.C} \qquad u(0,y,z) = 0,\ u(a,y,z) = 0 \qquad (130)$$

$$u(x,0,z) = 0,\ u(x,b,z) = 0$$

$$u(x,y,0) = 0,\ u(x,y,c) = f(x,y)$$

where $u = u(x, y, z)$ is the solution of Laplace's equation at any point located at the position (x, y, z) of a rectangular volume.

Following the steps used in the previous section, we first establish a set of fundamental solutions that will satisfy the partial differential equation and the homogeneous boundary conditions. Next we use the superposition principle to establish a general solution. The remaining constant of integration is determined by using the remaining inhomogeneous boundary condition.

The method of separation of variables assumes that $u(x, y, z)$ consists of the product of three distinct functions $F(x)$, $G(y)$, and $H(z)$ such that each function depends on one variable only. This means that we can set

$$u(x, y, z) = F(x)G(y)H(z). \tag{131}$$

Differentiating both sides of (131) twice with respect to x, y, and z and substituting into the PDE of (130) we find

$$F''(x)G(y)H(z) + F(x)G''(y)H(z) + F(x)G(y)H''(z) = 0. \tag{132}$$

Dividing both sides by $F(x)G(y)H(z)$ gives

$$\frac{F''(x)}{F(x)} = -\left(\frac{G''(y)}{G(y)} + \frac{H''(z)}{H(z)} \right). \tag{133}$$

It is obvious that the left hand side depends only on the variable x and the right hand side depends only on the variables y and z. The equality holds only if both sides are equal to the same constant. Accordingly, we set

$$\frac{F''(x)}{F(x)} = -\left(\frac{G''(y)}{G(y)} + \frac{H''(z)}{H(z)} \right) = -\lambda^2. \tag{134}$$

Equation (134) yields the second order ordinary differential equations

$$F''(x) + \lambda^2 F(x) = 0, \tag{135}$$
$$G''(y) + \mu^2 G(y) = 0, \tag{136}$$
$$H''(z) - (\lambda^2 + \mu^2)H(z) = 0, \tag{137}$$

where λ, and μ are constants.

Solving the second order differential equations (135) – (137) gives

$$F(x) = A\cos(\lambda x) + B\sin(\lambda x), \tag{138}$$
$$G(y) = \alpha\cos(\mu y) + \beta\sin(\mu y), \tag{139}$$
$$H(z) = \gamma\cosh(\nu z) + \delta\sinh(\nu z), \tag{140}$$

where

$$\nu = \sqrt{\lambda^2 + \mu^2}, \tag{141}$$

and $A, B, \alpha, \beta, \gamma$, and δ are constants.

Using the homogeneous boundary conditions gives

$$A = 0, \ \lambda_n = \frac{n\pi}{a}, n = 1, 2, 3 \cdots, \tag{142}$$

$$\alpha = 0, \ \mu_m = \frac{m\pi}{b}, m = 1, 2, 3, \cdots, \tag{143}$$

$$\gamma = 0, \ \nu_{nm} = \sqrt{(\frac{n\pi}{a})^2 + (\frac{m\pi}{b})^2}, \tag{144}$$

so that

$$F_n(x) = \sin(\frac{n\pi}{a}x), n = 1, 2, 3 \cdots, \tag{145}$$

$$G_m(y) = \sin(\frac{m\pi}{b}y), m = 1, 2, 3 \cdots, \tag{146}$$

$$H_{nm}(z) = \sinh\left(\sqrt{(\frac{n\pi}{a})^2 + (\frac{m\pi}{b})^2}\ z\right). \tag{147}$$

Combining (145) – (147) we obtain the fundamental set of solutions

$$u_n = \sin(\frac{n\pi}{a}x)\sin(\frac{m\pi}{b}y)\sinh\left(\sqrt{(\frac{n\pi}{a})^2 + (\frac{m\pi}{b})^2}\ z\right), n, m = 1, 2, \cdots. \tag{148}$$

Using the superposition principle we obtain

$$u = \sum_{m=1}^{\infty}\sum_{n=1}^{\infty} C_{nm}\sin(\frac{n\pi}{a}x)\sin(\frac{m\pi}{b}y)\sinh\left(\sqrt{(\frac{n\pi}{a})^2 + (\frac{m\pi}{b})^2}\ z\right), \tag{149}$$

where the constants C_{nm} are as yet undetermined. To determine the constants C_{nm}, we use the inhomogeneous boundary condition $u(x, y, c) = f(x, y)$ to find

$$\sum_{m=1}^{\infty}\sum_{n=1}^{\infty} C_{nm}\sin(\frac{n\pi}{a}x)\sin(\frac{m\pi}{b}y)\sinh\left(\sqrt{(\frac{n\pi}{a})^2 + (\frac{m\pi}{b})^2}\ c\right) = f(x, y). \tag{150}$$

The Fourier coefficients are then given by

$$C_{nm}\sinh\left(\sqrt{(\frac{n\pi}{a})^2 + (\frac{m\pi}{b})^2}\ c\right)$$
$$= \frac{4}{ab}\int_0^a\int_0^b \sin(\frac{n\pi}{a}x)\sin(\frac{m\pi}{a}y)f(x, y)\, dx\, dy. \tag{151}$$

Consequently, the solution of the Laplace's equation is given by (149) with C_{nm} defined by (151).

The method of separation of variables will be illustrated by discussing the following examples.

Example 5. Use the method of separation of variables to solve the boundary value problem

$$\text{PDE} \qquad u_{xx} + u_{yy} + u_{zz} = 0, \, 0 < x, y, z < \pi$$

$$\text{B.C} \qquad u(0, y, z) = 0, \, u(\pi, y, z) = 0 \tag{152}$$

$$u(x, 0, z) = 0, \, u(x, \pi, z) = 0$$

$$u(x, y, 0) = 0, \, u(x, y, \pi) = \sinh(\sqrt{2}\pi) \sin x \sin y$$

Solution.

Proceeding as discussed above, we obtain the second order ordinary differential equations

$$F''(x) + \lambda^2 F(x) = 0, \tag{153}$$
$$G''(y) + \mu^2 G(y) = 0, \tag{154}$$
$$H''(z) - (\lambda^2 + \mu^2) H(z) = 0, \tag{155}$$

so that

$$F(x) = A \cos(\lambda x) + B \sin(\lambda x), \tag{156}$$
$$G(y) = \alpha \cos(\mu y) + \beta \sin(\mu y), \tag{157}$$
$$H(z) = \gamma \cosh(\nu z) + \delta \sinh(\nu z), \tag{158}$$

where

$$\nu = \sqrt{\lambda^2 + \mu^2}, \tag{159}$$

and $A, B, \alpha, \beta, \gamma$, and δ are constants.

Using the homogeneous boundary conditions gives

$$A = 0, \, \lambda_n = n, n = 1, 2, 3 \cdots, \tag{160}$$
$$\alpha = 0, \, \mu_m = m, m = 1, 2, 3, \cdots, \tag{161}$$
$$\gamma = 0, \, \nu_{nm} = \sqrt{n^2 + m^2}, \tag{162}$$

so that

$$F_n(x) = \sin(nx), n = 1, 2, 3 \cdots, \tag{163}$$
$$G_m(y) = \sin(my), m = 1, 2, 3 \cdots, \tag{164}$$
$$H_{nm}(z) = \sinh\left(\sqrt{n^2 + m^2} \, z\right). \tag{165}$$

Consequently, we obtain the fundamental solutions

$$u_n(x, y, z) = \sin(nx) \sin(my) \sinh\left(\sqrt{n^2 + m^2} \, z\right), \, n, m = 1, 2, \cdots, \tag{166}$$

so that the general solution is

$$u(x, y, z) = \sum_{m=1}^{\infty} \sum_{n=1}^{\infty} C_{nm} \sin(nx) \sin(my) \sinh\left(\sqrt{n^2 + m^2}\, z\right). \qquad (167)$$

To determine the constants C_{nm}, we use the inhomogeneous boundary condition $u(x, y, \pi) = \sin x \sin y \sinh(\sqrt{2}\pi)$ to find

$$\sum_{m=1}^{\infty} \sum_{n=1}^{\infty} C_{nm} \sin(nx) \sin(my) \sinh\left(\sqrt{n^2 + m^2}\, \pi\right) = \sin x \sin y \sinh(\sqrt{2}\pi). \qquad (168)$$

Expanding the double series and equating the coefficients of like terms on both sides we find

$$C_{11} = 1, \text{ for } n = 1, m = 1, \ C_{nm} = 0, \text{ for } n \neq 1, m \neq 1. \qquad (169)$$

This gives the exact solution

$$u(x, y, z) = \sin x \sin y \sinh \sqrt{2}z, \qquad (170)$$

obtained upon substituting the (169) into (167).

Example 6. Use the method of separation of variables to solve the boundary value problem

$$\text{PDE} \quad u_{xx} + u_{yy} + u_{zz} = 0, \ 0 < x, y, z < \pi$$

$$\text{B.C} \quad u(0, y, z) = 0, \ u(\pi, y, z) = 0$$

$$u(x, 0, z) = 0, \ u(x, \pi, z) = 0 \qquad (171)$$

$$u(x, y, 0) = \sinh \sqrt{2}\pi \sin x \sin y, \ u(x, y, \pi) = 0$$

Solution.

Proceeding as before, we obtain

$$\begin{aligned}
F(x) &= A\cos(\lambda x) + B\sin(\lambda x), & (172) \\
G(y) &= \alpha\cos(\mu y) + \beta\sin(\mu y), & (173) \\
H(z) &= \gamma\cosh(\nu z) + \delta\sinh(\nu z), & (174)
\end{aligned}$$

where

$$\nu = \sqrt{\lambda^2 + \mu^2}, \qquad (175)$$

and $A, B, \alpha, \beta, \gamma,$ and δ are constants.

Because $H(\pi) = 0$, it is useful to rewrite (174) into the equivalent form

$$H(z) = C\sinh\nu(K - z), \qquad (176)$$

where K and C are constants. Using the homogeneous boundary conditions lead to

$$A \;=\; 0, \; \lambda_n = n, n = 1, 2, 3 \cdots, \tag{177}$$

$$\alpha \;=\; 0, \; \mu_m = m, m = 1, 2, 3, \cdots \tag{178}$$

$$K \;=\; \pi, \; \nu_{nm} = \sqrt{n^2 + m^2}, \tag{179}$$

so that

$$F_n(x) \;=\; \sin(nx), n = 1, 2, 3 \cdots, \tag{180}$$

$$G_m(y) \;=\; \sin(my), m = 1, 2, 3 \cdots, \tag{181}$$

$$H_{nm}(z) \;=\; \sinh\left(\sqrt{n^2 + m^2}(\pi - z)\right). \tag{182}$$

This gives

$$u_n = \sin(nx)\sin(my)\sinh\left(\sqrt{n^2 + m^2}(\pi - z)\right), n, m = 1, 2, \cdots. \tag{183}$$

Using the superposition principle gives

$$u(x, y, z) = \sum_{m=1}^{\infty}\sum_{n=1}^{\infty} C_{nm}\sin(nx)\sin(my)\sinh\left(\sqrt{n^2 + m^2}(\pi - z)\right). \tag{184}$$

To determine the constants C_{nm}, we substitute the inhomogeneous boundary condition $u(x, y, 0) = \sinh(\sqrt{2}\pi)\sin x \sin y$ into (184), and equate the coefficients of like terms on both sides, we find

$$C_{11} = 1, \text{ for } n = 1, m = 1, \; C_{nm} = 0, \text{ for } n \neq 1, m \neq 1. \tag{185}$$

This gives the exact solution

$$u(x, y, z) = \sin x \sin y \sinh \sqrt{2}(\pi - z), \tag{186}$$

obtained by combining (185) and (184).

Example 7. Use the method of separation of variables to solve the boundary value problem

$$\text{PDE} \qquad u_{xx} + u_{yy} + u_{zz} = 0, \, 0 < x, y, z < \pi$$

$$\text{B.C} \qquad u(0, y, z) = 0, \; u(\pi, y, z) = 0$$

$$u(x, 0, z) = 0, \; u(x, \pi, z) = 0 \tag{187}$$

$$u(x, y, 0) = 1, \; u(x, y, \pi) = 0$$

Because $u(x, y, 0) = 1$, it is necessary to use Fourier coefficients to determine the constants.

Solution.

Using the boundary conditions and following the last example we obtain

$$A = 0, \lambda_n = n, n = 1, 2, 3 \cdots, \tag{188}$$

$$\alpha = 0, \mu_m = m, m = 1, 2, 3, \cdots \tag{189}$$

$$K = \pi, \nu_{nm} = \sqrt{n^2 + m^2}, C \neq 0, \tag{190}$$

so that

$$F_n(x) = \sin(nx), n = 1, 2, 3 \cdots, \tag{191}$$

$$G_m(y) = \sin(my), m = 1, 2, 3 \cdots, \tag{192}$$

$$H_{nm}(z) = \sinh\left(\sqrt{n^2 + m^2}(\pi - z)\right). \tag{193}$$

Using the superposition principle gives

$$u(x, y, z) = \sum_{m=1}^{\infty} \sum_{n=1}^{\infty} C_{nm} \sin(nx) \sin(my) \sinh\left(\sqrt{n^2 + m^2}(\pi - z)\right). \tag{194}$$

To determine the constants C_{nm}, we use the inhomogeneous boundary condition $u(x, y, 0) = 1$ in (194) to find

$$\sum_{m=1}^{\infty} \sum_{n=1}^{\infty} C_{nm} \sin(nx) \sin(my) \sinh\left(\sqrt{n^2 + m)^2}\, \pi\right) = 1. \tag{195}$$

This gives

$$C_{nm} \sinh(\sqrt{n^2 + m^2}\, \pi) = \frac{4}{\pi^2} \int_0^{\pi} \int_0^{\pi} \sin(nx) \sin(my)\, dx\, dy,$$

$$= \begin{cases} \dfrac{8}{\pi^2 nm} & \text{for } n, m \text{ odd} \\[2mm] 0 & \text{for } n, m \text{ even} \end{cases} \tag{196}$$

The solution is given by

$$u(x, y, z) = \sum_{r=0}^{\infty} \sum_{s=0}^{\infty} \frac{8}{\pi^2(2r+1)(2s+1)} \sin(2r+1)x \sin(2s+1)y$$

$$\times \sinh\sqrt{(2r+1)^2 + (2s+1)^2}(\pi - z). \tag{197}$$

Example 8. Use the method of separation of variables to solve the boundary value problem

$$\begin{aligned} \text{PDE} \quad u_{xx} + u_{yy} + u_{zz} &= 0, \ 0 < x, y, z < \pi \\ \text{B.C} \quad u_x(0, y, z) &= 0, \ u_x(\pi, y, z) = 0 \\ u_y(x, 0, z) &= 0, \ u_y(x, \pi, z) = 0 \\ u_z(x, y, 0) &= 0, \ u_z(x, y, \pi) = \sqrt{2} \sinh(\sqrt{2}\pi) \cos x \cos y \end{aligned} \tag{198}$$

Solution.

Note that the boundary conditions are inhomogeneous, therefore the equation is a Neumann problem in three dimensions. Following the techniques used before gives

$$
\begin{align}
F_n(x) &= \cos(nx), n = 0, 1, 2, 3 \cdots, & (199)\\
G_m(y) &= \cos(my), m = 0, 1, 2, 3 \cdots, & (200)\\
H_{nm}(z) &= \cosh\left(\sqrt{n^2 + m^2}\, z\right). & (201)
\end{align}
$$

Using the superposition principle gives

$$
u(x, y, z) = C_0 + \sum_{m=1}^{\infty} \sum_{n=1}^{\infty} C_{nm} \sin(nx) \sin(my) \sinh\left(\sqrt{n^2 + m^2}\, z\right). \qquad (202)
$$

To determine the constants C_{nm}, we use the inhomogeneous boundary condition $u(x, y, \pi) = \sqrt{2}\sinh(\sqrt{2}\pi) \cos x \cos y$ to obtain

$$
C_{11} = 1, \text{for } n = 1, m = 1, C_{nm} = 0, \text{for } n \neq 1, m \neq 1. \qquad (203)
$$

Accordingly, the solution is given in the form

$$
u(x, y, z) = C_0 + \cos x \cos y \cosh(\sqrt{2}z). \qquad (204)
$$

Exercises 7.3.2
Use the method of separation of variables to solve the following three dimensional Laplace's equations:

1. $u_{xx} + y_{yy} + u_{zz} = 0, 0 < x, y, z < \pi$
    $u(0, y, z) = u(\pi, y, z) = 0$
    $u(x, 0, z) = u(x, \pi, z) = 0$
    $u(x, y, 0) = 0, u(x, y, \pi) = \sinh(\sqrt{5}\pi) \sin x \sin(2y)$

2. $u_{xx} + y_{yy} + u_{zz} = 0, 0 < x, y, z < \pi$
    $u(0, y, z) = u(\pi, y, z) = 0$
    $u(x, 0, z) = u(x, \pi, z) = 0$
    $u(x, y, 0) = 0, u(x, y, \pi) = \sinh(10\pi) \sin(6x) \sin(8y)$

3. $u_{xx} + y_{yy} + u_{zz} = 0, 0 < x, y, z < \pi$
    $u(0, y, z) = u(\pi, y, z) = 0$
    $u(x, 0, z) = u(x, \pi, z) = 0$
    $u(x, y, 0) = 0, u(x, y, \pi) = \sinh(\sqrt{8}\pi) \sin(2x) \sin(2y)$

4. $u_{xx} + y_{yy} + u_{zz} = 0, 0 < x, y, z < \pi$
    $u(0, y, z) = u(\pi, y, z) = 0$
    $u(x, 0, z) = u(x, \pi, z) = 0$
    $u(x, y, 0) = \sinh(\sqrt{5}\pi) \sin x \sin(2y), u(x, y, \pi) = 0$

5. $u_{xx} + y_{yy} + u_{zz} = 0,\ 0 < x, y, z < \pi$
   $u(0, y, z) = u(\pi, y, z) = 0$
   $u(x, 0, z) = u(x, \pi, z) = 0$
   $u(x, y, 0) = \sinh(5\pi)\sin(3x)\sin(4y),\ u(x, y, \pi) = 0$

6. $u_{xx} + y_{yy} + u_{zz} = 0,\ 0 < x, y, z < \pi$
   $u(0, y, z) = u(\pi, y, z) = 0$
   $u(x, 0, z) = u(x, \pi, z) = 0$
   $u(x, y, 0) = \sinh(13\pi)\sin(5x)\sin(12y),\ u(x, y, \pi) = 0$

7. $u_{xx} + y_{yy} + u_{zz} = 0,\ 0 < x, y, z < \pi$
   $u_x(0, y, z) = u_x(\pi, y, z) = 0$
   $u_y(x, 0, z) = u_y(x, \pi, z) = 0$
   $u_z(x, y, 0) = 0,\ u_z(x, y, \pi) = \sqrt{5}\sinh(\sqrt{5}\pi)\cos x \cos(2y)$

8. $u_{xx} + y_{yy} + u_{zz} = 0,\ 0 < x, y, z < \pi$
   $u_x(0, y, z) = u_x(\pi, y, z) = 0$
   $u_y(x, 0, z) = u_y(x, \pi, z) = 0$
   $u_z(x, y, 0) = 0,\ u_z(x, y, \pi) = 13\sinh(13\pi)\cos(5x)\cos(12y)$

9. $u_{xx} + y_{yy} + u_{zz} = 0,\ 0 < x, y, z < \pi$
   $u_x(0, y, z) = u_x(\pi, y, z) = 0$
   $u_y(x, 0, z) = u_y(x, \pi, z) = 0$
   $u_z(x, y, 0) = -5\sinh(5\pi)\cos(3x)\cos(4y),\ u_z(x, y, \pi) = 0$

10. $u_{xx} + y_{yy} + u_{zz} = 0,\ 0 < x, y, z < \pi$
    $u_x(0, y, z) = u_x(\pi, y, z) = 0$
    $u_y(x, 0, z) = u_y(x, \pi, z) = 0$
    $u_z(x, y, 0) = -\sqrt{8}\sinh(\sqrt{8}\pi)\cos(2x)\cos(2y),\ u_z(x, y, \pi) = 0$

11. $u_{xx} + y_{yy} + u_{zz} = 0,\ 0 < x, y, z < \pi$
    $u(0, y, z) = u(\pi, y, z) = 0$
    $u(x, 0, z) = u(x, \pi, z) = 0$
    $u(x, y, 0) = \sin 8x \sin 15y,\ u(x, y, \pi) = \cosh 17\pi \sin 8x \sin 15y$

12. $u_{xx} + y_{yy} + u_{zz} = 0,\ 0 < x, y, z < \pi$
    $u(0, y, z) = u(\pi, y, z) = 0$
    $u(x, 0, z) = u(x, \pi, z) = \sin 3x \sinh 5z$
    $u(x, y, 0) = 0,\ u(x, y, \pi) = \sinh 5\pi \sin 3x \cos 4y$

7.4 Laplace's Equation in Polar Coordinates

In the last two sections we have studied the two dimensional Dirichlet problem for a rectangle governed by the equation

$$u_{xx} + u_{yy} = 0. \tag{205}$$

The boundary conditions are specified on the boundary of a rectangle. However, if the domain of the solution $u(x, y)$ is a disc or a circular annulus, it is useful to study the two dimensional Laplace's equation in polar coordinates. It is well known that the polar coordinates (r, θ) of any point are related to its Cartesian coordinates (x, y) by the familiar formulas

$$x = r\cos\theta, \ y = r\sin\theta. \tag{206}$$

Using these formulas along with the chain rule, the Laplace's equation for a circular domain becomes

$$\frac{\partial^2 u}{\partial r^2} + \frac{1}{r}\frac{\partial u}{\partial r} + \frac{1}{r^2}\frac{\partial^2 u}{\partial \theta^2} = 0, \ 0 < r < a, \ 0 \le \theta \le 2\pi. \tag{207}$$

Recall that Laplace's equation is a boundary value problem. Accordingly, the boundary condition that describes the solution $u(r, \theta)$ at the circumference of a circular domain should be specified. Therefore, we set the boundary condition for a disc by

$$u(a, \theta) \ = \ f(\theta), \ 0 < r < a. \tag{208}$$

7.4.1   Laplace's Equation For A Disc

In this part, we will study Laplace's equation for a circular disc of radius a where the top and the bottom faces of the disc are insulated. The boundary condition at the circular edge is specified. The phenomenon that the temperature reaches a steady state inside the disc is governed by the Laplace equation in polar coordinates, and expressed by the boundary value problem

$$\begin{aligned} &\text{PDE} & &\frac{\partial^2 u}{\partial r^2} + \frac{1}{r}\frac{\partial u}{\partial r} + \frac{1}{r^2}\frac{\partial^2 u}{\partial \theta^2} = 0, \ 0 < r < a, \ 0 \le \theta \le 2\pi \\ &\text{BC} & &u(a, \theta) = f(\theta) \end{aligned} \tag{209}$$

Two important facts should be taken into consideration, namely:
1. the solution $u(r, \theta)$ should be **bounded** at $r = 0$.
2. the solution $u(r, \theta)$ should be **periodic** with period 2π. This means that $u(r, \theta + 2\pi) = u(r, \theta)$.
    It is interesting to point out that these facts are not boundary conditions. The solution $u(r, \theta)$ being bounded at $r = 0$ and the periodicity of the solution $u(r, \theta)$ play a major role in determining the solution. In addition, the boundary condition $f(\theta)$ must be periodic with period 2π. Also note that the coefficients $\frac{1}{r}$ and $\frac{1}{r^2}$ become an infinite at $r = 0$, hence $r = 0$ is excluded from the domain of the solution.
    To solve (209), we use the method of separation of variables, hence we set $u(r, \theta)$ in the form

$$u(r, \theta) = F(r)G(\theta). \tag{210}$$

This gives

$$
\begin{array}{rcl}
u_r(r, \theta) & = & F'(r)G(\theta), \\
u_{rr}(r, \theta) & = & F''(r)G(\theta), \\
u_{\theta\theta}(r, \theta) & = & F(r)G''(\theta).
\end{array}
\tag{211}
$$

Substituting (211) into the PDE of (209) gives

$$
F''(r)G(\theta) + \frac{1}{r}F'(r)G(\theta) + \frac{1}{r^2}F(r)G''(\theta) = 0.
\tag{212}
$$

Dividing both sides of (212) by $F(r)G(\theta)$ yields

$$
\frac{G''(\theta)}{G(\theta)} = -\left(r^2 \frac{F''(r)}{F(r)} + r\frac{F'(r)}{F(r)} \right).
\tag{213}
$$

It is well known now that the equality holds only if each side is equal to the same constant. Therefore, we set

$$
\frac{G''(\theta)}{G(\theta)} = -\lambda^2,
\tag{214}
$$

so that

$$
-\left(r^2 \frac{F''(r)}{F(r)} + r\frac{F'(r)}{F(r)} \right) = -\lambda^2.
\tag{215}
$$

This gives the second order differential equations

$$
G''(\theta) + \lambda^2 G(\theta) = 0,
\tag{216}
$$

and

$$
r^2 F''(r) + rF'(r) - \lambda^2 F(r) = 0.
\tag{217}
$$

Equation (216) gives the solution

$$
G(\theta) = A\cos(\lambda\theta) + B\sin(\lambda\theta).
\tag{218}
$$

As mentioned earlier, $u(r, \theta)$ is periodic, and hence, so is $G(\theta)$. Accordingly, the periodicity implies that

$$
\lambda_n = n, \; n = 0, 1, 2, \cdots.
\tag{219}
$$

Notice that $n = 0$ is included in our values for n since it results in a constant, which is also periodic. In view of (219), equation (218) becomes

$$
G_n(\theta) = A_n\cos(n\theta) + B_n\sin(n\theta), \; n = 0, 1, 2, \cdots.
\tag{220}
$$

Substituting $\lambda_n = n$ into (217) gives

$$
r^2 F''(r) + rF'(r) - n^2 F(r) = 0.
\tag{221}
$$

Equation (221) is the well-known second order Euler ordinary differential equation with general solution given by

$$F_0(r) = C_0 + D_0 \ln r, \text{ for } n = 0, \tag{222}$$

and

$$F_n(r) = C_n r^n + D_n r^{-n}, \text{ for } n = 1, 2, 3, \cdots. \tag{223}$$

It is important to recall that $u(r, \theta)$, and hence $F(r)$ should be bounded at $r = 0$. However, each of the components $\ln r$ in (222) and r^{-n} in (223) approaches infinity at $r = 0$. This means that we must set

$$D_0 = D_n = 0, \tag{224}$$

so that $u(r, \theta)$ becomes bounded at $r = 0$. Combining (222) – (224) gives the general solutions of Euler equation (221) by

$$F_n(r) = C_n r^n, \, n = 0, 1, 2, \cdots. \tag{225}$$

Using the superposition principle, and combining the results obtained above, we find

$$u(r, \theta) = C_0 + \sum_{n=1}^{\infty} r^n \left(A_n \cos(n\theta) + B_n \sin(n\theta) \right), \tag{226}$$

which is usually written in a more convenient equivalent form by

$$u(r, \theta) = \frac{a_0}{2} + \sum_{n=1}^{\infty} (\frac{r}{a})^n \left(a_n \cos(n\theta) + b_n \sin(n\theta) \right), \tag{227}$$

simply by modifying the constants.

To determine the constants $a_n, n \geq 0$ and $b_n, n \geq 1$, we use the boundary condition $u(a, \theta) = f(\theta)$. Setting $r = a$ in (227) gives

$$\frac{a_0}{2} + \sum_{n=1}^{\infty} \left(a_n \cos(n\theta) + b_n \sin(n\theta) \right) = f(\theta). \tag{228}$$

It is obvious that a_n and b_n are Fourier coefficients, and therefore can be determined by

$$a_n = \frac{1}{\pi} \int_{-\pi}^{\pi} f(\theta) \cos(n\theta) \, d\theta, \, n = 0, 1, 2, \cdots, \tag{229}$$

and

$$b_n = \frac{1}{\pi} \int_{-\pi}^{\pi} f(\theta) \sin(n\theta) \, d\theta, \, n = 1, 2, , \cdots. \tag{230}$$

To determine the constants a_n and b_n, Appendix A can be used to evaluate the integrals in (229) and in (230).

However, the constants a_n and b_n can also be determined by equating the coefficients of like terms if the boundary condition is given in terms of sines and cosines as discussed in previous chapters.

The technique discussed above will be illustrated by discussing the following examples.

Example 1. Use the method of separation of variables to solve the Dirichlet problem

PDE $\qquad \dfrac{\partial^2 u}{\partial r^2} + \dfrac{1}{r}\dfrac{\partial u}{\partial r} + \dfrac{1}{r^2}\dfrac{\partial^2 u}{\partial \theta^2} = 0,\ 0 < r < 1,\ 0 \leq \theta \leq 2\pi$

$$\tag{231}$$

BC $\qquad u(1,\theta) = \cos^2\theta$

Solution.

Using the method of separation of variables gives

$$G^{''}(\theta) + \lambda^2 G(\theta) = 0, \tag{232}$$

and

$$r^2 F^{''}(r) + r F^{'}(r) - \lambda^2 F(r) = 0. \tag{233}$$

The solution of (232) is given by

$$G(\theta) = A\cos(\lambda\theta) + B\sin(\lambda\theta). \tag{234}$$

The periodicity of $u(r,\theta)$ implies that

$$\lambda_n = n,\ n = 0, 1, 2, \cdots. \tag{235}$$

Equation (234) becomes

$$G_n(\theta) = A_n \cos(n\theta) + B_n \sin(n\theta),\ n = 0, 1, 2, \cdots. \tag{236}$$

Substituting $\lambda_n = n$ into (233) gives

$$r^2 F^{''}(r) + r F^{'}(r) - n^2 F(r) = 0, \tag{237}$$

a second order Euler differential equation with the general solution given by

$$F_0(r) = C_0 + D_0 \ln r,\ n = 0, \tag{238}$$

and

$$F_n(r) = C_n r^n + D_n r^{-n},\ n = 1, 2, 3, \cdots. \tag{239}$$

The essential fact that $u(r,\theta)$, and hence $F(r)$ should be bounded at $r = 0$ means that we must set

$$D_0 = D_n = 0, \tag{240}$$

This gives the general solutions of (237) by

$$F_n(r) = C_n r^n, \ n = 0, 1, 2, \cdots. \tag{241}$$

Using the superposition principle and proceeding as before, we find

$$u(r, \theta) = \frac{a_0}{2} + \sum_{n=1}^{\infty} r^n \left(a_n \cos(n\theta) + b_n \sin(n\theta) \right). \tag{242}$$

To determine the constants $a_n, n \geq 0$ and $b_n, n \geq 1$, set $r = 1$ in (242) and using the boundary condition, we obtain

$$\frac{a_0}{2} + \sum_{n=1}^{\infty} \left(a_n \cos(n\theta) + b_n \sin(n\theta) \right) = \frac{1}{2} + \frac{1}{2}\cos(2\theta), \tag{243}$$

where the identity

$$\cos^2\theta = \frac{1}{2} + \frac{1}{2}\cos(2\theta), \tag{244}$$

was used. Expanding the series in (243) and equating the coefficients of like terms in both sides we find

$$a_0 = 1, \tag{245}$$

$$a_2 = \frac{1}{2}, \ n = 2, \tag{246}$$

and

$$\begin{aligned} a_n &= 0, \text{for } n \neq 0, 2, \\ b_n &= 0, \text{for } n = 1, 2, 3, \cdots. \end{aligned} \tag{247}$$

This gives the particular solution by

$$u(r, \theta) = \frac{1}{2} + \frac{1}{2}r^2 \cos(2\theta). \tag{248}$$

Example 2. Use the method of separation of variables to solve the Dirichlet problem

$$\begin{aligned} \text{PDE} \quad & \frac{\partial^2 u}{\partial r^2} + \frac{1}{r}\frac{\partial u}{\partial r} + \frac{1}{r^2}\frac{\partial^2 u}{\partial \theta^2} = 0, \ 0 < r < 1, 0 \leq \theta \leq 2\pi \\ \text{BC} \quad & u(1, \theta) = 2\sin^2\theta + \sin\theta \end{aligned} \tag{249}$$

Solution.

We first set $u(r, \theta)$ in the form

$$u(r, \theta) = F(r)G(\theta). \tag{250}$$

Substituting (250) into the PDE of (249) and proceeding as before we obtain

$$u(r,\theta) = \frac{a_0}{2} + \sum_{n=1}^{\infty} r^n \left(a_n \cos(n\theta) + b_n \sin(n\theta)\right). \tag{251}$$

To determine the constants $a_n, n \geq 0$ and $b_n, n \geq 1$, we substitute $r = 1$ in (251) and we use the boundary condition, we obtain

$$\frac{a_0}{2} + \sum_{n=1}^{\infty} \left(a_n \cos(n\theta) + b_n \sin(n\theta)\right) = 1 - \cos(2\theta) + \sin\theta, \tag{252}$$

where the identity

$$\sin^2\theta = \frac{1}{2} - \frac{1}{2}\cos(2\theta), \tag{253}$$

was used. Expanding the series in (252) and equating the coefficients of like terms in both sides we find

$$a_0 = 2, a_2 = -1, b_1 = 1, \tag{254}$$

where each of the remaining coefficients is zero. Hence, the particular solution is given by

$$u(r,\theta) = 1 - r^2 \cos(2\theta) + r\sin\theta. \tag{255}$$

Example 3. Use the method of separation of variables to solve the Dirichlet problem

$$
\begin{array}{ll}
\text{PDE} & \dfrac{\partial^2 u}{\partial r^2} + \dfrac{1}{r}\dfrac{\partial u}{\partial r} + \dfrac{1}{r^2}\dfrac{\partial^2 u}{\partial \theta^2} = 0, \ 0 < r < 1, \ 0 \leq \theta \leq 2\pi \\[2mm]
\text{BC} & u(1,\theta) = \mid 2\theta \mid
\end{array} \tag{256}
$$

Solution.

We first set $u(r,\theta)$ in the form

$$u(r,\theta) = F(r)G(\theta). \tag{257}$$

Substituting (257) into the PDE of (256) and proceeding as before we obtain

$$u(r,\theta) = \frac{a_0}{2} + \sum_{n=1}^{\infty} r^n \left(a_n \cos(n\theta) + b_n \sin(n\theta)\right). \tag{258}$$

To determine the constants $a_n, n \geq 0$ and $b_n, n \geq 1$, we substitute $r = 1$ in (258) and by using the boundary condition, we obtain

$$\frac{a_0}{2} + \sum_{n=1}^{\infty} \left(a_n \cos(n\theta) + b_n \sin(n\theta)\right) = \mid 2\theta \mid . \tag{259}$$

The Fourier coefficients can be determined by using the formulas (229) and (230). For a_0, we find

$$
\begin{aligned}
a_0 &= \frac{1}{\pi} \int_{-\pi}^{\pi} |\, 2\theta \,| \, d\theta, \\
&= \frac{1}{\pi} \left(\int_{-\pi}^{0} -2\theta d\theta + \int_{0}^{\pi} 2\theta d\theta \right), \\
&= 2\pi,
\end{aligned}
\tag{260}
$$

$$
\begin{aligned}
a_n &= \frac{1}{\pi} \int_{-\pi}^{\pi} |\, 2\theta \,| \cos(n\theta) \, d\theta, \\
&= -\frac{8}{\pi n^2}, \text{ for } n \text{ odd, 0 otherwise,}
\end{aligned}
\tag{261}
$$

and

$$
\begin{aligned}
b_n &= \frac{1}{\pi} \int_{-\pi}^{\pi} |\, 2\theta) \,| \sin(n\theta) \, d\theta, \\
&= 0.
\end{aligned}
\tag{262}
$$

Therefore, the solution is given by

$$
u(r, \theta) = \pi - \frac{8}{\pi} \sum_{k=0}^{\infty} \frac{1}{(2k+1)^2} r^{2k+1} \cos(2k+1)\theta.
\tag{263}
$$

Example 4. Use the method of separation of variables to solve the Dirichlet problem

$$
\begin{array}{ll}
\text{PDE} & \dfrac{\partial^2 u}{\partial r^2} + \dfrac{1}{r}\dfrac{\partial u}{\partial r} + \dfrac{1}{r^2}\dfrac{\partial^2 u}{\partial \theta^2} = 0,\ 0 < r < 1,\ 0 \leq \theta \leq 2\pi \\[2mm]
\text{BC} & u_r(1, \theta) = 2\cos(2\theta)
\end{array}
\tag{264}
$$

Solution.

Note that a Neumann boundary condition is given. Recall that the solution will be determined up to an arbitrary constant, hence the solution is not unique. We first set $u(r, \theta)$ in the form

$$
u(r, \theta) = F(r)G(\theta).
\tag{265}
$$

Substituting (265) into the PDE of (264) and proceeding as before we obtain

$$
u(r, \theta) = \frac{a_0}{2} + \sum_{n=1}^{\infty} r^n \left(a_n \cos(n\theta) + b_n \sin(n\theta) \right).
\tag{266}
$$

To determine the constants $a_n, n \geq 0$ and $b_n, n \geq 1$, we set $r = 1$ in the derivative of (266) and by using the boundary condition, we obtain

$$
\sum_{n=1}^{\infty} n \left(a_n \cos(n\theta) + b_n \sin(n\theta) \right) = 2\cos(2\theta),
\tag{267}
$$

Expanding the series in (267) and equating the coefficients of like terms in both sides we find

$$a_2 = 1, n = 2, \tag{268}$$

where each of the remaining coefficients is zero. Hence, the particular solution is given by

$$u(r, \theta) = C_0 + r^2 \cos(2\theta), \tag{269}$$

where C_0 is an arbitrary constant.

Exercises 7.4.1

Use the method of separation of variables to solve the following Laplace's equations:

1. $\dfrac{\partial^2 u}{\partial r^2} + \dfrac{1}{r}\dfrac{\partial u}{\partial r} + \dfrac{1}{r^2}\dfrac{\partial^2 u}{\partial \theta^2} = 0,\ 0 < r < 1,\ 0 \le \theta \le 2\pi$

   $u(1, \theta) = 2 + 3\sin\theta + 4\cos\theta$

2. $\dfrac{\partial^2 u}{\partial r^2} + \dfrac{1}{r}\dfrac{\partial u}{\partial r} + \dfrac{1}{r^2}\dfrac{\partial^2 u}{\partial \theta^2} = 0,\ 0 < r < 1,\ 0 \le \theta \le 2\pi$

   $u(1, \theta) = 2\cos^2(2\theta)$

3. $\dfrac{\partial^2 u}{\partial r^2} + \dfrac{1}{r}\dfrac{\partial u}{\partial r} + \dfrac{1}{r^2}\dfrac{\partial^2 u}{\partial \theta^2} = 0,\ 0 < r < 1,\ 0 \le \theta \le 2\pi$

   $u(1, \theta) = 2\sin^2(3\theta)$

4. $\dfrac{\partial^2 u}{\partial r^2} + \dfrac{1}{r}\dfrac{\partial u}{\partial r} + \dfrac{1}{r^2}\dfrac{\partial^2 u}{\partial \theta^2} = 0,\ 0 < r < 1,\ 0 \le \theta \le 2\pi$

   $u(1, \theta) = \sin(2\theta) + \cos(2\theta)$

5. $\dfrac{\partial^2 u}{\partial r^2} + \dfrac{1}{r}\dfrac{\partial u}{\partial r} + \dfrac{1}{r^2}\dfrac{\partial^2 u}{\partial \theta^2} = 0,\ 0 < r < 1,\ 0 \le \theta \le 2\pi$

   $u_r(1, \theta) = 8\sin 4\theta$

6. $\dfrac{\partial^2 u}{\partial r^2} + \dfrac{1}{r}\dfrac{\partial u}{\partial r} + \dfrac{1}{r^2}\dfrac{\partial^2 u}{\partial \theta^2} = 0,\ 0 < r < 1,\ 0 \le \theta \le 2\pi$

   $u_r(1, \theta) = \sin\theta - \cos\theta$

7. $\dfrac{\partial^2 u}{\partial r^2} + \dfrac{1}{r}\dfrac{\partial u}{\partial r} + \dfrac{1}{r^2}\dfrac{\partial^2 u}{\partial \theta^2} = 0,\ 0 < r < 1,\ 0 \le \theta \le 2\pi$

   $u_r(1, \theta) = 2\sin(2\theta)$

8. $\dfrac{\partial^2 u}{\partial r^2} + \dfrac{1}{r}\dfrac{\partial u}{\partial r} + \dfrac{1}{r^2}\dfrac{\partial^2 u}{\partial \theta^2} = 0,\ 0 < r < 1,\ 0 \le \theta \le 2\pi$

   $u_r(1, \theta) = 3\cos(3\theta)$

9. $\dfrac{\partial^2 u}{\partial r^2} + \dfrac{1}{r}\dfrac{\partial u}{\partial r} + \dfrac{1}{r^2}\dfrac{\partial^2 u}{\partial \theta^2} = 0,\ 0 < r < 1,\ 0 \le \theta \le 2\pi$

$u_r(1,\theta) = 2\sin(2\theta) + 3\cos(3\theta)$

10. $\dfrac{\partial^2 u}{\partial r^2} + \dfrac{1}{r}\dfrac{\partial u}{\partial r} + \dfrac{1}{r^2}\dfrac{\partial^2 u}{\partial \theta^2} = 0,\ 0 < r < 1,\ 0 \le \theta \le 2\pi$

$u_r(1,\theta) = 2\cos(2\theta)$

7.4.2   Laplace's Equation For An Annulus

In this part, we will study the Laplace's equation in the domain lying between the concentric circles K_1 and K_2 of radii a and b where $0 < a < b$. In other words, the domain of the solution includes two circular edges, an interior boundary with radius $r_1 = a$ and an exterior boundary with radius $r_2 = b$. As a result, two boundary conditions should be specified in this case.

The Laplace's equation for an annulus is therefore defined by the boundary value problem

$$\text{PDE} \qquad \frac{\partial^2 u}{\partial r^2} + \frac{1}{r}\frac{\partial u}{\partial r} + \frac{1}{r^2}\frac{\partial^2 u}{\partial \theta^2} = 0,\ 0 < a < r < b,\ 0 \le \theta \le 2\pi \tag{270}$$

$$\text{BC} \qquad u(a,\theta) = f(\theta),\ u(b,\theta) = g(\theta)$$

It is essential to note that $u(r,\theta)$ must be periodic with period 2π. This means that $u(r,\theta + 2\pi) = u(r,\theta)$. Accordingly, the boundary conditions $f(\theta)$ and $g(\theta)$ must be periodic with period 2π. Moreover, we do not claim that $u(r,\theta)$ is bounded at $r = 0$. This is due to the fact that $a < r < b$, and r will never be 0.

We begin our analysis by seeking a solution expressed as a product of two distinct functions, where each function depends on one variable only. The method of separation of variables suggests that

$$u(r,\theta) = F(r)G(\theta). \tag{271}$$

Substituting in (270) gives

$$\frac{G''(\theta)}{G(\theta)} = -\left(r^2\frac{F''(r)}{F(r)} + r\frac{F'(r)}{F(r)} \right) = -\lambda^2. \tag{272}$$

This gives the differential equations

$$G''(\theta) + \lambda^2 G(\theta) = 0, \tag{273}$$

and

$$r^2 F''(r) + r F'(r) - \lambda^2 F(r) = 0. \tag{274}$$

Equation (273) gives the solution

$$G(\theta) = A\cos(\lambda\theta) + B\sin(\lambda\theta). \qquad (275)$$

As mentioned earlier, $u(r,\theta)$ is periodic, and hence, so is $G(\theta)$. Accordingly, the periodicity implies that

$$\lambda_n = n, \; n = 0, 1, 2, \cdots. \qquad (276)$$

Notice that $n = 0$ is included in our values for n since it results in a constant, which is also periodic. In view of (276), equation (275) becomes

$$G_n(\theta) = A_n\cos(n\theta) + B_n\sin(n\theta), \; n = 0, 1, 2, \cdots. \qquad (277)$$

Substituting $\lambda_n = n$ into (274) gives

$$r^2 F''(r) + r F'(r) - n^2 F(r) = 0. \qquad (278)$$

Equation (278) is the well-known second order Euler ordinary differential equation with general solution given by

$$F_0(r) = \frac{1}{2}(C_0 + D_0 \ln r), \text{ for } n = 0, \qquad (279)$$

and

$$F_n(r) = C_n r^n + D_n r^{-n}, \text{ for } n = 1, 2, 3, \cdots. \qquad (280)$$

Using the superposition principle, the general solution is given by

$$u(r,\theta) = \tfrac{1}{2}(a_0 + b_0 \ln r)$$

$$\sum_{n=1}^{\infty} \left((a_n r^n + b_n r^{-n})\cos(n\theta) + (c_n r^n + d_n r^{-n})\sin(n\theta) \right), \qquad (281)$$

where a_n, b_n, c_n and d_n are constants.

Using the boundary conditions, we first set $r = b$ into (281) to obtain

$$\tfrac{1}{2}(a_0 + b_0 \ln b)$$

$$+ \sum_{n=1}^{\infty} \left((a_n b^n + b_n b^{-n})\cos(n\theta) + (c_n b^n + d_n b^{-n})\sin(n\theta) \right) = g(\theta). \qquad (282)$$

The Fourier coefficients are thus given by

$$a_0 + b_0 \ln b = \frac{1}{\pi} \int_{-\pi}^{\pi} g(\theta)\, d\theta, \qquad (283)$$

$$a_n b^n + b_n b^{-n} = \frac{1}{\pi} \int_{-\pi}^{\pi} g(\theta)\cos(n\theta)\, d\theta, \qquad (284)$$

$$c_n b^n + d_n b^{-n} = \frac{1}{\pi} \int_{-\pi}^{\pi} g(\theta)\sin(n\theta)\, d\theta. \qquad (285)$$

Substituting $r = a$ into (281) gives

$$\frac{1}{2}(a_0 + b_0 \ln a)$$

$$+ \sum_{n=1}^{\infty} \left((a_n a^n + b_n b a^{-n}) \cos(n\theta) + (c_n a^n + d_n a^{-n}) \sin(n\theta) \right) = f(\theta), \qquad (286)$$

so that the Fourier coefficients are given by

$$a_0 + b_0 \ln a \;\; = \;\; \frac{1}{\pi} \int_{-\pi}^{\pi} f(\theta)\, d\theta, \qquad (287)$$

$$a_n a^n + b_n a^{-n} \;\; = \;\; \frac{1}{\pi} \int_{-\pi}^{\pi} f(\theta) \cos(n\theta)\, d\theta, \qquad (288)$$

$$c_n a^n + d_n a^{-n} \;\; = \;\; \frac{1}{\pi} \int_{-\pi}^{\pi} f(\theta) \sin(n\theta)\, d\theta. \qquad (289)$$

Solving (283) and (287) for a_0 and b_0, (284) and (288) for a_n and b_n, and (285) and (289) for c_n and d_n completes the determination of the constants a_0, b_0, a_n, b_n, c_n, and d_n. This gives the formal solution of Laplace's equation for a circular annulus.

For simplicity reasons, the following illustrative examples will include the boundary conditions in terms of sines and cosines so as to equate the coefficients of both sides as applied before.

Example 5. Use the method of separation of variables to solve the following Dirichlet problem for an annulus:

PDE $\qquad \dfrac{\partial^2 u}{\partial r^2} + \dfrac{1}{r}\dfrac{\partial u}{\partial r} + \dfrac{1}{r^2}\dfrac{\partial^2 u}{\partial \theta^2} = 0,\; 1 < r < 2,\, 0 \le \theta \le 2\pi$

BC $\qquad u(1, \theta) = \frac{1}{2} + \sin\theta,\; u(2, \theta) = \frac{1}{2} + \frac{1}{2}\ln 2 + \cos\theta$ $\qquad (290)$

Solution.

Following the procedure outlined above, we set

$$u(r, \theta) = F(r)G(\theta). \qquad (291)$$

Substituting in the PDE of (290) and using the periodicity condition give

$$u(r, \theta) = \frac{1}{2}(a_0 + b_0 \ln r)$$

$$+ \sum_{n=1}^{\infty} \left((a_n r^n + b_n r^{-n}) \cos(n\theta) + (c_n r^n + d_n r^{-n}) \sin(n\theta) \right). \qquad (292)$$

We first set $r = 1$ into (292) and use the related boundary condition give

$$\frac{1}{2}a_0 + \sum_{n=1}^{\infty} \left((a_n + b_n)\cos(n\theta) + (c_n + d_n)\sin(n\theta) \right) = \frac{1}{2} + \sin\theta. \qquad (293)$$

Expanding the series at the left side, and equating the coefficients of like terms on both sides give

$$a_0 = 1, \tag{294}$$
$$a_n + b_n = 0, \text{ for } n \geq 1, \tag{295}$$
$$c_1 + d_1 = 1, c_n + d_n = 0, \text{ for } n > 1. \tag{296}$$

We next set $r = 2$ into (292), using the related boundary condition give

$$\frac{1}{2}a_0 + \frac{1}{2}b_0 \ln 2 + \sum_{n=1}^{\infty} \left((a_n 2^n + b_n 2^{-n}) \cos(n\theta) + (c_n 2^n + d_n 2^{-n}) \sin(n\theta) \right) \tag{297}$$

$$= \frac{1}{2} + \frac{1}{2} \ln 2 + \cos \theta$$

Expanding the series at the left side, and equating the coefficients of like terms on both sides we obtain

$$b_0 = 1, \tag{298}$$
$$2a_1 + \frac{1}{2}b_1 = 1, 2^n a_n + 2^{-n} b_n = 0, \text{ for } n > 1, \tag{299}$$
$$2^n c_n + 2^{-n} d_n = 0, \text{ for } n \geq 1. \tag{300}$$

Equations (294) and (298) give

$$a_0 = 1, b_0 = 1. \tag{301}$$

Equations (295) and (299) give

$$a_1 = \frac{2}{3}, b_1 = -\frac{2}{3}. \tag{302}$$

Equations (296) and (300) give

$$c_1 = -\frac{1}{3}, d_1 = \frac{4}{3}, \tag{303}$$

where all other coefficients vanish. Accordingly, the solution is given by

$$u(r,\theta) = \frac{1}{2} + \frac{1}{2} \ln r + (\frac{2}{3}r - \frac{2}{3}r^{-1}) \cos \theta + (-\frac{1}{3}r + \frac{4}{3}r^{-1}) \sin \theta, \tag{304}$$

obtained upon substituting (301) – (303) into (292).

Example 6. Use the method of separation of variables to solve the following Dirichlet problem for an annulus:

PDE $\quad \dfrac{\partial^2 u}{\partial r^2} + \dfrac{1}{r}\dfrac{\partial u}{\partial r} + \dfrac{1}{r^2}\dfrac{\partial^2 u}{\partial \theta^2} = 0, \ 1 < r < e, \ 0 \leq \theta \leq 2\pi$

BC $\quad u(1,\theta) = \frac{1}{2} + 4\cos\theta \tag{305}$

$\quad\quad u(e,\theta) = 1 + 4\cosh 1 \cos\theta + 4\sinh 1 \sin\theta.$

Solution.

Following the procedure outlined above, we find

$$u(r, \theta) = \tfrac{1}{2}(a_0 + b_0 \ln r)$$

$$+ \sum_{n=1}^{\infty} \left((a_n r^n + b_n r^{-n}) \cos(n\theta) + (c_n r^n + d_n r^{-n}) \sin(n\theta) \right). \tag{306}$$

Substituting $r = 1$ into (306) and using the related boundary condition give

$$\frac{1}{2}a_0 + \sum_{n=1}^{\infty} \left((a_n + b_n) \cos(n\theta) + (c_n + d_n) \sin(n\theta) \right) = \frac{1}{2} + 4 \cos \theta. \tag{307}$$

Expanding the series at the left side, and equating the coefficients of like terms on both sides we obtain

$$a_0 = 1, \tag{308}$$
$$a_1 + b_1 = 4, a_n + b_n = 0, \text{ for } n > 1, \tag{309}$$
$$c_n + d_n = 0, \text{ for } n \geq 1, \tag{310}$$

Substituting $r = e$ into (306) and using the related boundary condition give

$$\tfrac{1}{2}a_0 + \tfrac{1}{2}b_0 + \sum_{n=1}^{\infty} \left((a_n e^n + b_n e^{-n}) \cos(n\theta) + (c_n e^n + d_n e^{-n}) \sin(n\theta) \right)$$

$$= 1 + \cosh 1 \cos \theta + \cosh 1 \sin \theta. \tag{311}$$

Expanding the series at the left side, and equating the coefficients of like terms on both sides we obtain

$$\frac{1}{2}a_0 + \frac{1}{2}b_0 = 1, \tag{312}$$
$$ea_1 + e^{-1}b_1 = 2(e + e^{-1}), ea_n + e^{-1}b_n = 0, n > 1, \tag{313}$$
$$ec_1 + e^{-1}d_1 = 2(e + e^{-1}), ec_n + e^{-1}d_n = 0, n > 1. \tag{314}$$

Equations (308) and (312) give

$$a_0 = 1, b_0 = 1. \tag{315}$$

Equations (309) and (313) give

$$a_1 = 2, b_1 = 2. \tag{316}$$

Equations (310) and (314) give

$$c_1 = 2, d_1 = -2, \tag{317}$$

and all other coefficients vanish. Accordingly, the solution is given by

$$u(r, \theta) = \frac{1}{2} + \frac{1}{2} \ln r + 2(r + r^{-1}) \cos \theta + 2(r - r^{-1}) \sin \theta. \tag{318}$$

Example 7. Use the method of separation of variables to solve the following Neumann problem for an annulus:

PDE $\quad \dfrac{\partial^2 u}{\partial r^2} + \dfrac{1}{r} \dfrac{\partial u}{\partial r} + \dfrac{1}{r^2} \dfrac{\partial^2 u}{\partial \theta^2} = 0,\ 1 < r < 2,\ 0 \le \theta \le 2\pi$

BC $\quad u_r(1, \theta) = 1,\ u_r(2, \theta) = \frac{1}{2} + \frac{3}{4} \cos \theta$

$$\tag{319}$$

Solution.

Note that this is a Neumann problem for an annulus. Following the procedure outlined above, we find

$$u(r, \theta) = \frac{1}{2}(a_0 + b_0 \ln r)$$

$$+ \sum_{n=1}^{\infty} \left((a_n r^n + b_n r^{-n}) \cos(n\theta) + (c_n r^n + d_n r^{-n}) \sin(n\theta) \right). \tag{320}$$

We first set $r = 1$ into the derivative of (320) and use the related boundary condition to obtain

$$\frac{1}{2} b_0 + \sum_{n=1}^{\infty} \left((na_n - nb_n) \cos(n\theta) + (nc_n - nd_n) \sin(n\theta) \right) = 1. \tag{321}$$

Expanding the series at the left side, and equating the coefficients of like terms on both sides we obtain

$$b_0 \ =\ 2, \tag{322}$$
$$n(a_n - b_n) \ =\ 0,\ n \ge 1, \tag{323}$$
$$n(c_n - d_n) \ =\ 0,\ n \ge 1. \tag{324}$$

We next set $r = 2$ into the derivative of (320), using the related boundary condition, and equating the coefficients of like terms on both sides we obtain

$$b_0 \ =\ 2, \tag{325}$$
$$a_1 - \frac{1}{4} b_1 \ =\ \frac{3}{4},\ n > 1, \tag{326}$$
$$2^{n-1} c_n - 2^{-n-1} d_n \ =\ 0,\ n \ge 1, \tag{327}$$

Proceeding as before we find

$$b_0 \ =\ 2, \tag{328}$$
$$a_1 \ =\ 1,\ b_1 = 1.$$

Note that other coefficients vanish. Accordingly, the solution is given by

$$u(r, \theta) = C_0 + \ln r + (r + r^{-1}) \cos \theta. \tag{329}$$

Example 8. Use the method of separation of variables to solve the following Neumann problem for an annulus:

PDE $\qquad \dfrac{\partial^2 u}{\partial r^2} + \dfrac{1}{r}\dfrac{\partial u}{\partial r} + \dfrac{1}{r^2}\dfrac{\partial^2 u}{\partial \theta^2} = 0,\ 1 < r < 2,\ 0 \le \theta \le 2\pi$

$\qquad\qquad\qquad\qquad\qquad\qquad\qquad\qquad\qquad\qquad\qquad\qquad\qquad\quad$ (330)

BC $\qquad u_r(1, \theta) = 0,\ u_r(2, \theta) = \frac{3}{4}\cos\theta + \frac{3}{4}\sin\theta.$

Solution.

Following the procedure outlined above, we find

$$u(r, \theta) = \tfrac{1}{2}(a_0 + b_0 \ln r)$$

$$+ \sum_{n=1}^{\infty}\left((a_n r^n + b_n r^{-n})\cos(n\theta) + (c_n r^n + d_n r^{-n})\sin(n\theta)\right), \tag{331}$$

Substituting $r = 1$ into the derivative of (331) and using the related boundary condition give

$$\frac{1}{2}b_0 + \sum_{n=1}^{\infty}\left(n(a_n - b_n)\cos(n\theta) + n(c_n - d_n)\sin(n\theta)\right) = 0. \tag{332}$$

Expanding the series at the left side, and equating the coefficients of like terms on both sides we obtain

$$\begin{aligned}
b_0 &= 0, & (333)\\
a_n - b_n &= 0,\ n \ge 1, & (334)\\
c_n - d_n &= 0,\ n \ge 1, & (335)
\end{aligned}$$

Substituting $r = 2$ into the derivative of (331), using the related boundary condition, and equating the coefficients of like terms on both sides we obtain

$$\begin{aligned}
b_0 &= 0, & (336)\\
a_1 - \frac{1}{4}b_1 &= \frac{3}{4}, & (337)\\
c_1 - \frac{1}{4}d_1 &= \frac{3}{4}. & (338)
\end{aligned}$$

Equations (333) and (336) give

$$b_0 = 0. \tag{339}$$

Equations (334) and (337) give

$$a_1 = 1,\ b_1 = 1. \tag{340}$$

Equations (335) and (338) give

$$c_1 = 1, d_1 = 1, \tag{341}$$

and all other coefficients vanish. Accordingly, the solution is given by

$$u(r,\theta) = C_0 + (r + r^{-1})\cos\theta + (r + r^{-1})\sin\theta. \tag{342}$$

Exercises 7.4.2

Use the method of separation of variables to solve the following Laplace's equations:

1. $\dfrac{\partial^2 u}{\partial r^2} + \dfrac{1}{r}\dfrac{\partial u}{\partial r} + \dfrac{1}{r^2}\dfrac{\partial^2 u}{\partial \theta^2} = 0, \ 1 < r < e, \ 0 \le \theta \le 2\pi$

   $u(1,\theta) = 1$

   $u(e,\theta) = 2 + 2\sinh 1(\cos\theta + \sin\theta)$

2. $\dfrac{\partial^2 u}{\partial r^2} + \dfrac{1}{r}\dfrac{\partial u}{\partial r} + \dfrac{1}{r^2}\dfrac{\partial^2 u}{\partial \theta^2} = 0, \ 1 < r < e, \ 0 \le \theta \le 2\pi$

   $u(1,\theta) = 1 + \cos\theta + \sin\theta$

   $u(e,\theta) = 2 + e(\cos\theta + \sin\theta)$

3. $\dfrac{\partial^2 u}{\partial r^2} + \dfrac{1}{r}\dfrac{\partial u}{\partial r} + \dfrac{1}{r^2}\dfrac{\partial^2 u}{\partial \theta^2} = 0, \ 1 < r < 2, \ 0 \le \theta \le 2\pi$

   $u(1,\theta) = 1$

   $u(2,\theta) = 1 + 1.5\sin\theta$

4. $\dfrac{\partial^2 u}{\partial r^2} + \dfrac{1}{r}\dfrac{\partial u}{\partial r} + \dfrac{1}{r^2}\dfrac{\partial^2 u}{\partial \theta^2} = 0, \ 1 < r < 2, \ 0 \le \theta \le 2\pi$

   $u(1,\theta) = 1 - \cos\theta - \sin\theta$

   $u(2,\theta) = 1 + \cos\theta + \sin\theta$

5. $\dfrac{\partial^2 u}{\partial r^2} + \dfrac{1}{r}\dfrac{\partial u}{\partial r} + \dfrac{1}{r^2}\dfrac{\partial^2 u}{\partial \theta^2} = 0, \ 1 < r < e^2, \ 0 \le \theta \le 2\pi$

   $u(1,\theta) = 1$

   $u(e^2,\theta) = 3 + 2\sinh 2\cos\theta$

6. $\dfrac{\partial^2 u}{\partial r^2} + \dfrac{1}{r}\dfrac{\partial u}{\partial r} + \dfrac{1}{r^2}\dfrac{\partial^2 u}{\partial \theta^2} = 0, \ 1 < r < e^2, \ 0 \le \theta \le 2\pi$

$u(1, \theta) = 1$

$u(e^2, \theta) = 3 + 2 \sinh 2 \sin \theta$

7. $\dfrac{\partial^2 u}{\partial r^2} + \dfrac{1}{r} \dfrac{\partial u}{\partial r} + \dfrac{1}{r^2} \dfrac{\partial^2 u}{\partial \theta^2} = 0,\ \dfrac{1}{2} < r < 1,\ 0 \leq \theta \leq 2\pi$

$u_r\left(\tfrac{1}{2}, \theta\right) = 2 - 3 \sin \theta$

$u_r(1, \theta) = 1$

8. $\dfrac{\partial^2 u}{\partial r^2} + \dfrac{1}{r} \dfrac{\partial u}{\partial r} + \dfrac{1}{r^2} \dfrac{\partial^2 u}{\partial \theta^2} = 0,\ \dfrac{1}{2} < r < 1,\ 0 \leq \theta \leq 2\pi$

$u_r\left(\tfrac{1}{2}, \theta\right) = 5 \cos \theta$

$u_r(1, \theta) = 2 \cos \theta$

9. $\dfrac{\partial^2 u}{\partial r^2} + \dfrac{1}{r} \dfrac{\partial u}{\partial r} + \dfrac{1}{r^2} \dfrac{\partial^2 u}{\partial \theta^2} = 0,\ 1 < r < 2,\ 0 \leq \theta \leq 2\pi$

$u_r(1, \theta) = \sin \theta$

$u_r(2, \theta) = 2.5 \sin \theta$

10. $\dfrac{\partial^2 u}{\partial r^2} + \dfrac{1}{r} \dfrac{\partial u}{\partial r} + \dfrac{1}{r^2} \dfrac{\partial^2 u}{\partial \theta^2} = 0,\ 1 < r < 2,\ 0 \leq \theta \leq 2\pi$

$u_r(1, \theta) = 1$

$u_r(2, \theta) = \tfrac{1}{2} + \tfrac{3}{4} \cos \theta + \tfrac{3}{4} \sin \theta$

11. $\dfrac{\partial^2 u}{\partial r^2} + \dfrac{1}{r} \dfrac{\partial u}{\partial r} + \dfrac{1}{r^2} \dfrac{\partial^2 u}{\partial \theta^2} = 0,\ 1 < r < 2,\ 0 \leq \theta \leq 2\pi$

$u_r(1, \theta) = 4 \cos \theta$

$u_r(2, \theta) = 2.5 \cos \theta$

12. $\dfrac{\partial^2 u}{\partial r^2} + \dfrac{1}{r} \dfrac{\partial u}{\partial r} + \dfrac{1}{r^2} \dfrac{\partial^2 u}{\partial \theta^2} = 0,\ 1 < r < 2,\ 0 \leq \theta \leq 2\pi$

$u_r(1, \theta) = \sin \theta$

$u_r(2, \theta) = 2.5 \sin \theta$

Chapter 8

Nonlinear Partial Differential Equations

8.1 Introduction

So far in this text we have been mainly concerned with studying classic methods and the newly developed Adomian decomposition method in discussing first order and second order linear partial differential equations. In this chapter, we will focus our study on the nonlinear partial differential equations. The nonlinear partial differential equations arise in a wide variety of physical problems such as fluid dynamics, plasma physics, solid mechanics and quantum field theory. Systems of nonlinear partial differential equations have been also noticed to arise in chemical and biological applications. The nonlinear wave equations and the solitons concept have introduced remarkable achievements in the field of applied sciences. The solutions obtained from nonlinear wave equations are different from the solutions of the linear wave equations.

Recently, a special type of KdV equation has been under a thorough investigation and new phenomenon was observed. It was discovered that when the wave dispersion is purely nonlinear, some features may be observed which is the existence of the so-called compactons:solitons with finite wave length [72,78,79,100–103,135]. As will be discussed in Chapter 9, solitons appear as a result of balance between week nonlinearity and dispersion. The characteristics of the solitons and the compactons concepts will be addressed in Chapter 11.

It is important to note that several traditional methods, such as the method of characteristics and the variational principle, are among the methods that are used to handle the nonlinear partial differential equations. Moreover, nonlinear partial differential equations are not easy to handle especially if the questions of uniqueness and stability of solutions are to be discussed. It is interesting to point out that the superposition principle, that we used for the linear partial differential equations, cannot be applied for the nonlinear partial differential equations. For this reason numerical solutions are usually established for nonlinear partial differential equations.

It is well known that a general method for determining analytical solutions for partial differential equations has not been found among traditional methods. However, we believe that the Adomian decomposition method, the noise terms phenomenon, and the related modification presented in previous chapters provide an effective, reliable, and powerful tool for handling nonlinear partial differential equations.

In Chapter 2, a detailed outline about the works conducted on Adomian's method, the implementation of this method to many scientific models and frontier physics problems, and the comparisons of this method with existing techniques was introduced. The convergence concept of the decomposition series was thoroughly investigated by many researchers to confirm the rapid convergence of the resulting series. Cherruault examined the convergence of Adomian's method in [35,36]. Cherruault *et. al* [36] discussed the convergence of Adomian's method when applied to integral equations. In addition, Cherruault and Adomian presented a new proof of convergence of the method in [37]. Moreover, Abbaoui *et. al* [1] formally proved the convergence of Adomian's method when applied to differential equations in general.

Rach *et. al.* [97] used Adomian's method to solve differential equations with singular coefficients such as Legendre's equation, Bessel's equation, and Hermite's equation. Moreover, in [104], a suitable definition of the operator was used to overcome the difficulty of singular points of Lane-Emden equation. In [127], a transformation formula was introduced to overcome the singularity concept for the Lane-Emden type of equations. Recently, a useful study was presented in [33] to implement the decomposition method for differential equations with discontinuities.

However, Adomian decomposition method does not assure [91], on its own, existence and uniqueness of the solution. In fact, it can be safely applied when a fixed point theorem holds [99]. A theorem developed by Re'paci [99] indicates that the decomposition method can be used as an algorithm for the approximation of the dynamical response in a sequence of time intervals $[0, t_1), [t_1, t_2), \cdots, [t_{n-1}, T)$ such that the condition at t_p is taken as initial condition in the interval $[t_p, t_{p-1})$ which follows.

Unlike the preceding chapters, we will apply only the Adomian decomposition method and all related phenomena in discussing the topic of nonlinear partial differential equations. This is due to the fact that the method is efficient in that it provides the solution in a rapidly convergent series and reduces the volume of computational work.

The nonlinear partial differential equation was defined in Chapter 1. The first order nonlinear partial differential equation in two independent variables x and y can be generally expressed in the form

$$F(x, y, u, u_x, u_y) = f, \tag{1}$$

where f is a function of one or two of the independent variables x and y. Similarly, the second order nonlinear partial differential equation in two independent variables x and y can be expressed by

$$F(x, y, u, u_x, u_y, u_{xx}, u_{xy}, u_{yy}) = f. \tag{2}$$

The nonlinear partial differential equation is called *homogeneous* if $f = 0$. On the other hand, the nonlinear partial differential equation is called *inhomogeneous* if $f \neq 0$. Examples of the first order nonlinear partial differential equations are given by

$$u_t + 2uu_x = 0, \tag{3}$$
$$u_x - u^2 u_y = 0. \tag{4}$$
$$u_x + uu_y = 6x, \tag{5}$$
$$u_t + uu_x = \sin x, \tag{6}$$

Note that equations (3) and (4) are homogeneous equations. On the other hand, equations (5) and (6) are inhomogeneous equations. Examples of second order nonlinear partial differential equations are given by

$$u_t + uu_x - \nu u_{xx} = 0, \tag{7}$$
$$u_{tt} - c^2 u_{xx} + \sin u = 0. \tag{8}$$

On the other hand, the modified Korteweg-de Vries equation

$$u_t - 6u^2 u_x + u_{xxx} = 0, \tag{9}$$

is an example of a third order nonlinear homogeneous partial differential equation.

In the following section, the Adomian decomposition method will be presented for finding analytical solutions of nonlinear partial differential equations, homogeneous or inhomogeneous.

8.2 Adomian Decomposition Method

The Adomian decomposition method has been outlined before in previous chapters and has been applied to a wide class of linear partial differential equations. The method has been applied directly and in a straightforward manner to a homogeneous and inhomogeneous problems without any restrictive assumptions or linearization. The method usually decomposes the unknown function u into an infinite sum of components that will be determined recursively through iterations as discussed before.

The Adomian decomposition method will be applied in this chapter and in the coming chapters to handle nonlinear partial differential equations. An important remark should be made here concerning the representation of the nonlinear terms that appear in the equation. Although the linear term u is expressed as an infinite series of components, the Adomian decomposition method [5,7,129] requires a special representation for the nonlinear terms such as $u^2, u^3, u^4, \sin u, e^u, uu_x, u_x^2$, etc. that appear in the equation. The method introduces a formal algorithm to establish a proper representation for all forms of nonlinear terms. The representation of the nonlinear terms is necessary to handle the nonlinear equations in an effective and successful way.

In the following, the Adomian scheme for calculating representation of nonlinear terms will be introduced in details. The discussion will be supported by several illustrative examples that will cover a wide variety of forms of nonlinearity. In a like manner, an alternative algorithm for for calculating Adomian's polynomials will be outlined in details supported by illustrative examples. For further readings, see [5,7,11,129]

8.2.1   Calculation of Adomian Polynomials

It is well known now that Adomian decomposition method suggests that the unknown linear function u may be represented by the decomposition series

$$u = \sum_{n=0}^{\infty} u_n, \tag{10}$$

where the components $u_n, n \geq 0$ can be elegantly computed in a recursive way. However, the nonlinear term $F(u)$, such as $u^2, u^3, u^4, \sin u, e^u, uu_x, u_x^2$, etc. can be expressed by an infinite series of the so-called Adomian polynomials A_n given in the form

$$F(u) = \sum_{n=0}^{\infty} A_n(u_0, u_1, u_2, \cdots, u_n), \tag{11}$$

where the so-called Adomian polynomials A_n can be evaluated for all forms of nonlinearity. Several schemes have been introduced in the literature by researchers to calculate Adomian polynomials. Adomian introduced a scheme in [5,7] for the calculation of Adomian polynomials that was formally justified. The scheme is considered by many as simple and practical. The method is based on specific steps that require specific formula for each polynomial. Later, an alternative reliable method that is based on algebraic and trigonometric identities and on Taylor series has been established in [129]. We believe that the alternative method, that will be presented later, introduces a more simplified and more practical approach. The reason for simplicity is that the alternative scheme employs only elementary operations and does not require specific formulas.

In the following, we will present the general Adomian algorithm for the calculation of the so-called Adomian polynomials followed by a summary of the necessary steps to calculate the first few Adomian polynomials.

The Adomian polynomials A_n for the nonlinear term $F(u)$ can be evaluated by using the following expression

$$A_n = \left(\frac{1}{n!}\right)\left(\frac{d^n}{d\lambda^n}\right)F(u(\lambda))|_{\lambda=0}. \tag{12}$$

The general formula (12) can be simplified as follows. Assuming that the nonlinear function is $F(u)$, therefore by using (12), Adomian polynomials are given by

$$A_0 = F(u_0),$$

$$A_1 = u_1 F'(u_0),$$

$$A_2 = u_2 F'(u_0) + \frac{1}{2!}u_1^2 F''(u_0),$$

$$A_3 = u_3 F'(u_0) + u_1 u_2 F''(u_0) + \frac{1}{3!}u_1^3 F'''(u_0),$$

$$A_4 = u_4 F'(u_0) + (\frac{1}{2!}u_2^2 + u_1 u_3)F''(u_0) + \frac{1}{2!}u_1^2 u_2 F'''(u_0) + \frac{1}{4!}u_1^4 F^{(iv)}(u_0),$$

$$A_5 = u_5 F'(u_0) + (u_2 u_3 + u_1 u_4)F''(u_0) + (\frac{1}{2!}u_1 u_2^2 + \frac{1}{2!}u_1^2 u_3)F'''(u_0)$$

$$+ \frac{1}{3!}u_1^3 u_2 F^{(iv)}(u_0) + \frac{1}{5!}u_1^5 F^{(v)}(u_0).$$

$$\tag{13}$$

Other polynomials can be generated in a similar manner.

Two important observations can be made here. First, A_0 depends only on u_0, A_1 depends only on u_0 and u_1, A_2 depends only on u_0, u_1, and u_2, and so on. Second, substituting (13) into (11) gives

$$\begin{aligned}
F(u) &= A_0 + A_1 + A_2 + A_3 + \cdots \\
&= F(u_0) + (u_1 + u_2 + u_3 + \cdots)F'(u_0) \\
&\quad + \frac{1}{2!}(u_1^2 + 2u_1 u_2 + 2u_1 u_3 + u_2^2 + \cdots)F''(u_0) + \cdots \\
&\quad + \frac{1}{3!}(u_1^3 + 3u_1^2 u_2 + 3u_1^2 u_3 + 6u_1 u_2 u_3 + \cdots)F'''(u_0) + \cdots \\
&= F(u_0) + (u - u_0)F'(u_0) + \frac{1}{2!}(u - u_0)^2 F''(u_0) + \cdots.
\end{aligned}$$

The last expansion confirms a fact that the series in A_n polynomials is a Taylor series about a function u_0 and not about a point as is usually used. The few Adomian polynomials given above in (13) clearly show that the sum of the subscripts of the components of u of each term of A_n is equal to n. As stated before, it is clear that A_0 depends only on u_0, A_1 depends only u_0 and u_1, A_2 depends only on u_0, u_1 and u_2. The same conclusion holds for other polynomials.

In the following, we will calculate Adomian polynomials for several forms of nonlinearity that may arise in nonlinear ordinary or partial differential equations.

Calculation of Adomian Polynomials A_n

I. Nonlinear Polynomials

Case 1. $F(u) = u^2$

The polynomials can be obtained as follows:

$A_0 = F(u_0) = u_0^2,$

$A_1 = u_1 F'(u_0) = 2u_0 u_1,$

$A_2 = u_2 F'(u_0) + \frac{1}{2!} u_1^2 F''(u_0) = 2u_0 u_2 + u_1^2,$

$A_3 = u_3 F'(u_0) + u_1 u_2 F''(u_0) + \frac{1}{3!} u_1^3 F'''(u_0) = 2u_0 u_3 + 2u_1 u_2.$

Case 2. $F(u) = u^3$

The polynomials are given by

$A_0 = F(u_0) = u_0^3,$

$A_1 = u_1 F'(u_0) = 3u_0^2 u_1,$

$A_2 = u_2 F'(u_0) + \frac{1}{2!} u_1^2 F''(u_0) = 3u_0^2 u_2 + 3u_0 u_1^2,$

$A_3 = u_3 F'(u_0) + u_1 u_2 F''(u_0) + \frac{1}{3!} u_1^3 F'''(u_0) = 3u_0^2 u_3 + 6u_0 u_1 u_2 + u_1^3.$

Case 3. $F(u) = u^4$

Proceeding as before we find

$A_0 = u_0^4,$

$A_1 = 4u_0^3 u_1,$

$A_2 = 4u_0^3 u_2 + 6u_0^2 u_1^2,$

$A_3 = 4u_0^3 u_3 + 4u_1^3 u_0 + 12u_0^2 u_1 u_2.$

II. Nonlinear Derivatives

Case 1. $F(u) = u_x^2$

Proceeding as before we find

$A_0 = u_{0_x}^2,$

$A_1 = 2u_{0_x} u_{1_x},$

$A_2 = 2u_{0_x} u_{2_x} + u_{1_x}^2,$

$A_3 = 2u_{0_x} u_{3_x} + 2u_{1_x} u_{2_x}.$

Case 2. $F(u) = u_x^3$

The Adomian polynomials are given by

$A_0 = u_{0_x}^3,$

$A_1 = 3u_{0_x}^2 u_{1_x},$

$A_2 = 3u_{0_x}^2 u_{2_x} + 3u_{0_x} u_{1_x}^2,$

$A_3 = 3u_{0_x}^2 u_{3_x} + 6u_{0_x} u_{1_x} u_{2_x} + u_{1_x}^3.$

Case 3. $F(u) = uu_x = \frac{1}{2}L_x(u^2)$

The Adomian polynomials for this nonlinearity are given by

$A_0 = F(u_0) = u_0 u_{0_x},$

$A_1 = \frac{1}{2}L_x(2u_0 u_1) = u_{0_x} u_1 + u_0 u_{1_x},$

$A_2 = \frac{1}{2}L_x(2u_0 u_2 + u_1^2) = u_{0_x} u_2 + u_{1_x} u_1 + u_{2_x} u_0,$

$A_3 = \frac{1}{2}L_x(2u_0 u_3 + 2u_1 u_2) = u_{0_x} u_3 + u_{1_x} u_2 + u_{2_x} u_1 + u_{3_x} u_0.$

II. Trigonometric Nonlinearity

Case 1. $F(u) = \sin u$

The Adomian polynomials for this form of nonlinearity are given by

$A_0 = \sin u_0,$

$A_1 = u_1 \cos u_0,$

$A_2 = u_2 \cos u_0 - \frac{1}{2!}u_1^2 \sin u_0,$

$A_3 = u_3 \cos u_0 - u_1 u_2 \sin u_0 - \frac{1}{3!}u_1^3 \cos u_0.$

Case 2. $F(u) = \cos u$

Proceeding as before gives

$A_0 = \cos u_0,$

$A_1 = -u_1 \sin u_0,$

$A_2 = -u_2 \sin u_0 - \frac{1}{2!}u_1^2 \cos u_0,$

$A_3 = -u_3 \sin u_0 - u_1 u_2 \cos u_0 + \frac{1}{3!}u_1^3 \sin u_0.$

III. Hyperbolic Nonlinearity

Case 1. $F(u) = \sinh u$

The A_n polynomials for this form of nonlinearity are given by

$A_0 = \sinh u_0,$

$A_1 = u_1 \cosh u_0,$

$A_2 = u_2 \cosh u_0 + \frac{1}{2!}u_1^2 \sinh u_0,$

$A_3 = u_3 \cosh u_0 + u_1 u_2 \sinh u_0 + \frac{1}{3!}u_1^3 \cosh u_0.$

Case 2. $F(u) = \cosh u$

The Adomian polynomials are given by $A_0 = \cosh u_0$,

$A_1 = u_1 \sinh u_0$,

$A_2 = u_2 \sinh u_0 + \frac{1}{2!} u_1^2 \cosh u_0$,

$A_3 = u_3 \sinh u_0 + u_1 u_2 \cosh u_0 + \frac{1}{3!} u_1^3 \sinh u_0$.

III. Exponential Nonlinearity

Case 1. $F(u) = e^u$

The Adomian polynomials for this form of nonlinearity are given by

$A_0 = e^{u_0}$,

$A_1 = u_1 e^{u_0}$,

$A_2 = (u_2 + \frac{1}{2!} u_1^2) e^{u_0}$,

$A_3 = (u_3 + u_1 u_2 + \frac{1}{3!} u_1^3) e^{u_0}$.

Case 2. $F(u) = e^{-u}$

Proceeding as before gives

$A_0 = e^{-u_0}$,

$A_1 = -u_1 e^{-u_0}$,

$A_2 = (-u_2 + \frac{1}{2!} u_1^2) e^{-u_0}$,

$A_3 = (-u_3 + u_1 u_2 - \frac{1}{3!} u_1^3) e^{-u_0}$.

IV. Logarithmic Nonlinearity

Case 1. $F(u) = \ln u, \ u > 0$

The A_n polynomials for logarithmic nonlinearity are give by

$A_0 = \ln u_0$,

$A_1 = \frac{u_1}{u_0}$,

$A_2 = \frac{u_2}{u_0} - \frac{1}{2} \frac{u_1^2}{u_0^2}$,

$A_3 = \frac{u_3}{u_0} - \frac{u_1 u_2}{u_0^2} + \frac{1}{3} \frac{u_1^3}{u_0^3}$.

Case 2. $F(u) = \ln(1 + u), \ -1 < u \le 1$

The A_n polynomials are give by

$A_0 = \ln(1 + u_0)$,

$A_1 = \frac{u_1}{1 + u_0}$,

$$A_2 = \frac{u_2}{1+u_0} - \frac{1}{2}\frac{u_1^2}{(1+u_0)^2},$$

$$A_3 = \frac{u_3}{1+u_0} - \frac{u_1 u_2}{(1+u_0)^2} + \frac{1}{3}\frac{u_1^3}{(1+u_0)^3}.$$

8.2.2 Alternative Algorithm for Calculating Adomian Polynomials

It is worth noting that a considerable amount of research work has been invested to develop an alternative method to Adomian algorithm for calculating Adomian polynomials A_n. The aim was to develop a practical technique that will calculate Adomian polynomials in a practical way without any need to the formulae introduced before. However, the methods developed so far in this regard are identical to that used by Adomian.

We believe that a simple and reliable technique can be established to make the calculations less dependable on the formulae presented before.

In the following, we will introduce an alternative algorithm that can be used to calculate Adomian polynomials for nonlinear terms in an easy way. The newly developed method in [129] depends mainly on algebraic and trigonometric identities, and on Taylor expansions as well. Moreover, we should use the fact that the sum of subscripts of the components of u in each term of the polynomial A_n is equal to n.

The alternative algorithm suggests that we substitute u as a sum of components $u_n, n \geq 0$ as defined by the decomposition method. It is clear that A_0 is always determined independent of the other polynomials $A_n, n \geq 1$, where A_0 is defined by

$$A_0 = F(u_0). \tag{14}$$

The alternative method assumes that we first separate $A_0 = F(u_0)$ for every non-linear term $F(u)$. With this separation done, the remaining components of $F(u)$ can be expanded by using algebraic operations, trigonometric identities, and Taylor series as well. We next collect all terms of the expansion obtained such that the sum of the subscripts of the components of u in each term is the same. Having collected these terms, the calculation of the Adomian polynomials is thus completed. Several examples have been tested, and the obtained results have shown that Adomian polynomials can be elegantly computed without any need to the formulas established by Adomian. The technique will be explained as follows.

Adomian Polynomials by Using the Alternative Method

I. Nonlinear Polynomials

Case 1. $F(u) = u^2$

We first set

$$u = \sum_{n=0}^{\infty} u_n. \tag{15}$$

Substituting (15) into $F(u) = u^2$ gives

$$F(u) = (u_0 + u_1 + u_2 + u_3 + u_4 + u_5 + \cdots)^2. \tag{16}$$

Expanding the expression at the right hand side gives

$$F(u) = u_0^2 + 2u_0u_1 + 2u_0u_2 + u_1^2 + 2u_0u_3 + 2u_1u_2 + \cdots. \tag{17}$$

The expansion in (17) can be rearranged by grouping all terms with the sum of the subscripts is the same. This means that we can rewrite (17) as

$$F(u) \;=\; \underbrace{u_0^2}_{A_0} + \underbrace{2u_0u_1}_{A_1} + \underbrace{2u_0u_2 + u_1^2}_{A_2} + \underbrace{2u_0u_3 + 2u_1u_2}_{A_3}$$
$$+ \underbrace{2u_0u_4 + 2u_1u_3 + u_2^2}_{A_4} + \underbrace{2u_0u_5 + 2u_1u_4 + 2u_2u_3}_{A_5} + \cdots. \tag{18}$$

This completes the determination of Adomian polynomials given by

$A_0 = u_0^2,$

$A_1 = 2u_0u_1,$

$A_2 = 2u_0u_2 + u_1^2,$

$A_3 = 2u_0u_3 + 2u_1u_2,$

$A_4 = 2u_0u_4 + 2u_1u_3 + u_2^2,$

$A_5 = 2u_0u_5 + 2u_1u_4 + 2u_2u_3.$

This is consistent with the results obtained before by using Adomian's algorithm.

Case 2. $F(u) = u^3$

Proceeding as before, we set

$$u = \sum_{n=0}^{\infty} u_n. \tag{19}$$

Substituting (19) into $F(u) = u^3$ gives

$$F(u) = (u_0 + u_1 + u_2 + u_3 + u_4 + u_5 + \cdots)^3. \tag{20}$$

Expanding the right hand side yields

$$F(u) \;=\; u_0^3 + 3u_0^2u_1 + 3u_0^2u_2 + 3u_0u_1^2 + 3u_0^2u_3 + 6u_0u_1u_2 + u_1^3$$
$$+ 3u_0^2u_4 + 3u_1^2u_2 + 3u_2^2u_0 + 6u_0u_1u_3 \cdots. \tag{21}$$

The expansion in (21) can be rearranged by grouping all terms with the sum of the subscripts is the same. This means that we can rewrite (21) as

$$
\begin{aligned}
F(u) \;=\; & \underbrace{u_0^3}_{A_0} + \underbrace{3u_0^2 u_1}_{A_1} + \underbrace{3u_0^2 u_2 + 3u_0 u_1^2}_{A_2} + \underbrace{3u_0^2 u_3 + 6u_0 u_1 u_2 + u_1^3}_{A_3} \\
& + \underbrace{3u_0^2 u_4 + 3u_1^2 u_2 + 3u_2^2 u_0 + 6u_0 u_1 u_3}_{A_4} + \cdots .
\end{aligned}
\tag{22}
$$

Consequently, Adomian polynomials can be written by

$A_0 = u_0^3,$

$A_1 = 3u_0^2 u_1,$

$A_2 = 3u_0^2 u_2 + 3u_0 u_1^2,$

$A_3 = 3u_0^2 u_3 + 6u_0 u_1 u_2 + u_1^3,$

$A_4 = 3u_0^2 u_4 + 3u_1^2 u_2 + 3u_2^2 u_0 + 6u_0 u_1 u_3.$

II. Nonlinear Derivatives

Case 1. $F(u) = u_x^2$

We first set

$$
u_x = \sum_{n=0}^{\infty} u_{n_x}.
\tag{23}
$$

Substituting (23) into $F(u) = u_x^2$ gives

$$
F(u) = (u_{0_x} + u_{1_x} + u_{2_x} + u_{3_x} + u_{4_x} + \cdots)^2.
\tag{24}
$$

Squaring the right side gives

$$
F(u) = u_{0_x}^2 + 2u_{0_x} u_{1_x} + 2u_{0_x} u_{2_x} + u_{1_x}^2 + 2u_{0_x} u_{3_x} + 2u_{1_x} u_{2_x} + \cdots .
\tag{25}
$$

Grouping the terms as discussed above we find

$$
\begin{aligned}
F(u) \;=\; & \underbrace{u_{0_x}^2}_{A_0} + \underbrace{2u_{0_x} u_{1_x}}_{A_1} + \underbrace{2u_{0_x} u_{2_x} + u_{1_x}^2}_{A_2} \\
& + \underbrace{2u_{0_x} u_{3_x} + 2u_{1_x} u_{2_x}}_{A_3} + \underbrace{u_{2_x}^2 + 2u_{0_x} u_{4_x} + 2u_{1_x} u_{3_x}}_{A_4} + \cdots .
\end{aligned}
\tag{26}
$$

Adomian polynomials are given by

$A_0 = u_{0_x}^2,$

$A_1 = 2u_{0_x} u_{1_x},$

$$A_2 = 2u_{0_x} u_{2_x} + u_{1_x}^2,$$

$$A_3 = 2u_{0_x} u_{3_x} + 2u_{1_x} u_{2_x},$$

$$A_4 = 2u_{0_x} u_{4_x} + 2u_{1_x} u_{3_x} + u_{2_x}^2.$$

Case 2. $F(u) = uu_x$

We first set

$$u = \sum_{n=0}^{\infty} u_n, \tag{27}$$

$$u_x = \sum_{n=0}^{\infty} u_{n_x}.$$

Substituting (27) into $F(u) = uu_x$ yields

$$F(u) = (u_0 + u_1 + u_2 + u + 3 + u_4 + \cdots) \times \tag{28}$$

$$(u_{0_x} + u_{1_x} + u_{2_x} + u_{3_x} + u_{4_x} + \cdots).$$

Multiplying the two factors gives

$$F(u) = u_0 u_{0_x} + u_{0_x} u_1 + u_0 u_{1_x} + u_{0_x} u_2 + u_{1_x} u_1 + u_{2_x} u_0 + u_{0_x} u_3$$

$$+ u_{1_x} u_2 + u_{2_x} u_1 + u_{3_x} u_0 + u_{0_x} u_4 + u_0 u_{4_x} + u_{1_x} u_3 \tag{29}$$

$$+ u_1 u_{3_x} + u_2 u_{2_x} + \cdots.$$

Proceeding with grouping the terms we obtain

$$F(u) = \underbrace{u_{0_x} u_0}_{A_0} + \underbrace{u_{0_x} u_1 + u_0 u_{1_x}}_{A_1} + \underbrace{u_{0_x} u_2 + u_{1_x} u_1 + u_{2_x} u_0}_{A_2}$$

$$+ \underbrace{u_{0_x} u_3 + u_{1_x} u_2 + u_{2_x} u_1 + u_{3_x} u_0}_{A_3} \tag{30}$$

$$+ \underbrace{u_{0_x} u_4 + u_{1_x} u_3 + u_{2_x} u_2 + u_{3_x} u_1 + u_{4_x} u_0}_{A_4} + \cdots.$$

It then follows that Adomian polynomials are given by

$$A_0 = u_{0_x} u_0,$$

$$A_1 = u_{0_x} u_1 + u_0 u_{1_x},$$

$$A_2 = u_{0_x} u_2 + u_{1_x} u_1 + u_{2_x} u_0,$$

$$A_3 = u_{0_x} u_3 + u_{1_x} u_2 + u_{2_x} u_1 + u_{3_x} u_0,$$

$$A_4 = u_{0_x} u_4 + u_0 u_{4_x} + u_{1_x} u_3 + u_1 u_{3_x} + u_2 u_{2_x}.$$

II. Trigonometric Nonlinearity

Case 1. $F(u) = \sin u$

Note that algebraic operations cannot be applied here. Therefore, our main aim is to separate $A_0 = F(u_0)$ from other terms. To achieve this goal, we first substitute

$$u = \sum_{n=0}^{\infty} u_n, \tag{31}$$

into $F(u) = \sin u$ to obtain

$$F(u) = \sin[u_0 + (u_1 + u_2 + u_3 + u_4 + \cdots)]. \tag{32}$$

To calculate A_0, recall the trigonometric identity

$$\sin(\theta + \phi) = \sin\theta \cos\phi + \cos\theta \sin\phi. \tag{33}$$

Accordingly, Eq. (32) becomes

$$\begin{aligned}
F(u) &= \sin u_0 \cos(u_1 + u_2 + u_3 + u_4 + \cdots) \\
&\quad + \cos u_0 \sin(u_1 + u_2 + u_3 + u_4 + \cdots).
\end{aligned} \tag{34}$$

Separating $F(u_0) = \sin u_0$ from other factors and using Taylor expansions for $\cos(u_1 + u_2 \cdots)$ and $\sin(u_1 + u_2 + \cdots)$ give

$$\begin{aligned}
F(u) &= \sin u_0 \left(1 - \tfrac{1}{2!}(u_1 + u_2 + \cdots)^2 + \tfrac{1}{4!}(u_1 + u_2 + \cdots)^4 - \cdots\right) \\
&\quad + \cos u_0 \left((u_1 + u_2 + \cdots) - \tfrac{1}{3!}(u_1 + u_2 + \cdots)^3 + \cdots\right),
\end{aligned} \tag{35}$$

so that

$$\begin{aligned}
F(u) &= \sin u_0 \left(1 - \tfrac{1}{2!}(u_1^2 + 2u_1 u_2 + \cdots)\right) \\
&\quad + \cos u_0 \left((u_1 + u_2 + \cdots) - \tfrac{1}{3!}u_1^3 + \cdots\right).
\end{aligned} \tag{36}$$

Note that we expanded the algebraic terms; then few terms of each expansion are listed. The last expansion can be rearranged by grouping all terms with the same sum of subscripts. This means that Eq. (36) can be rewritten in the form

$$\begin{aligned}
F(u) &= \underbrace{\sin u_0}_{A_0} + \underbrace{u_1 \cos u_0}_{A_1} + \underbrace{u_2 \cos u_0 - \frac{1}{2!}u_1^2 \sin u_0}_{A_2} \\
&\quad + \underbrace{u_3 \cos u_0 - u_1 u_2 \sin u_0 - \frac{1}{3!}u_1^3 \cos u_0}_{A_3} + \cdots
\end{aligned} \tag{37}$$

Case 2. $F(u) = \cos u$

Proceeding as before we obtain

$$F(u) = \underbrace{\cos u_0}_{A_0} - \underbrace{u_1 \sin u_0}_{A_1} + \underbrace{(-u_2 \sin u_0 - \frac{1}{2!}u_1^2 \cos u_0)}_{A_2}$$

$$+ \underbrace{(-u_3 \sin u_0 - u_1 u_2 \cos u_0 + \frac{1}{3!}u_1^3 + \cdots)}_{A_3} \tag{38}$$

III. Hyperbolic Nonlinearity

Case 1. $F(u) = \sinh u$

To calculate the A_n polynomials for $F(u) = \sinh u$, we first substitute

$$u = \sum_{n=0}^{\infty} u_n, \tag{39}$$

into $F(u) = \sinh u$ to obtain

$$F(u) = \sinh[u_0 + (u_1 + u_2 + u_3 + u_4 + \cdots)]. \tag{40}$$

To calculate A_0, recall the hyperbolic identity

$$\sinh(\theta + \phi) = \sinh\theta \cosh\phi + \cosh\theta \sinh\phi. \tag{41}$$

Accordingly, Eq. (40) becomes

$$F(u) = \sinh u_0 \cosh(u_1 + u_2 + u_3 + u_4 + \cdots)$$

$$+ \cosh u_0 \sinh(u_1 + u_2 + u_3 + u_4 + \cdots). \tag{42}$$

Separating $F(u_0) = \sinh u_0$ from other factors and using Taylor expansions for $\cosh(u_1 + u_2 + \cdots)$ and $\sinh(u_1 + u_2 + \cdots)$ give

$$F(u) = \sinh u_0 \times$$

$$\left(1 + \tfrac{1}{2!}(u_1 + u_2 + \cdots)^2 + \tfrac{1}{4!}(u_1 + u_2 + \cdots)^4 + \cdots\right)$$

$$+ \cosh u_0 \left((u_1 + u_2 + \cdots) - \tfrac{1}{3!}(u_1 + u_2 + \cdots)^3 + \cdots\right)$$

$$= \sinh u_0 \left(1 + \tfrac{1}{2!}(u_1^2 + 2u_1 u_2 + \cdots)\right)$$

$$+ \cosh u_0 \left((u_1 + u_2 + \cdots) - \tfrac{1}{3!}u_1^3 + \cdots\right).$$

By grouping all terms with the same sum of subscripts we find

$$
\begin{aligned}
F(u) \;=\; & \underbrace{\sinh u_0}_{A_0} + \underbrace{u_1 \cosh u_0}_{A_1} + \underbrace{u_2 \cosh u_0 + \frac{1}{2!}u_1^2 \sinh u_0}_{A_2} \\
& + \underbrace{u_3 \cosh u_0 + u_1 u_2 \sinh u_0 + \frac{1}{3!}u_1^3 \cosh u_0 + \cdots}_{A_3}.
\end{aligned}
\tag{43}
$$

Case 2. $F(u) = \cosh u$

Proceeding as in $\sinh x$ we find

$$
\begin{aligned}
F(u) \;=\; & \underbrace{\cosh u_0}_{A_0} + \underbrace{u_1 \sinh u_0}_{A_1} + \underbrace{u_2 \sinh u_0 + \frac{1}{2!}u_1^2 \cosh u_0}_{A_2} \\
& + \underbrace{u_3 \sinh u_0 + u_1 u_2 \cosh u_0 + \frac{1}{3!}u_1^3 \sinh u_0 + \cdots}_{A_3}.
\end{aligned}
\tag{44}
$$

III. Exponential Nonlinearity

Case 1. $F(u) = e^u$

Substituting

$$
u = \sum_{n=0}^{\infty} u_n,
\tag{45}
$$

into $F(u) = e^u$ gives

$$
F(u) = e^{(u_0 + u_1 + u_2 + u_3 + \cdots)},
\tag{46}
$$

or equivalently

$$
F(u) = e^{u_0} e^{(u_1 + u_2 + u_3 + \cdots)}.
\tag{47}
$$

Keeping the term e^{u_0} and using the Taylor expansion for the other factor we obtain

$$
\begin{aligned}
F(u) = e^{u_0} \times \\
\left(1 + (u_1 + u_2 + u_3 + \cdots) + \tfrac{1}{2!}(u_1 + u_2 + u_3 + \cdots)^2 + \cdots\right).
\end{aligned}
\tag{48}
$$

By grouping all terms with identical sum of subscripts we find

$$
\begin{aligned}
F(u) \;=\; & \underbrace{e^{u_0}}_{A_0} + \underbrace{u_1 e^{u_0}}_{A_1} + \underbrace{(u_2 + \frac{1}{2!}u_1^2)e^{u_0}}_{A_2} + \underbrace{(u_3 + u_1 u_2 + \frac{1}{3!}u_1^3)e^{u_0}}_{A_3} \\
& + \underbrace{(u_4 + u_1 u_3 + \frac{1}{2!}u_2^2 + \frac{1}{2!}u_1^2 u_2 + \frac{1}{4!}u_1^4)e^{u_0} + \cdots}_{A_4}.
\end{aligned}
\tag{49}
$$

Case 2. $F(u) = e^{-u}$

Proceeding as before we find

$$
F(u) = \underbrace{e^{-u_0}}_{A_0} + \underbrace{-u_1 e^{-u_0}}_{A_1} + \underbrace{(-u_2 + \frac{1}{2!}u_1^2)e^{-u_0}}_{A_2}
$$

$$
+ \underbrace{(-u_3 + u_1 u_2 - \frac{1}{3!}u_1^3)e^{-u_0}}_{A_3}
\tag{50}
$$

$$
+ \underbrace{(-u_4 + u_1 u_3 + \frac{1}{2!}u_2^2 - \frac{1}{2!}u_1^2 u_2 + \frac{1}{4!}u_1^4)e^{-u_0}}_{A_4} + \cdots .
$$

IV. Logarithmic Nonlinearity

Case 1. $F(u) = \ln u,\ u > 0$

Substituting

$$
u = \sum_{n=0}^{\infty} u_n,
\tag{51}
$$

into $F(u) = \ln u$ gives

$$
F(u) = \ln(u_0 + u_1 + u_2 + u_3 + \cdots).
\tag{52}
$$

Equation (52) can be written as

$$
F(u) = \ln\left(u_0(1 + \frac{u_1}{u_0} + \frac{u_2}{u_0} + \frac{u_3}{u_0} + \cdots) \right).
\tag{53}
$$

Using the fact that $\ln(\alpha\beta) = \ln\alpha + \ln\beta$, Eq. (53) becomes

$$
F(u) = \ln u_0 + \ln(1 + \frac{u_1}{u_0} + \frac{u_2}{u_0} + \frac{u_3}{u_0} + \cdots).
\tag{54}
$$

Separating $F(u_0) = \ln u_0$ and using the Taylor expansion for the remaining term we obtain

$$
F(u) = \ln u_0 + \{(\frac{u_1}{u_0} + \frac{u_2}{u_0} + \frac{u_3}{u_0} + \cdots) - \frac{1}{2}(\frac{u_1}{u_0} + \frac{u_2}{u_0} + \frac{u_3}{u_0} + \cdots)^2
$$

$$
+ \frac{1}{3}(\frac{u_1}{u_0} + \frac{u_2}{u_0} + \frac{u_3}{u_0} + \cdots)^3 - \frac{1}{4}(\frac{u_1}{u_0} + \frac{u_2}{u_0} + \frac{u_3}{u_0} + \cdots)^4 + \cdots \}.
\tag{55}
$$

Proceeding as before, Eq. (55) can be written as

$$
F(u) = \underbrace{\ln u_0}_{A_0} + \underbrace{\frac{u_1}{u_0}}_{A_1} + \underbrace{\frac{u_2}{u_0} - \frac{1}{2}\frac{u_1^2}{u_0^2}}_{A_2} + \underbrace{\frac{u_3}{u_0} - \frac{u_1 u_2}{u_0^2} + \frac{1}{3}\frac{u_1^3}{u_0^3}}_{A_3} + \cdots .
\tag{56}
$$

Case 2. $F(u) = \ln(1+u)$, $-1 < u \le 1$

In a like manner we obtain

$$
F(u) = \underbrace{\ln(1+u_0)}_{A_0} + \underbrace{\frac{u_1}{1+u_0}}_{A_1} + \underbrace{\frac{u_2}{1+u_0} - \frac{1}{2}\frac{u_1^2}{(1+u_0)^2}}_{A_2}
$$
$$
+ \underbrace{\frac{u_3}{1+u_0} - \frac{u_1 u_2}{(1+u_0)^2} + \frac{1}{3}\frac{u_1^3}{(1+u_0)^3}}_{A_3} + \cdots .
$$

(57)

As stated before, there are other methods that can be used to evaluate Adomian polynomials. However, these methods suffer from the huge size of calculations. For this reason, the most commonly used methods are presented in this text.

Exercises 8.2

Use Adomian algorithm or the alternative method to calculate the first four Adomian polynomials of the following nonlinear terms:

1. $F(u) = u^4$

2. $F(u) = u^2 + u^3$

3. $F(u) = \cos 2u$

4. $F(u) = \sinh 2u$

5. $F(u) = e^{2u}$

6. $F(u) = u^2 u_x$

7. $F(u) = u u_x^2$

8. $F(u) = u e^u$

9. $F(u) = u \sin u$

10. $F(u) = u \cosh u$

11. $F(u) = u^2 + \sin u$

12. $F(u) = u + \cos u$

13. $F(u) = u \ln u$, $u > 0$

14. $F(u) = u + \ln u$, $u > 0$

15. $F(u) = u^{\frac{1}{2}}$, $u > 0$

16. $F(u) = u^{-1}$, $u > 0$

8.3   Nonlinear Ordinary Differential Equations

Although this book is devoted to investigate partial differential equations, it seems useful to employ the Adomian decomposition method first to nonlinear ordinary differential equations. It is well known that nonlinear ordinary differential equations are, in general, difficult to handle. The Adomian decomposition method will be applied in a direct manner as discussed in previous chapters except that nonlinear terms should be represented by the so called Adomian polynomials. It is interesting to point out that the modified decomposition method and the noise terms phenomenon that were introduced in Chapter 2 will be used here at proper places.

Recall that in solving differential or integral equations, solutions are usually obtained as exact solutions defined in closed form expressions, or as series solutions normally obtained from concrete problems.

To apply the Adomian decomposition method for solving nonlinear ordinary differential equations, we consider the equation

$$Ly + Ry + F(y) = g(x), \tag{58}$$

where the differential operator L may be considered as the highest order derivative in the equation, R is the remainder of the differential operator, $F(y)$ expresses the nonlinear terms, and $g(x)$ is an inhomogeneous term. If L is a first order operator defined by

$$L = \frac{d}{dx}, \tag{59}$$

then, we assume that L is invertible and the inverse operator L^{-1} is given by

$$L^{-1}(.) = \int_0^x (.)dx, \tag{60}$$

so that

$$L^{-1}Ly = y(x) - y(0). \tag{61}$$

However, if L is a second order differential operator given by

$$L = \frac{d^2}{dx^2}, \tag{62}$$

so that the inverse operator L^{-1} is regarded a two-fold integration operator defined by

$$L^{-1}(.) = \int_0^x \int_0^x (.)dx\,dx, \tag{63}$$

which means that

$$L^{-1}Ly = y(x) - y(0) - xy'(0). \tag{64}$$

In a parallel manner, if L is a third order differential operator, we can easily show that

$$L^{-1}Ly = y(x) - y(0) - xy'(0) - \frac{1}{2!}x^2y''(0). \tag{65}$$

For higher order operators we can easily define the related inverse operators in a similar way.

Applying L^{-1} to both sides of (58) gives

$$y(x) = \psi_0 - L^{-1}g(x) - L^{-1}Ry - L^{-1}F(y), \tag{66}$$

where

$$\psi_0 = \begin{cases} y(0) & \text{for } L = \frac{d}{dx}, \\[2mm] y(0) - xy'(0) & \text{for } L = \frac{d^2}{dx^2}, \\[2mm] y(0) - xy'(0) - \frac{1}{2!}x^2y''(0) & \text{for } L = \frac{d^3}{dx^3}, \\[2mm] y(0) - xy'(0) - \frac{1}{2!}x^2y''(0) - \frac{1}{3!}x^3y'''(0) & \text{for } L = \frac{d^4}{dx^4}, \\[2mm] y(0) - xy'(0) - \frac{1}{2!}x^2y''(0) - \frac{1}{3!}x^3y'''(0) - \frac{1}{4!}x^4y^{(iv)}(0) & \text{for } L = \frac{d^5}{dx^5}, \end{cases} \tag{67}$$

and so on. The Adomian decomposition method admits the decomposition of y into an infinite series of components

$$y(x) = \sum_{n=0}^{\infty} y_n, \tag{68}$$

and the nonlinear term $F(y)$ be equated to an infinite series of polynomials

$$F(y) = \sum_{n=0}^{\infty} A_n, \tag{69}$$

where A_n are the Adomian polynomials. Substituting (68) and (69) into (66) gives

$$\sum_{n=0}^{\infty} y_n = \psi_0 - L^{-1}g(x) - L^{-1}R\left(\sum_{n=0}^{\infty} y_n\right) - L^{-1}\left(\sum_{n=0}^{\infty} A_n\right). \tag{70}$$

The various components y_n of the solution y can be easily determined by using the recursive relation

$$\begin{aligned} y_0 &= \psi_0 - L^{-1}(g(x)), \\ y_{k+1} &= L^{-1}(Ry_k) - L^{-1}(A_k), k \geq 0. \end{aligned} \tag{71}$$

Consequently, the first few components can be written as

$$
\begin{aligned}
y_0 &= \psi_0 - L^{-1}g(x), \\
y_1 &= -L^{-1}(Ry_0) - L^{-1}(A_0), \\
y_2 &= -L^{-1}(Ry_1) - L^{-1}(A_1), \\
y_3 &= -L^{-1}(Ry_2) - L^{-1}(A_2), \\
y_4 &= -L^{-1}(Ry_3) - L^{-1}(A_3).
\end{aligned}
\tag{72}
$$

Having determined the components y_n, $n \geq 0$, the solution y in a series form follows immediately. As stated before, the series may be summed to provide the solution in a closed form. However, for concrete problems, the $n-$term partial sum

$$
\phi_n = \sum_{k=0}^{n-1} y_k,
\tag{73}
$$

may be used to give the approximate solution.

In the following, several examples will be discussed for illustration.

Example 1. Solve the first order nonlinear ordinary differential equation

$$
y' - y^2 = 1, \; y(0) = 0.
\tag{74}
$$

Solution.

In an operator form, Eq. (74) can be written as

$$
Ly = 1 + y^2, \; y(0) = 0,
\tag{75}
$$

where L is a first order differential operator. It is clear that L^{-1} is invertible and given by

$$
L^{-1}(.) = \int_0^x (.) \, dx.
\tag{76}
$$

Applying L^{-1} to both sides of (75) and using the initial condition give

$$
y = x + L^{-1}(y^2).
\tag{77}
$$

The decomposition method suggests that the solution $y(x)$ be expressed by the decomposition series

$$
y(x) = \sum_{n=0}^{\infty} y_n(x),
\tag{78}
$$

and the nonlinear terms y^2 be equated to

$$y^2 = \sum_{n=0}^{\infty} A_n, \tag{79}$$

where $y_n(x), n \geq 0$ are the components of $y(x)$ that will be determined recursively, and A_n, $n \geq 0$ are the Adomian polynomials that represent the nonlinear term y^2.
Inserting (78) and (79) into (77) yields

$$\sum_{n=0}^{\infty} y_n(x) = x + L^{-1}\left(\sum_{n=0}^{\infty} A_n\right). \tag{80}$$

The zeroth component y_0 is usually defined by all terms that are not included under the operator L^{-1}. The remaining components can be determined recurrently such that each term is determined by using the previous component. Consequently, the components of $y(x)$ can be elegantly determined by using the recursive relation

$$\begin{aligned} y_0(x) &= x, \\ y_{k+1}(x) &= L^{-1}(A_k), k \geq 0. \end{aligned} \tag{81}$$

Note that the Adomian polynomials A_n for the nonlinear term y^2 were determined before by using Adomian algorithm and by using the alternative method where we found

$A_0 = y_0^2$,
$A_1 = 2y_0 y_1$,
$A_2 = 2y_0 y_2 + y_1^2$,
$A_3 = 2y_0 y_3 + 2y_1 y_2$,
$A_4 = 2y_0 y_4 + 2y_1 y_3 + y_2^2$,

and so on. Using these polynomials into (81), the first few components can be determined recursively by

$$\begin{aligned} y_0(x) &= x, \\ y_1(x) &= L^{-1}A_0 = L^{-1}(y_0^2) = \tfrac{1}{3}x^3, \\ y_2(x) &= L^{-1}A_1 = L^{-1}(2y_0 y_1) = \tfrac{2}{15}x^5, \\ y_3(x) &= L^{-1}A_2 = L^{-1}(2y_0 y_2 + y_1^2) = \tfrac{17}{315}x^7, \end{aligned} \tag{82}$$

Consequently, the solution in a series form is given by

$$y(x) = x + \frac{1}{3}x^3 + \frac{2}{15}x^5 + \frac{17}{315}x^7 + \cdots, \tag{83}$$

and in a closed form by

$$y(x) = \tan x. \tag{84}$$

We point out that the ordinary differential equation (74) can be solved as a separable differential equation.

Example 2. Solve the first order nonlinear ordinary differential equation

$$y' + y^2 = 1, \; y(0) = 0. \tag{85}$$

Solution.

Operating with L^{-1} we obtain

$$y = x - L^{-1}(y^2). \tag{86}$$

Using the decomposition series for the solution $y(x)$ and the polynomial representation for y^2 give

$$\sum_{n=0}^{\infty} y_n(x) = x - L^{-1}\left(\sum_{n=0}^{\infty} A_n\right). \tag{87}$$

This leads to the recursive relation

$$\begin{aligned} y_0 &= x, \\ y_{k+1} &= -L^{-1}(A_k), k \geq 0. \end{aligned} \tag{88}$$

The Adomian polynomials A_n for the nonlinear term y^2 were used in the previous example, hence, the first few components can be determined recursively as

$$\begin{aligned} y_0 &= x, \\ y_1 &= -L^{-1}A_0 = L^{-1}(y_0^2) = -\tfrac{1}{3}x^3, \\ y_2 &= -L^{-1}A_1 = -L^{-1}(2y_0y_1) = \tfrac{2}{15}x^5, \\ y_3 &= -L^{-1}A_2 = -L^{-1}(2y_0y_2 + y_1^2) = -\tfrac{17}{315}x^7, \end{aligned} \tag{89}$$

and so on. Consequently, the solution in a series form is given by

$$y(x) = x - \frac{1}{3}x^3 + \frac{2}{15}x^5 - \frac{17}{315}x^7 + \cdots, \tag{90}$$

and in a closed form by

$$y(x) = \tanh x. \tag{91}$$

We point out that the ordinary differential equation (85) can be solved as a separable differential equation where partial fractions should be used.

Example 3. Use the modified decomposition method to solve the Riccati differential equation

$$y' = 1 - x^2 + y^2, \ y(0) = 0. \tag{92}$$

Solution.

Applying the inverse operator L^{-1} we obtain

$$y = x - \frac{1}{3}x^3 + L^{-1}(y^2). \tag{93}$$

Using the decomposition series $y(x)$ and the polynomial representation for y^2 give

$$\sum_{n=0}^{\infty} y_n(x) = x - \frac{1}{3}x^3 + L^{-1}\left(\sum_{n=0}^{\infty} A_n\right). \tag{94}$$

It is important to point out that the modified decomposition method is recommended here. In this approach we split the polynomial $x - \frac{1}{3}x^3$ into two parts, namely, x will be assigned to the zeroth component y_0, and $-\frac{1}{3}x^3$ that will be assigned to the component y_1 among other terms. In this case, we use a modified recursive relation to accelerate the convergence of the solution. The modified recursive relation is defined by

$$
\begin{aligned}
y_0 &= x, \\
y_1 &= -\tfrac{1}{3}x^3 + L^{-1}(A_0), \\
y_{k+2} &= -L^{-1}(A_{k+1}), k \geq 0.
\end{aligned} \tag{95}
$$

Consequently, the first few components are given by

$$
\begin{aligned}
y_0 &= x, \\
y_1 &= -\tfrac{1}{3}x^3 + L^{-1}A_0 = -\tfrac{1}{3}x^3 + L^{-1}(y_0^2) = 0, \\
y_{k+2} &= 0, k \geq 0.
\end{aligned} \tag{96}
$$

The exact solution is given by

$$y(x) = x. \tag{97}$$

We next consider a first order nonlinear differential equation where a closed form solution is not easily observed.

Example 4. Solve the first order nonlinear differential equation

$$y' = -y + y^2, \ y(0) = 2. \tag{98}$$

Solution.

Applying the inverse operator L^{-1} and using the initial condition give

$$y(x) = 2 - L^{-1}(y) + L^{-1}(y^2). \tag{99}$$

We next represent the linear term $y(x)$ by the decomposition series of components y_n, $n \geq 0$, and equate the nonlinear term y^2 by the series of Adomian polynomials A_n, $n \geq 0$, to find

$$\sum_{n=0}^{\infty} y_n(x) = 2 - L^{-1}\left(\sum_{n=0}^{\infty} y_n(x)\right) + L^{-1}\left(\sum_{n=0}^{\infty} A_n\right). \tag{100}$$

The Adomian polynomials A_n for y^2 have been derived and used before. Following the decomposition method we set the recursive relation

$$\begin{aligned} y_0 &= 2, \\ y_{k+1} &= -L^{-1}(y_k) + L^{-1}(A_k), k \geq 0. \end{aligned} \tag{101}$$

Consequently, the first few components of the solution are given by

$$\begin{aligned} y_0 &= 2, \\ y_1 &= -L^{-1}(y_0) + L^{-1}A_0 = -L^{-1}(2) + L^{-1}(4) = 2x, \\ y_2 &= -L^{-1}(y_1) + L^{-1}A_1 = -L^{-1}(2x) + L^{-1}(8x) = 3x^2, \\ y_3 &= -L^{-1}(y_2) + L^{-1}A_2 = -L^{-1}(3x^2) + L^{-1}(16x^2) = \tfrac{13}{3}x^3, \end{aligned} \tag{102}$$

and so on. Based on these calculations, the solution in a series form is given by

$$y(x) = 2 + 2x + 3x^2 + \frac{13}{3}x^3 + \cdots, \quad x < \ln 2. \tag{103}$$

It is clear that a closed form solution is not easily observed. However, the closed form solution is given by

$$y(x) = \frac{2}{2 - e^x}. \tag{104}$$

For numerical purposes, additional terms can be easily evaluated to enhance the accuracy of the approximate solution in (103). As will be discussed in Chapter 10, it will be shown that few terms only can lead to a high accuracy level with minimum error.

Example 5. Solve the first order nonlinear differential equation

$$y' = \frac{y^2}{1 - xy}, \quad y(0) = 1. \tag{105}$$

Solution.

We first rewrite the equation by

$$y' = xyy' + y^2, \ y(0) = 1. \tag{106}$$

It is useful to note that the differential equation (106) contains two nonlinear terms yy' and y^2. The Adomian polynomials for both terms have been derived before; hence will be implemented directly. Applying the inverse operator L^{-1} and using the initial condition give

$$y(x) = 1 + L^{-1}(xyy') + L^{-1}(y^2). \tag{107}$$

We next represent the linear term $y(x)$ by the decomposition series of components y_n, $n \geq 0$, equate the nonlinear term yy' by the Adomian polynomials A_n, $n \geq 0$, and equate the nonlinear term y^2 by the series of Adomian polynomials B_n, $n \geq 0$, to find

$$\sum_{n=0}^{\infty} y_n(x) = 1 + L^{-1}\left(\sum_{n=0}^{\infty} x A_n\right) + L^{-1}\left(\sum_{n=0}^{\infty} B_n\right). \tag{108}$$

Identifying the zeroth component y_0, and following the decomposition method we set the recursive relation

$$\begin{aligned} y_0(x) &= 1, \\ y_{k+1}(x) &= L^{-1}(xA_k) + L^{-1}(B_k), k \geq 0. \end{aligned} \tag{109}$$

This relation leads to the component-by-component identification

$$\begin{aligned} y_0 &= 1, \\ y_1 &= L^{-1}(A_0) + L^{-1}(B_0) = L^{-1}(0) + L^{-1}(1) = x, \\ y_2 &= L^{-1}(A_1) + L^{-1}(B_1) = L^{-1}(x) + L^{-1}(2x) = \tfrac{3}{2}x^2, \\ y_3 &= L^{-1}(A_2) + L^{-1}(B_2) = L^{-1}(4x^2) + L^{-1}(4x^2) = \tfrac{8}{3}x^3, \end{aligned} \tag{110}$$

and so on. Based on these calculations, the solution in a series form is given by

$$y(x) = 1 + x + \frac{3}{2}x^2 + \frac{8}{3}x^3 + \cdots. \tag{111}$$

It is clear that a closed form solution where y is expressed explicitly in terms of x cannot be found. However, the exact solution can be expressed in the implicit expression

$$y = e^{xy}. \tag{112}$$

In the following example, the ordinary differential equation contains an exponential nonlinearity. The Adomian polynomials for this form of nonlinearity have been calculated before.

Example 6. Solve the first order nonlinear differential equation

$$y' - e^y = 0, \; y(0) = 1. \tag{113}$$

Solution.

Applying the inverse operator L^{-1} and using the initial condition give

$$y(x) = 1 + L^{-1}(e^y). \tag{114}$$

Equating the linear term $y(x)$ by an infinite series of components y_n, $n \geq 0$, and representing the nonlinear term e^y by an infinite series of Adomian polynomials A_n, $n \geq 0$, we obtain

$$\sum_{n=0}^{\infty} y_n(x) = 1 + L^{-1}\left(\sum_{n=0}^{\infty} A_n\right). \tag{115}$$

The Adomian polynomials A_n for e^y have been calculated before and given by

$$
\begin{aligned}
A_0 &= e^{y_0}, \\[4pt]
A_1 &= y_1 e^{y_0}, \\[4pt]
A_2 &= (y_2 + \tfrac{1}{2!}y_1^2)e^{y_0}, \\[4pt]
A_3 &= (\tfrac{1}{3!}y_1^3 + y_1 y_2 + y_3)e^{y_0}, \\[4pt]
A_4 &= (y_4 + y_1 y_3 + \tfrac{1}{2!}y_2^2 + \tfrac{1}{2!}y_1^2 y_2 + \tfrac{1}{4!}y_1^4)e^{y_0}.
\end{aligned}
\tag{116}
$$

Identifying $y_0 = 1$ and applying the decomposition method give the recursive relation

$$
\begin{aligned}
y_0 &= 1, \\[4pt]
y_{k+1} &= L^{-1}(A_k), k \geq 0.
\end{aligned}
\tag{117}
$$

This gives

$$
\begin{aligned}
y_0 &= 1, \\[4pt]
y_1 &= L^{-1}A_0 = L^{-1}(e) = ex, \\[4pt]
y_2 &= L^{-1}A_1 = L^{-1}(e^2 x) = \tfrac{e^2}{2}x^2, \\[4pt]
y_3 &= L^{-1}A_2 = L^{-1}(e^3 x^2) = \tfrac{e^3}{3}x^3,
\end{aligned}
\tag{118}
$$

and so on. The solution in a series form is given by

$$y(x) = 1 + ex + \frac{1}{2}(ex)^2 + \frac{1}{3}(ex)^3 + \cdots, \quad -1 \le ex < 1, \tag{119}$$

and in a closed form by

$$y(x) = 1 - \ln(1 - ex), \quad -1 \le ex < 1. \tag{120}$$

Example 7. Use the noise terms phenomenon to solve the second order nonlinear differential equation

$$y'' + (y')^2 + y^2 = 1 - \sin x, \; y(0) = 0, \; y'(0) = 1. \tag{121}$$

Solution.

Applying the two-fold integral operator L^{-1} to both sides of Eq. (121) gives

$$y(x) = \sin x + \frac{1}{2}x^2 - L^{-1}\left((y')^2 + y^2\right). \tag{122}$$

Using the assumptions of the decomposition method yields

$$\sum_{n=0}^{\infty} y_n(x) = \sin x + \frac{1}{2}x^2 - L^{-1}\left(\sum_{n=0}^{\infty} A_n\right). \tag{123}$$

This leads to the recursive relation

$$
\begin{aligned}
y_0 &= \sin x + \tfrac{1}{2}x^2, \\
y_{k+1} &= -L^{-1}(A_k), k \ge 0.
\end{aligned}
\tag{124}
$$

This relation leads to the identification

$$
\begin{aligned}
y_0 &= \sin x + \tfrac{1}{2}x^2, \\
y_1 &= -L^{-1}((y_0')^2 + y_0^2) = -\tfrac{1}{2!}x^2 + \cdots.
\end{aligned}
\tag{125}
$$

The zeroth component contains the trigonometric function $\sin x$, therefore it is recommended that the noise terms phenomenon be used here. By canceling the noise terms $\frac{1}{2}x^2$ and $-\frac{1}{2}x^2$ between y_0 and y_1, and justifying that the remaining non-canceled term of y_0 satisfies the differential equation leads to the exact solution given by

$$y(x) = \sin x. \tag{126}$$

This example shows that the noise terms phenomenon, if exists, works effectively for inhomogeneous ordinary differential equations and for inhomogeneous partial differential equations as well.

Exercises 8.3

In Exercises 1 – 6, use the Adomian scheme to find the exact solution of each of the
following nonlinear ordinary differential equations:

1. $y' - 3y^2 = 3$, $y(0) = 0$

2. $y' + 4y^2 = 4$, $y(0) = 0$

3. $y' - y^2 = -2x - x^2$, $y(0) = 1$

4. $y' + e^y = 0$, $y(0) = 1$

In Exercises 5 – 8, use the Adomian decomposition method to find the exact solution
of the following Riccati differential equations:

5. $y' = (2x - 3) - xy + y^2$, $y(0) = -2$

6. $y' = 1 - 2y + y^2$, $y(0) = 2$

7. $y' = 1 - x^2 + y^2$, $y(0) = 0$

8. $y' = -1 + xy + y^2$, $y(0) = 0$

In Exercises 9 – 12, use the Adomian decomposition method to find the first four
terms of the series solution of the following first order nonlinear ordinary differential
equations:

9. $y' + y = \sin y$, $y(0) = \frac{\pi}{2}$

10. $y' = x^2 + y^2$, $y(0) = 1$

11. $y' + y = y^2$, $y(0) = 2$

12. $y' = -\frac{y^2}{1+xy}$, $y(0) = 1$

In Exercises 13 – 16, use the Adomian decomposition method to find the exact
solution of the following second order nonlinear ordinary differential equations:

13. $y'' + (y')^2 + y^2 = 1 - \cos x$, $y(0) = 1$, $y'(0) = 0$

14. $y'' - 2yy' = 0$, $y(0) = 0$, $y'(0) = 1$

15. $y'' + 2yy' = 0$, $y(0) = 0$, $y'(0) = 1$

16. $y'' + yy' + (y')^2 = 0$, $y(0) = 0$, $y'(0) = 1$

In Exercises 17 – 20, use the Adomian decomposition method to find the first
four terms of the series solution of the following second order nonlinear ordinary
differential equations:

17. $y'' + y^2 = 0$, $y(0) = 1$, $y'(0) = 0$

18. $y'' - y^3 = 0$, $y(0) = 1$, $y'(0) = 0$

19. $y'' - \sin y = 0$, $y(0) = \frac{\pi}{2}$, $y'(0) = 0$

20. $y'' - ye^y = 0$, $y(0) = 1$, $y'(0) = 0$

In Exercises 20 – 24, use the Adomian decomposition method to find the exact solution of the following third order nonlinear ordinary differential equations:

21. $y''' + (y')^2 - 12y' = 6$, $y(0) = y'(0) = y''(0) = 0$

22. $y''' + (y'')^2 + (y')^2 = 2 + \cos x$, $y(0) = 0$, $y'(0) = 2$, $y''(0) = 0$

23. $y''' + (y'')^2 + (y')^2 = 1 - \sin x$, $y(0) = 0$, $y'(0) = 0$, $y''(0) = 1$

24. $y''' - (y'')^2 + (y')^2 = 1 + \cosh x$, $y(0) = 0$, $y'(0) = 1$, $y''(0) = 0$

8.4 Nonlinear Partial Differential Equations

It was indicated before that nonlinear partial differential equations arise in different areas of physics, engineering, and applied mathematics such as fluid mechanics, condensed matter physics, soliton physics and quantum field theory. A considerable amount of research work has been invested in the study of numerous problems modeled by nonlinear partial differential equations. In this section, a straightforward implementation of the decomposition method to nonlinear partial differential equations will be carried out in general. However, a wide variety of physically significant problems modeled by nonlinear partial differential equations, such as the advection problem, the KdV equation, the modified KdV equation, the KP equation, Boussinesq equation, will be investigated in details in Chapter 9. Moreover, a comparative study will be conducted in Chapter 11 to show the physical behavior of the solitons concept and the recently discovered compactons: solitons with the absence of infinite wings.

An important note worth mentioning is that there is no general method that can be employed for obtaining analytical solutions for nonlinear partial differential equations. Several methods are usually used and numerical solutions are often obtained. Further, transformation methods are sometimes used to convert a nonlinear equation to an ordinary equation or to a system of ordinary differential equations. Furthermore, perturbation techniques and discretization methods, that require a massive size of computational work, are also used for some types of equations.

However, we have stated before that Adomian decomposition method can be used generally for all types of differential and integral equations. The method can be applied in a straightforward manner and it provides a rapidly convergent series solution.

The description of the decomposition method has been presented in details in the preceding chapters. However, in the following we will discuss a general description of the method that will be used for nonlinear partial differential equations. We first consider the nonlinear partial differential equation given in an operator form

$$L_x u(x, y) + L_y u(x, y) + R u(x, y) + F(u(x, y)) = g(x, y), \qquad (127)$$

where L_x is the highest order differential in x, L_y is the highest order differential in y, R contains the remaining linear terms of lower derivatives, $F(u(x, y))$ is an analytic nonlinear term, and $g(x, y)$ is an inhomogeneous or forcing term.

The solutions for $u(x, y)$ obtained from the operator equations $L_x u$ and $L_y u$ are called partial solutions. It has been shown before that these partial solutions are equivalent and each converges to the exact solution. However, the decision as to which operator L_x or L_y should be used to solve the problem depends mainly on two bases:

(i) The operator of lowest order should be selected to minimize the size of computational work.

(ii) The selected operator of lowest order should be of best known conditions to accelerate the evaluation of the components of the solution.

Assuming that the operator L_x meets the two bases of selection, therefore we set

$$L_x u(x, y) = g(x, y) - L_y u(x, y) - R u(x, y) - F(u(x, y)). \qquad (128)$$

Applying L_x^{-1} to both sides of (128) gives

$$
\begin{aligned}
u(x, y) \;=\; & \Phi_0 - L_x^{-1} g(x, y) - L_x^{-1} L_y u(x, y) - L_x^{-1} R u(x, y) \\
& - L_x^{-1} F(u(x, y)),
\end{aligned}
\qquad (129)
$$

where

$$
\Phi_0 =
\begin{cases}
u(0, y) & \text{for } L = \frac{\partial}{\partial x}, \\[2mm]
u(0, y) - x u_x(0, y) & \text{for } L = \frac{\partial^2}{\partial x^2}, \\[2mm]
u(0, y) - x u_x(0, y) - \frac{1}{2!} x^2 u_{xx}(0, y) & \text{for } L = \frac{\partial^3}{\partial x^3}, \\[2mm]
u(0, y) - x u_x(0, y) - \frac{1}{2!} x^2 u_{xx}(0, y) - \frac{1}{3!} x^3 u_{xxx}(0, y) & \text{for } L = \frac{\partial^4}{\partial x^4},
\end{cases}
$$

We proceed in exactly the same manner by calculating the solution $u(x, y)$ in a series form

$$u(x, y) = \sum_{n=0}^{\infty} u_n(x, y), \qquad (130)$$

and the nonlinear term $F(u(x, y))$ by

$$F(u(x, y)) = \sum_{n=0}^{\infty} A_n, \qquad (131)$$

where A_n are Adomian polynomials that can be generated for all forms of nonlinearity. Based on these assumptions, Eq. (129) becomes

$$\sum_{n=0}^{\infty} u_n(x,y) = \Phi_0 - L_x^{-1}g(x,y) - L_x^{-1}L_y\left(\sum_{n=0}^{\infty} u_n(x,y)\right)$$
$$- L_x^{-1}\left(\sum_{n=0}^{\infty} u_n(x,y)\right) - L_x^{-1}\left(\sum_{n=0}^{\infty} A_n\right). \tag{132}$$

The components $u_n(x,y), n \geq 0$ of the solution $u(x,y)$ can be recursively determined by using the relation

$$u_0(x,y) = \Phi_0 - L_x^{-1}g(x,y),$$
$$u_{k+1}(x,y) = -L_x^{-1}L_y u_k - L_x^{-1}Ru_k - L_x^{-1}(A_k), \ k \geq 0. \tag{133}$$

Using the algorithms described before for calculating A_n for the nonlinear term $F(u)$, the first few components can be identified by

$$u_0(x,y) = \Phi_0 - L_x^{-1}g(x,y),$$
$$u_1(x,y) = -L_x^{-1}L_y u_0(x,y) - L_x^{-1}Ru_0(x,y) - L_x^{-1}A_0,$$
$$u_2(x,y) = -L_x^{-1}L_y u_1(x,y) - L_x^{-1}Ru_1(x,y) - L_x^{-1}A_1,$$
$$u_3(x,y) = -L_x^{-1}L_y u_2(x,y) - L_x^{-1}Ru_2(x,y) - L_x^{-1}A_2,$$
$$u_4(x,y) = -L_x^{-1}L_y u_3(x,y) - L_x^{-1}Ru_3(x,y) - L_x^{-1}A_3,$$

where each component can be determined by using the preceding component. Having calculated the components $u_n(x,y), n \geq 0$, the solution in a series form is readily obtained.

In the following, several distinct nonlinear partial differential equations will be discussed to illustrate the procedure outlined above.

Example 1. Solve the nonlinear partial differential equation

$$u_t + uu_x = 0, \ u(x,0) = x, \ x \in R, t > 0, \tag{134}$$

where $u = u(x,t)$.

Solution.

In an operator form, Eq. (134) becomes

$$L_t u(x,t) = -uu_x, \tag{135}$$

where L_t is defined by

$$L_t = \frac{\partial}{\partial t}. \tag{136}$$

The inverse operator L_t^{-1} is identified by

$$L_t^{-1}(.) = \int_0^t (.)dt. \tag{137}$$

Applying L_t^{-1} to both sides of (135) and using the initial condition we obtain

$$u(x,t) = x - L_t^{-1}uu_x. \tag{138}$$

Substituting

$$u(x,t) = \sum_{n=0}^{\infty} u_n(x,t), \tag{139}$$

and the nonlinear term by

$$uu_x = \sum_{n=0}^{\infty} A_n, \tag{140}$$

into (138) gives

$$\sum_{n=0}^{\infty} u_n(x,t) = x - L_t^{-1}\left(\sum_{n=0}^{\infty} A_n\right). \tag{141}$$

This gives the recursive relation

$$\begin{aligned} u_0(x,t) &= x, \\ u_{k+1}(x,t) &= -L_t^{-1}(A_k), \ k \geq 0. \end{aligned} \tag{142}$$

The first few components are given by

$$\begin{aligned} u_0(x,t) &= x, \\ u_1(x,t) &= -L_t^{-1}A_0 = -L_t^{-1}(x) = -xt, \\ u_2(x,t) &= -L_t^{-1}A_1 = -L_t^{-1}(-2xt) = xt^2, \\ u_3(x,t) &= -L_t^{-1}A_2 = -L_t^{-1}(3xt^2) = -xt^3, \end{aligned} \tag{143}$$

where additional terms can be easily computed. Combining the results obtained above, the solution in a series form is given by

$$u(x,t) = x(1 - t + t^2 - t^3 + \cdots), \tag{144}$$

and in a closed form by

$$u(x,t) = \frac{x}{1+t}, \ |t| < 1. \tag{145}$$

Example 2. Use the modified decomposition method to solve the nonlinear partial differential equation

$$u_t + uu_x = x + xt^2, \ u(x,0) = 0, \ x \in R, t > 0, \tag{146}$$

where $u = u(x,t)$.

Solution.

Note that the equation is an inhomogeneous equation. Proceeding as in Example 1, Eq. (146) becomes

$$L_t u(x,t) = x + xt^2 - uu_x. \tag{147}$$

Applying the inverse operator L_t^{-1} to both sides of (147) and using the initial condition we find

$$u(x,t) = xt + \frac{1}{3}xt^3 - L_t^{-1}uu_x. \tag{148}$$

Using the decomposition assumptions for the linear term $u(x,t)$ and for the nonlinear term uu_x defined by

$$u(x,t) = \sum_{n=0}^{\infty} u_n(x,t), \tag{149}$$

and

$$uu_x = \sum_{n=0}^{\infty} A_n, \tag{150}$$

into (148) gives

$$\sum_{n=0}^{\infty} u_n(x,t) = xt + \frac{1}{3}xt^3 - L_t^{-1}\left(\sum_{n=0}^{\infty} A_n\right), \tag{151}$$

where

$$A_0 = u_0 u_{0_x}.$$

The modified decomposition method admits the use of a modified recursive relation given by

$$\begin{aligned}
u_0(x,t) &= xt, \\
u_1(x,t) &= \tfrac{1}{3}xt^3 - L_t^{-1}(A_0), \\
u_{k+2}(x,t) &= -L_t^{-1}A_{k+1}, \ k \geq 0.
\end{aligned} \tag{152}$$

Consequently, we obtain

$$\begin{aligned}
u_0(x,t) &= xt, \\
u_1(x,t) &= \tfrac{1}{3}xt^3 - L_t^{-1}(xt^2) = 0, \\
u_{k+2}(x,t) &= 0, k \geq 0.
\end{aligned} \tag{153}$$

In view of (153), the exact solution is given by

$$u(x,t) = xt. \tag{154}$$

Example 3. Solve the nonlinear partial differential equation

$$u_t = x^2 + \frac{1}{4}u_x^2, \quad u(x,0) = 0, \tag{155}$$

where $u = u(x,t)$.

Solution.

Operating with L_t^{-1} we find

$$u(x,t) = x^2 t + \frac{1}{4}L_t^{-1}u_x^2. \tag{156}$$

The decomposition method suggests that $u(x,t)$ can be defined by

$$u(x,t) = \sum_{n=0}^{\infty} u_n(x,t), \tag{157}$$

and the nonlinear term u_x^2 by

$$u_x^2 = \sum_{n=0}^{\infty} A_n, \tag{158}$$

where $A_n, n \geq 0$, are Adomian polynomials. Using these assumptions gives

$$\sum_{n=0}^{\infty} u_n(x,t) = x^2 t + \frac{1}{4}L_t^{-1}\left(\sum_{n=0}^{\infty} A_n\right). \tag{159}$$

This gives the recursive relation

$$\begin{aligned} u_0(x,t) &= x^2 t, \\ u_{k+1}(x,t) &= \tfrac{1}{4}L_t^{-1}A_k, \ k \geq 0. \end{aligned} \tag{160}$$

The Adomian polynomials A_n for this form of nonlinearity are given by

$A_0 = u_{0_x}^2,$
$A_1 = 2u_{0_x}u_{1_x},$
$A_2 = 2u_{0_x}u_{2_x} + u_{1_x}^2,$
$A_3 = 2u_{0_x}u_{3_x} + 2u_{1_x}u_{2_x},$

and so on. The first few components are determined as follows:

$$\begin{aligned} u_0(x,t) &= x^2 t, \\ u_1(x,t) &= \tfrac{1}{4}L_t^{-1}A_0 = \tfrac{1}{4}L_t^{-1}(4x^2 t^2) = \tfrac{1}{3}x^3 t^3, \\ u_2(x,t) &= \tfrac{1}{4}L_t^{-1}A_1 = \tfrac{1}{4}L_t^{-1}(\tfrac{8}{3}x^2 t^4) = \tfrac{2}{15}x^2 t^5, \\ u_3(x,t) &= \tfrac{1}{4}L_t^{-1}A_2 = \tfrac{1}{4}L_t^{-1}(\tfrac{68}{45}x^2 t^6) = \tfrac{17}{315}x^3 t^7, \end{aligned} \tag{161}$$

and so on. Combining the results obtained for the components, the solution in a series form is given by

$$u(x,t) = x^2 \left(t + \frac{1}{3}t^3 + \frac{2}{15}t^5 + \frac{17}{315}t^7 + \cdots \right), \tag{162}$$

and in a closed form by

$$u(x,t) = x^2 \tan t. \tag{163}$$

Example 4. Solve the nonlinear partial differential equation by the modified decomposition method

$$u_{xx} - u_x u_{yy} = -x + u, \ u(0,y) = \sin y, \ u_x(0,y) = 1, \tag{164}$$

where $u = u(x,y)$.

Solution.

Note that the equation is an inhomogeneous equation. We first write Eq. (164) in an operator form

$$L_x u = -x + u + u_x u_{yy}, \tag{165}$$

where L_x is a second order partial differential operator given by

$$L_x = \frac{\partial^2}{\partial x^2}. \tag{166}$$

Assuming that L_x^{-1} is invertible, and the inverse operator L_x^{-1} is a two-fold integral operator defined by

$$L_x^{-1}(.) = \int_0^x \int_0^x (.)dx\,dx, \tag{167}$$

so that

$$\begin{aligned} L_x^{-1}L_x u &= \int_0^x \int_0^x u_{xx}dx\,dx, \\ &= u(x,y) - u(0,y) - x u_x(0,y). \end{aligned} \tag{168}$$

Proceeding as before we find

$$u(x,y) = \sin y + x - \frac{1}{3!}x^3 + L_x^{-1}(u + u_x u_{yy}) \tag{169}$$

Following Adomian method we obtain

$$\sum_{n=0}^{\infty} u_n(x,y) = \sin y + x - \frac{1}{3!}x^3 + L_x^{-1}\left(\sum_{n=0}^{\infty} u_n(x,y) + \sum_{n=0}^{\infty} A_n \right), \tag{170}$$

where A_n are the Adomian polynomials that represent the nonlinear term $u_x u_{yy}$.

To use the modified decomposition method, we identify the component u_0 by $u_0(x, y) = \sin y + x$, and the remaining term $-\frac{1}{2}x^2$ will be assigned to $u_1(x, y)$ among other terms. Consequently, we obtain the recursive relation

$$
\begin{aligned}
u_0(x, y) &= \sin y + x \\
u_1(x, y) &= -\tfrac{1}{3!}x^3 + L_x^{-1}(u_0 + A_0), \\
u_{k+2}(x, y) &= L_x^{-1}(u_{k+1} + A_{k+1}), k \geq 0.
\end{aligned}
\tag{171}
$$

Consequently, we obtain

$$
\begin{aligned}
u_0(x, y) &= \sin y + x, \\
u_1(x, y) &= -\tfrac{1}{3!}x^3 + L_x^{-1}(u_0 + A_0) = -\tfrac{1}{3!}x^3 + L_x^{-1}(x) = 0,
\end{aligned}
\tag{172}
$$

The exact solution is therefore given by

$$
u(x, y) = x + \sin y
\tag{173}
$$

Example 5. Solve the nonlinear partial differential equation

$$
u_{xx} + \frac{1}{4}u_y^2 = u, \; u(0, y) = 1 + y^2, \; u_x(0, y) = 1,
\tag{174}
$$

where $u = u(x, y)$.

Solution.

Operating with L_x^{-1} on (174) and using the given conditions we find

$$
u(x, y) = y^2 + 1 + x + L_x^{-1}\left(u(x, y) - \frac{1}{4}u_y^2\right).
\tag{175}
$$

Proceeding as before we obtain

$$
\sum_{n=0}^{\infty} u_n(x, y) = y^2 + 1 + x + L_x^{-1}\left(\sum_{n=0}^{\infty} u_n(x, y) - \frac{1}{4}\sum_{n=0}^{\infty} A_n\right),
\tag{176}
$$

where A_n are the Adomian polynomials that represent the nonlinear term u_y^2. The decomposition method admits the use of the recursive relation

$$
\begin{aligned}
u_0(x, y) &= y^2 + 1 + x \\
u_{k+1}(x, y) &= L_x^{-1}(u_k - \tfrac{1}{4}A_k), k \geq 0.
\end{aligned}
\tag{177}
$$

The Adomian polynomials are given by

$$A_0 = u_{0_y}^2,$$
$$A_1 = 2u_{0_y} u_{1_y},$$
$$A_2 = 2u_{0_y} u_{2_y} + u_{1_y}^2,$$
$$A_3 = 2u_{0_y} u_{3_y} + 2u_{1_y} u_{2_y},$$

and so on. The first few components of the solution $u(x,y)$ are given by

$$
\begin{aligned}
u_0(x,y) &= y^2 + 1 + x, \\
u_1(x,y) &= L_x^{-1}(u_0 - \tfrac{1}{4}A_0) = L_x^{-1}(1+x) - \tfrac{1}{2!}x^2 + \tfrac{1}{3!}x^3, \\
u_2(x,y) &= L_x^{-1}(u_1 - \tfrac{1}{4}A_1) = L_x^{-1}(\tfrac{1}{2!}x^2 + \tfrac{1}{3!}x^3) = \tfrac{1}{4!}x^4 + \tfrac{1}{5!}x^5,
\end{aligned}
\tag{178}
$$

and so on for other components. Consequently, the solution in a series form is given by

$$u(x,y) = y^2 + (1 + x + \frac{1}{2!}x^2 + \frac{1}{3}x^3 + \frac{1}{4!}x^4 + \cdots), \tag{179}$$

which gives the solution in a closed form by

$$u(x,y) = y^2 + e^x. \tag{180}$$

Example 6. Solve the nonlinear partial differential equation

$$u_{xx} + u^2 - u_y^2 = 0, \ u(0,y) = 0, \ u_x(0,y) = e^y, \tag{181}$$

where $u = u(x,y)$.

Solution.

We first write Eq. (181) in an operator form by

$$L_x u = u_y^2 - u^2, \tag{182}$$

where L_x is a second order partial differential operator. Operating with L_x^{-1} gives

$$u(x,y) = xe^y + L_x^{-1}\left(u_y^2 - u^2\right), \tag{183}$$

so that

$$\sum_{n=0}^{\infty} u_n(x,y) = xe^y + L_x^{-1}\left(\sum_{n=0}^{\infty} A_n - \sum_{n=0}^{\infty} B_n\right), \tag{184}$$

where A_n and B_n are the Adomian polynomials that represent the nonlinear terms u_y^2 and u^2 respectively. We next set the recursive relation

$$
\begin{aligned}
u_0(x,y) &= xe^y \\
u_{k+1}(x,y) &= L_x^{-1}(A_k - B_k), k \geq 0.
\end{aligned}
\tag{185}
$$

The first few components of the solution $u(x,y)$ are given by

$$
\begin{aligned}
u_0(x,y) &= xe^y, \\
u_1(x,y) &= L_x^{-1}(A_0 - B_0) = 0,
\end{aligned}
\tag{186}
$$

and therefore other components vanish. The exact solution is given by

$$
u(x,y) = xe^y.
\tag{187}
$$

Example 7. Solve the nonlinear partial differential equation

$$
u_{xx} + u^2 - u_{yy}^2 = 0, \ u(0,y) = 0, \ u_x(0,y) = \cos y,
\tag{188}
$$

where $u = u(x,y)$.

Solution.

Operating with the two-fold integral operator L_x^{-1} on (188) leads to

$$
u(x,y) = x\cos y + L_x^{-1}\left(u_{yy}^2 - u^2\right).
\tag{189}
$$

Following Adomian decomposition method we obtain

$$
\sum_{n=0}^{\infty} u_n(x,y) = x\cos y + L_x^{-1}\left(\sum_{n=0}^{\infty} A_n - \sum_{n=0}^{\infty} B_n\right),
\tag{190}
$$

where A_n and B_n are the Adomian polynomials that represent the nonlinear terms $(u_{yy})^2$ and u^2 respectively. This gives the recursive relation

$$
\begin{aligned}
u_0(x,y) &= x\cos y \\
u_{k+1}(x,y) &= L_x^{-1}(A_k - B_k), k \geq 0.
\end{aligned}
\tag{191}
$$

The first few components of the solution $u(x,y)$ are given by

$$
\begin{aligned}
u_0(x,y) &= x\cos y, \\
u_1(x,y) &= L_x^{-1}(A_0 - B_0) = 0,
\end{aligned}
\tag{192}
$$

and other components vanish as well. Consequently, the exact solution is given by

$$
u(x,y) = x\cos y.
\tag{193}
$$

Example 8. Solve the nonlinear partial differential equation

$$
u_t + \frac{1}{36}xu_{xx}^2 = x^3, \ u(x,0) = 0.
\tag{194}
$$

Solution.

Using the integral operator L_t^{-1} on (194) and using the given condition we obtain

$$u(x,t) = x^3 t - \frac{1}{36} L_t^{-1}(x u_{xx}^2). \tag{195}$$

Following the analysis presented before we obtain

$$\sum_{n=0}^{\infty} u_n(x,t) = x^3 t - \frac{1}{36} L_t^{-1}\left(x \sum_{n=0}^{\infty} A_n\right), \tag{196}$$

where A_n are the Adomian polynomials that represent the nonlinear terms u_{xx}^2. This gives the recursive relation

$$\begin{aligned}
u_0(x,t) &= x^3 t \\
u_{k+1}(x,t) &= -\frac{1}{36} L_t^{-1}(A_k), k \geq 0.
\end{aligned} \tag{197}$$

The Adomian polynomials are given by

$A_0 = u_{0_{xx}}^2,$
$A_1 = 2 u_{0_{xx}} u_{1_{xx}},$
$A_2 = 2 u_{0_{xx}} u_{2_{xx}} + u_{1_{xx}}^2,$

and so on. The first few components of the solution $u(x,y)$ are given by

$$\begin{aligned}
u_0(x,t) &= x^3 t, \\
u_1(x,t) &= -\frac{1}{36} L_t^{-1}(36 x^3 t^2) = -\frac{1}{3} x^3 t^3, \\
u_2(x,t) &= -\frac{1}{36} L_t^{-1}(-24 x^3 t^4) = \frac{2}{15} x^3 t^5, \\
u_3(x,t) &= -\frac{1}{36} L_t^{-1}\left(\frac{68}{5} x^2 t^6\right) = -\frac{17}{315} x^3 t^7,
\end{aligned} \tag{198}$$

and so on. Consequently, the solution in a series form is given by

$$u(x,t) = x^3\left(t - \frac{1}{3} t^3 + \frac{2}{15} t^5 - \frac{17}{315} t^7 + \cdots\right), \tag{199}$$

and in a closed form by

$$u(x,t) = x^3 \tanh t. \tag{200}$$

Example 9. Solve the nonlinear partial differential equation

$$u_t + u^2 u_x = 0, \ u(x,0) = 2x, \ x \in R, \ t > 0 \tag{201}$$

where $u = u(x,t)$.

Solution.

Applying L_t^{-1} on (201) and using the given condition we obtain

$$u(x,t) = 2x - L_t^{-1}(u^2 u_x). \qquad (202)$$

It follows that

$$\sum_{n=0}^{\infty} u_n(x,t) = 2x - L_t^{-1}\left(\sum_{n=0}^{\infty} A_n\right), \qquad (203)$$

where A_n are the Adomian polynomials that represent the nonlinear terms $u^2 u_x$. This gives the recursive relation

$$\begin{aligned} u_0(x,t) &= 2x \\ u_{k+1}(x,t) &= -L_t^{-1}(A_k), k \geq 0. \end{aligned} \qquad (204)$$

This gives the first few components of $u(x,y)$ by

$$\begin{aligned} u_0(x,t) &= 2x, \\ u_1(x,t) &= -L_t^{-1}(A_0) = -L_t^{-1}(8x^2) = -8x^2 t, \\ u_2(x,t) &= -L_t^{-1}(A_1) = -L_t^{-1}(-128x^3 t) = 64x^3 t^2 \\ u_3(x,t) &= -L_t^{-1}(A_2) = -L_t^{-1}(1920x^4 t^2) = -640x^4 t^3, \end{aligned} \qquad (205)$$

and so on. It follows that the solution in a series form is given by

$$u(x,t) = 2x - 8x^2 t + 64x^3 t^2 - 640x^4 t^3 + \cdots. \qquad (206)$$

Two observations can be made here. First, we can easily observe that

$$u(x,t) = 2x, \text{ for } t = 0. \qquad (207)$$

We next observe that for $t > 0$, the series solution in (206) can be formally expressed in a closed form by

$$u(x,t) = \frac{1}{4t}\left(\sqrt{1 + 16xt} - 1\right). \qquad (208)$$

Combining (206) and (207) gives the solution in the form

$$u(x,t) = \begin{cases} 2x & \text{for } t = 0 \\ \frac{1}{4t}\left(\sqrt{1 + 16xt} - 1\right) & \text{for } t > 0 \end{cases} \qquad (209)$$

Example 10. Solve the nonlinear partial differential equation

$$u_t + uu_x = 0, \ u(x,0) = \sin x, \ x \in R, \ t > 0 \tag{210}$$

where $u = u(x,t)$.

Solution.

Eq. (210) can be written in the form

$$L_t u = -uu_x. \tag{211}$$

Operating with L_t^{-1} on (211) gives

$$u(x,t) = \sin x - L_t^{-1}(uu_x). \tag{212}$$

Using the decomposition assumptions for the linear and the nonlinear terms we find

$$\sum_{n=0}^{\infty} u_n(x,t) = \sin x - L_t^{-1}\left(\sum_{n=0}^{\infty} A_n\right), \tag{213}$$

where A_n are the Adomian polynomials that represent the nonlinear terms uu_x. The following recursive relation

$$\begin{aligned} u_0(x,t) &= \sin x \\ u_{k+1}(x,t) &= -L_t^{-1}(A_k), k \geq 0. \end{aligned} \tag{214}$$

follows immediately. This gives the first few components of $u(x,y)$ by

$$\begin{aligned} u_0(x,t) &= \sin x, \\ u_1(x,t) &= -L_t^{-1}(A_0) = -t\sin x \cos x, \\ u_2(x,t) &= -L_t^{-1}(A_1) = (\sin x \cos^2 x - \tfrac{1}{2}\sin^3 x)t^2 \end{aligned} \tag{215}$$

and so on. It is clear that the solution obtained will be in terms of a series of functions rather than a power series. It then follows that

$$u(x,t) = \sin x - t\sin x \cos x + (\sin x\cos^2 x - \frac{1}{2}\sin^3 x)t^2 + \cdots. \tag{216}$$

However, by using the traditional method of characteristics, we can show that the solution can be expressed in the parametric form

$$\begin{aligned} u(x,t) &= \sin\xi, \\ \xi &= x - t\sin\xi. \end{aligned} \tag{217}$$

For numerical approximations, the series solution obtained above is more effective and practical compared to the parametric form solution given in (217).

In closing this section, it is important to note that the well-known nonlinear models that characterize physical models will be examined in details in Chapter 9. The aim of Chapter 8 is to introduce the algorithm in a general way so that it can be applied in scientific applications as will be seen in the coming chapters. For numerical purposes, the decomposition series solution will be combined with the powerful Padé approximants to handle the boundary condition at infinity in particular.

Exercises 8.4
In Exercises 1 – 12, use the Adomian decomposition method to find the exact solution of the following nonlinear partial differential equations:

1. $u_{xx} + u_y u_{yy} = 2$, $u(0, y) = 0$, $u_x(0, y) = y$

2. $u_{yy} + u_x u_{xx} = 2$, $u(x, 0) = 0$, $u_y(x, 0) = x$

3. $u_t + u u_x = 1 + x + t$, $u(x, 0) = x$

4. $u_t + u u_x = x + t + xt^2$, $u(x, 0) = 1$

5. $u_t - u u_x = 0$, $u(x, 0) = x$

6. $u_t + u u_x = 1 + t \cos x + \sin x \cos x$, $u(x, 0) = \sin x$

7. $u_t = 2x^2 - \frac{1}{8} u_x^2$, $u(x, 0) = 0$

8. $u_t = x^3 + \frac{1}{36} x u_{xx}^2$, $u(x, 0) = 0$

9. $u_t + u^2 u_x = 0$, $u(x, 0) = 3x$

10. $u_x + u_y u_{yy} = \frac{1}{1+x^2}$, $u(0, y) = 2y$

11. $u_{xx} + 2u_x(u - t) = 0$, $u(0, t) = t$, $u_x(0, t) = 1$

12. $u_{xx} - 2u_x(u - t) = 0$, $u(0, t) = t$, $u_x(0, t) = 1$

In Exercises 13 – 24, use the modified decomposition method to find the exact solution of the following nonlinear partial differential equations:

13. $u_{xx} + u u_y = 2y^2 + 2x^4 y^3$, $u(0, y) = 0$, $u_x(0, y) = 0$

14. $u_{xx} + u u_y = 2u - y$, $u(0, y) = 1 + y$, $u_x(0, y) = 1$

15. $u_{xx} + u u_x = x + \ln y$, $u(0, y) = \ln y$, $u_x(0, y) = 1$, $y > 0$

16. $u_{yy} + u u_y = y + \ln x$, $u(x, 0) = \ln x$, $u_y(x, 0) = 1$, $x > 0$

17. $u_{yy} + u_x^2 - u^2 = 0$, $u(x, 0) = 0$, $u_y(x, 0) = e^{-x}$

18. $u_{yy} - u_{xx}^2 + u^2 = 0$, $u(x,0) = 0$, $u_y(x,0) = \cos x$

19. $u_{xx} + uu_x = x + \ln(1+y)$, $u(0,y) = \ln(1+y)$,
    $u_x(0,y) = 1$, $y > -1$

20. $u_{xx} + yuu_x = 1 + xy$, $u(0,y) = \frac{1}{y}$, $u_x(0,y) = 1$, $y > 0$

21. $u_{yy} + u_x u_y = \frac{1}{1+x^2}$, $u(x,0) = \arctan x$, $u_y(x,0) = 1$

22. $u_{xx} + uu_t = -t$, $u(0,t) = t$, $u_x(0,t) = 1$

23. $u_{xx} + uu_t = t$, $u(0,t) = t$, $u_x(0,t) = 1$

24. $u_{xx} + uu_t = t$, $u(0,t) = t$, $u_x(0,t) = 0$

In Exercises 25 – 30, use the Adomian decomposition method to find the series solution of the following nonlinear partial differential equations:

25. $u_t + uu_x = 0$, $u(x,0) = \sinh x$

26. $u_t + uu_x = 0$, $u(x,0) = \cos x$

27. $u_t + uu_{xx} = x^2$, $u(x,0) = 0$

28. $u_t + u_x^2 = 0$, $u(x,0) = x$

29. $u_t + u^2 u_x = 0$, $u(x,0) = x$

30. $u_t + uu_x^2 = 0$, $u(x,0) = x$

8.5 Systems of Nonlinear PDEs

In this section, systems of nonlinear partial differential equations will be examined by using Adomian decomposition method. Systems of nonlinear partial differential equations arise in many scientific models such as the propagation of shallow water waves and the Brusselator model of chemical reaction-diffusion model. To achieve our goal in handling systems of nonlinear partial differential equations, we write a system in an operator form by

$$
\begin{aligned}
L_t u + L_x v + N_1(u,v) &= g_1, \\
L_t v + L_x u + N_2(u,v) &= g_2,
\end{aligned}
\tag{218}
$$

with initial data

$$
\begin{aligned}
u(x,0) &= f_1(x), \\
v(x,0) &= f_2(x),
\end{aligned}
\tag{219}
$$

where L_t and L_x are considered, without loss of generality, first order partial differential operators, N_1 and N_2 are nonlinear operators, and g_1 and g_2 are source terms. Operating with the integral operator L_t^{-1} to the system (218) and using the initial data (219) yields

$$
\begin{aligned}
u(x,t) &= f_1(x) + L_t^{-1}g_1 - L_t^{-1}L_x v - L_t^{-1}N_1(u,v), \\
v(x,t) &= f_2(x) + L_t^{-1}g_2 - L_t^{-1}L_x u - L_t^{-1}N_2(u,v).
\end{aligned}
\tag{220}
$$

The linear unknown functions $u(x,t)$ and $v(x,t)$ can be decomposed by infinite series of components

$$
\begin{aligned}
u(x,t) &= \sum_{n=0}^{\infty} u_n(x,t), \\
v(x,t) &= \sum_{n=0}^{\infty} v_n(x,t).
\end{aligned}
\tag{221}
$$

However, the nonlinear operators $N_1(u,v)$ and $N_2(u,v)$ should be represented by using the infinite series of the so-called Adomian polynomials A_n and B_n as follows:

$$
\begin{aligned}
N_1(u,v) &= \sum_{n=0}^{\infty} A_n, \\
N_2(u,v) &= \sum_{n=0}^{\infty} B_n,
\end{aligned}
\tag{222}
$$

where $u_n(x,t)$ and $v_n(x,t), n \geq 0$ are the components of $u(x,t)$ and $v(x,t)$ respectively that will be recurrently determined, and A_n and $B_n, n \geq 0$ are Adomian polynomials that can be generated for all forms of nonlinearity. The algorithms for calculating Adomian polynomials were introduced in Sections 8.2 and 8.3. Substituting (221) and (222) into (220) gives

$$
\begin{aligned}
\sum_{n=0}^{\infty} u_n(x,t) &= f_1(x) + L_t^{-1}g_1 - L_t^{-1}L_x\left(\sum_{n=0}^{\infty} v_n\right) - L_t^{-1}\left(\sum_{n=0}^{\infty} A_n\right), \\
\sum_{n=0}^{\infty} v_n(x,t) &= f_2(x) + L_t^{-1}g_2 - L_t^{-1}L_x\left(\sum_{n=0}^{\infty} u_n\right) - L_t^{-1}\left(\sum_{n=0}^{\infty} B_n\right).
\end{aligned}
\tag{223}
$$

Two recursive relations can be constructed from (223) given by

$$
\begin{aligned}
u_0(x,t) &= f_1(x) + L_t^{-1}g_1, \\
u_{k+1}(x,t) &= -L_t^{-1}\left(L_x v_k\right) - L_t^{-1}\left(A_k\right), \; k \geq 0,
\end{aligned}
\tag{224}
$$

and

$$
\begin{aligned}
v_0(x,t) &= f_2(x) + L_t^{-1}g_2, \\
v_{k+1}(x,t) &= -L_t^{-1}\left(L_x u_k\right) - L_t^{-1}\left(B_k\right), \; k \geq 0.
\end{aligned}
\tag{225}
$$

It is an essential feature of the decomposition method that the zeroth components $u_0(x,t)$ and $v_0(x,t)$ are defined always by all terms that arise from initial data and from integrating the source terms. Having defined the zeroth pair (u_0, v_0), the remaining pair $(u_k, v_k), k \geq 1$ can be obtained in a recurrent manner by using (224) and (225). Additional pairs for the decomposition series solutions normally account for higher accuracy. Having determined the components of $u(x,t)$ and $v(x,t)$, the solution (u, v) of the system follows immediately in the form of a power series expansion upon using (221).

To give a clear overview of the analysis introduced above, two illustrative systems of nonlinear partial differential equations have been selected to demonstrate the efficiency of the method.

Example 1. Consider the nonlinear system:

$$
\begin{aligned}
u_t + vu_x + u &= 1, \\
v_t - uv_x - v &= 1,
\end{aligned}
\tag{226}
$$

with the conditions

$$
u(x,0) = e^x, \quad v(x,0) = e^{-x}.
\tag{227}
$$

Solution.

Operating with L_t^{-1} on (226) we obtain

$$
\begin{aligned}
u(x,t) &= e^x + t - L_t^{-1}(vu_x + u), \\
v(x,t) &= e^{-x} + t + L_t^{-1}(uv_x + v).
\end{aligned}
\tag{228}
$$

The linear terms $u(x,t)$ and $v(x,t)$ can be represented by the decomposition series

$$
\begin{aligned}
u(x,t) &= \sum_{n=0}^{\infty} u_n(x,t), \\
v(x,t) &= \sum_{n=0}^{\infty} v_n(x,t),
\end{aligned}
\tag{229}
$$

and the nonlinear terms vu_x and uv_x by an infinite series of polynomials

$$
\begin{aligned}
vu_x &= \sum_{n=0}^{\infty} A_n, \\
uv_x &= \sum_{n=0}^{\infty} B_n,
\end{aligned}
\tag{230}
$$

where A_n and B_n are the Adomian polynomials that can be generated for any form of nonlinearity. Substituting (229) and (230) into (228) gives

$$
\begin{aligned}
\sum_{n=0}^{\infty} u_n(x,t) &= e^x + t - L_t^{-1}\left(\sum_{n=0}^{\infty} A_n + \sum_{n=0}^{\infty} u_n\right), \\
\sum_{n=0}^{\infty} v_n(x,t) &= e^{-x} + t + L_t^{-1}\left(\sum_{n=0}^{\infty} B_n + \sum_{n=0}^{\infty} v_n\right).
\end{aligned}
\tag{231}
$$

To accelerate the convergence of the solution, the modified decomposition method of will be applied here. The modified decomposition method defines the recursive relations in the form

$$u_0(x,t) \;=\; e^x,$$

$$u_1(x,t) \;=\; t - L_t^{-1}\left(\sum_{n=0}^{\infty} A_0 + u_0\right),$$ (232)

$$u_{k+1}(x,t) \;=\; -L_t^{-1}\left(\sum_{n=0}^{\infty} A_k + u_k\right),\, k \geq 1,$$

and

$$v_0(x,t) \;=\; e^{-x},$$

$$v_1(x,t) \;=\; t + L_t^{-1}\left(\sum_{n=0}^{\infty} B_0 + v_0\right),$$ (233)

$$v_{k+1}(x,t) \;=\; L_t^{-1}\left(\sum_{n=0}^{\infty} B_k + v_k\right),\, k \geq 1.$$

The Adomian polynomials for the nonlinear term vu_x are given by

$A_0 = v_0 u_{0_x},$
$A_1 = v_1 u_{0_x} + v_0 u_{1_x},$
$A_2 = v_2 u_{0_x} + v_1 u_{1_x} + v_0 u_{2_x},$
$A_3 = v_3 u_{0_x} + v_2 u_{1_x} + v_1 u_{2_x} + v_0 u_{3_x},$

and for the nonlinear term uv_x by

$B_0 = u_0 v_{0_x},$
$B_1 = u_1 v_{0_x} + u_0 v_{1_x},$
$B_2 = u_2 v_{0_x} + u_1 v_{1_x} + u_0 v_{2_x},$
$B_3 = u_3 v_{0_x} + u_2 v_{1_x} + u_1 v_{2_x} + u_0 v_{3_x}.$

Using the derived Adomian polynomials into (232) and (233), we obtain the following pairs of components

$$(u_0, v_0) \;=\; (e^x, e^{-x}),$$

$$(u_1, v_1) \;=\; (-te^x, te^{-x}),$$ (234)

$$(u_2, v_2) \;=\; (\tfrac{t^2}{2!}e^x, \tfrac{t^2}{2!}e^{-x}),$$

$$(u_3, v_3) \;=\; (-\tfrac{t^3}{3!}e^x, \tfrac{t^3}{3!}e^{-x}).$$

Accordingly, the solution of the system in a series form is given by

$$(u,v) = \left(e^x(1 - t + \frac{t^2}{2!} - \frac{t^3}{3!} + \cdots). e^{-x}(1 + t + \frac{t^2}{2!} + \frac{t^3}{3!} + \cdots)\right).$$ (235)

and in a closed form by

$$(u, v) = \left(e^{x-t}, e^{-x+t}\right).\qquad(236)$$

In what follows, a system of three nonlinear partial differential equations in three unknown functions $u(x, y, t), v(x, y, t)$ and $w(x, y, t)$ will be studied. It is worth noting that handling this system by traditional methods is quiet complicated.

Example 2. Consider the following nonlinear system:

$$
\begin{aligned}
u_t - v_x w_y &= 1, \\
v_t - w_x u_y &= 5, \qquad(237) \\
w_t - u_x v_y &= 5,
\end{aligned}
$$

with the initial conditions

$$u(x, y, 0) = x + 2y, \; v(x, y, 0) = x - 2y, \; w(x, y, 0) = -x + 2y.\qquad(238)$$

Solution.

Following the analysis presented above we obtain

$$
\begin{aligned}
u(x, y, t) &= (x + 2y + t) + L_t^{-1}(v_x w_y), \\
v(x, y, t) &= (x - 2y + 5t) + L_t^{-1}(w_x u_y), \qquad(239) \\
w(x, y, t) &= (-x + 2y + 5t) + L_t^{-1}(u_x v_y).
\end{aligned}
$$

Substituting the decomposition representations for linear and nonlinear terms into (239) yields

$$
\begin{aligned}
\sum_{n=0}^{\infty} u_n(x, y, t) &= (x + 2y + t) + L_t^{-1}\left(\sum_{n=0}^{\infty} A_n\right), \\
\sum_{n=0}^{\infty} v_n(x, y, t) &= (x - 2y + 5t) + L_t^{-1}\left(\sum_{n=0}^{\infty} B_n\right), \qquad(240) \\
\sum_{n=0}^{\infty} w_n(x, y, t) &= (-x + 2y + 5t) + L_t^{-1}\left(\sum_{n=0}^{\infty} C_n\right),
\end{aligned}
$$

where A_n, B_n, and C_n, are Adomian polynomials for the nonlinear terms $v_x w_y, w_x u_y$, and $u_x v_y$ respectively. For brevity, we list the first three Adomian polynomials for A_n, B_n, and C_n as follows:

For $v_x w_y$, we find

$$
\begin{aligned}
A_0 &= v_{0_x} w_{0_y}, \\
A_1 &= v_{1_x} w_{0_y} + v_{0_x} w_{1_y}, \\
A_2 &= v_{2_x} w_{0_y} + v_{1_x} w_{1_y} + v_{0_x} w_{2_y},
\end{aligned}
$$

and for $w_x u_y$ we find

$$B_0 = w_{0_x} u_{0_y},$$
$$B_1 = w_{1_x} u_{0_y} + w_{0_x} u_{1_y},$$
$$B_2 = w_{2_x} u_{0_y} + w_{1_x} u_{1_y} + w_{0_x} u_{2_y}.$$

and for $u_x v_y$ we find

$$C_0 = u_{0_x} v_{0_y},$$
$$C_1 = u_{1_x} v_{0_y} + u_{0_x} v_{1_y},$$
$$C_2 = u_{2_x} v_{0_y} + u_{1_x} v_{1_y} + u_{0_x} v_{2_y}.$$

Substituting these polynomials into the appropriate recursive relations we find

$$
\begin{aligned}
(u_0, v_0, w_0) &= (x + 2y + t,\ x - 2y + 5t,\ -x + 2y + 5t), \\
(u_1, v_1, w_1) &= (2t,\ -2t,\ -2t), \\
(u_k, v_k, w_k) &= (0, 0, 0),\ k \geq 2.
\end{aligned}
\tag{241}
$$

Consequently, the exact solution of the system of nonlinear partial differential equations is given by

$$(u, v, w) = (x + 2y + 3t,\ x - 2y + 3t,\ -x + 2y + 3t). \tag{242}$$

Exercises 8.5

Use Adomian decomposition method to solve the following systems of nonlinear partial differential equations:

1. $u_t + u_x v_x = 2,\ v_t + u_x v_x = 0,$
   with the initial data: $u(x, 0) = x,\ v(x, 0) = x$

2. $u_t - v u_x - u = 1,\ v_t + u v_x + v = 1,$
   with the initial data: $u(x, 0) = e^{-x},\ v(x, 0) = e^x.$

3. $u_t + 2 v u_x - u = 2,\ v_t - 3 u v_x + v = 3,$
   with the initial data: $u(x, 0) = e^x,\ v(x, 0) = e^{-x}.$

4. $u_t + v u_x - 3u = 2,\ v_t - u v_x + 3v = 2,$
   with the initial data: $u(x, 0) = e^{2x},\ v(x, 0) = e^{-2x}.$

5. $u_t + u_x v_x - w_y = 1,\ v_t + v_x w_x + u_y = 1,\ w_t + w_x u_x - v_y = 1$
   $u(x, y, 0) = x + y,\ v(x, y, 0) = x - y,\ w(x, y, 0) = -x + y.$

6. $u_t + v_x w_y - v_y w_x = -u, \ v_t + w_x u_y + w_y u_x = v,$
$w_t + u_x v_y + u_y v_x = w$

$u(x, y, 0) = e^{x+y}, \ v(x, y, 0) = e^{x-y}, \ w(x, y, 0) = e^{-x+y}.$

7. $u_t + u_x v_x - u_y v_y + u = 0, \ v_t + v_x w_x - v_y w_y - v = 0,$
$w_t + w_x u_x + w_y u_y - w = 0$

$u(x, y, 0) = e^{x+y}, \ v(x, y, 0) = e^{x-y}, \ w(x, y, 0) = e^{-x+y}.$

8. $u_t + u_y v_x = 1 + e^t, \ v_t + v_y w_x = 1 - e^{-t},$
$w_t + w_y u_y = 1 - e^{-t}$

$u(x, y, 0) = 1 + x + y, \ v(x, y, 0) = 1 + x - y, \ w(x, y, 0) = 1 - x + y.$

Chapter 9

Linear and Nonlinear Physical Models

9.1 Introduction

This chapter is devoted to treatments of linear and nonlinear particular applications that appear in applied sciences. A wide variety of physically significant problems modeled by linear and nonlinear partial differential equations has been the focus of extensive studies for the last decades. A huge size of research and investigation has been invested in these scientific applications. Several approaches have been used such as the characteristics method, spectral methods and perturbation techniques to examine these problems.

Nonlinear PDEs have undergone remarkable developments. Nonlinear problems arise in different areas including gravitation, chemical reaction, fluid dynamics, dispersion, nonlinear optics, plasma physics, acoustics, and inviscid fluids. Nonlinear wave propagation problems have provided solutions of different physical structures than solutions of linear wave equations.

It is well known that many physical, chemical and biological problems are characterized by the interaction of convection and diffusion and by the interaction of diffusion and reaction processes. Burgers' equation is considered as a model equation that describes the interaction of convection and diffusion, whereas Fisher's equation is an appropriate model that describes the process of interaction between diffusion and reaction.

In this chapter, Adomian decomposition method, the modified decomposition method, and the self-canceling noise-terms phenomenon will be employed in the treatments of these models. The linear and nonlinear models will be approached directly and in a like manner. The series representation of the linear term u, and the representation of the nonlinear term $F(u)$ by a series of Adomian polynomials will be used in a like manner to that used in Chapter 8; hence details will be skipped.

9.2   The Nonlinear Advection Problem

The nonlinear partial differential equation of the advection problem is of the form

$$u_t + uu_x = f(x,t), \; u(x,0) = g(x). \tag{1}$$

The problem has been handled by using the characteristic method, and recently by applying numerical methods such as Fourier series and Runge-Kutta method.

In this section, we approach the advection problem by utilizing the decomposition method to find a rapidly convergent power series solution. The phenomenon of self-canceling noise terms will be used where appropriate.

In an operator form, Eq. (1) can be rewritten as

$$L_t u + \frac{1}{2} L_x(u^2) = f(x,t), \; u(x,0) = g(x). \tag{2}$$

Operating with L_t^{-1} yields

$$u(x,t) = g(x) + L_t^{-1}(f(x,t)) - \frac{1}{2} L_t^{-1} L_x(u^2). \tag{3}$$

Substituting the linear term $u(x,t)$ by the series

$$u(x,t) = \sum_{n=0}^{\infty} u_n(x,t), \tag{4}$$

and the nonlinear term u^2 by a series of Adomian polynomials

$$u^2(x,t) = \sum_{n=0}^{\infty} A_n, \tag{5}$$

into (3) gives

$$\sum_{n=0}^{\infty} u_n(x,t) = g(x) + L_t^{-1}(f(x,t)) - \frac{1}{2} L_t^{-1} L_x \left(\sum_{n=0}^{\infty} A_n \right). \tag{6}$$

Following Adomian approach, we obtain the recursive relation

$$\begin{aligned} u_0(x,t) &= g(x) + L_t^{-1}(f(x,t)), \\ u_{k+1}(x,t) &= -\tfrac{1}{2} L_t^{-1} L_x(A_k), \; k \geq 0. \end{aligned} \tag{7}$$

In view of (7), the components u_n, $n \geq 0$ can be easily computed, and the series solution can be formally constructed.

As stated in Chapter 2, the phenomenon of the self-canceling noise terms may appear for inhomogeneous problems, whereas homogeneous problems do not produce the noise terms in $u_0(x, t)$ and $u_1(x, t)$. The self-canceling noise terms will play a major role in accelerating the convergence.

To give a clear overview of this analysis, the following illustrative homogeneous and inhomogeneous examples will be discussed.

Example 1. Solve the inhomogeneous advection problem

$$u_t + \frac{1}{2}(u^2)_x = e^x + t^2 e^{2x}, \; u(x, 0) = 0. \tag{8}$$

Solution.

Operating with L_t^{-1}, Eq. (8) becomes

$$u(x, t) = te^x + \frac{1}{3}t^3 e^{2x} - \frac{1}{2}L_t^{-1}L_x(u^2). \tag{9}$$

Substituting the decomposition representation for the linear term $u(x, t)$ and for the nonlinear term $u^2(x, t)$ gives

$$\sum_{n=0}^{\infty} u_n(x, t) = te^x + \frac{1}{3}t^3 e^{2x} - \frac{1}{2}L_t^{-1}L_x\left(\sum_{n=0}^{\infty} A_n\right), \tag{10}$$

where A_n are the Adomian polynomials for the nonlinear term u^2. This gives the recursive relation

$$
\begin{aligned}
u_0(x, t) &= te^x + \frac{1}{3}t^3 e^{2x}, \\
u_{k+1}(x, t) &= -\frac{1}{2}L_t^{-1}L_x(A_k), \; k \geq 0,
\end{aligned}
\tag{11}
$$

so that the first two components are given by

$$
\begin{aligned}
u_0(x, t) &= te^x + \frac{1}{3}t^3 e^{2x}, \\
u_1(x, t) &= -\frac{1}{2}L_t^{-1}L_x(A_0), \\
&= -\frac{1}{3}t^3 e^{2x} - \frac{1}{5}t^5 e^{3x} - \frac{2}{63}t^7 e^{4x}.
\end{aligned}
\tag{12}
$$

The noise terms phenomenon suggests that if terms in u_0 are canceled by terms in u_1, even though u_1 contains further terms, then the remaining non-canceled terms of u_0 may provide the exact solution of the problem. This should be justified through substitution. Thus by canceling the term $-\frac{1}{3}t^2 e^{2x}$ in u_0, and by justifying that the remaining non-canceled term of u_0 satisfies the equation, it then follows that the exact solution is given by

$$u(x, t) = te^x. \tag{13}$$

Example 2. Solve the inhomogeneous advection problem

$$u_t + \frac{1}{2}(u^2)_x = -\sin(x+t) - \frac{1}{2}\sin 2(x+t), \ u(x,0) = \cos x. \tag{14}$$

Solution.

Applying L_t^{-1} to both sides of (14) gives

$$u(x,t) = \cos(x+t) + \frac{1}{4}\cos 2(x+t) - \frac{1}{4}\cos 2x - \frac{1}{2}L_t^{-1}L_x(u^2). \tag{15}$$

Using the decomposition representation for the linear and nonlinear terms yields

$$\sum_{n=0}^{\infty} u_n(x,t) = \cos(x+t) + \frac{1}{4}\cos 2(x+t) - \frac{1}{4}\cos 2x - \frac{1}{2}L_t^{-1}L_x\left(\sum_{n=0}^{\infty} A_n\right), \tag{16}$$

where A_n are the Adomian polynomials for the nonlinear term u^2. Following Adomian analysis, the recursive relation

$$\begin{aligned} u_0(x,t) &= \cos(x+t) + \tfrac{1}{4}\cos 2(x+t) - \tfrac{1}{4}\cos 2x, \\ u_{k+1}(x,t) &= -\tfrac{1}{2}L_t^{-1}L_x(A_k), \ k \geq 0, \end{aligned} \tag{17}$$

follows immediately. The first two components are given by

$$\begin{aligned} u_0(x,t) &= \cos(x+t) + \tfrac{1}{4}\cos 2(x+t) - \tfrac{1}{4}\cos 2x, \\ u_1(x,t) &= -\tfrac{1}{2}L_t^{-1}L_x(A_0) = -\tfrac{1}{4}\cos 2(x+t) + \tfrac{1}{4}\cos 2x - \cdots. \end{aligned} \tag{18}$$

It is easily observed that two noise terms appear in the components $u_0(x,t)$ and $u_1(x,t)$. By canceling these terms from u_0, the remaining non-canceled term of u_0 may provide the exact solution. It follows that the exact solution is given by

$$u(x,t) = \cos(x+t), \tag{19}$$

that can be justified by substitution.

It is interesting to note that other noise terms appear between other components. Such noise terms will vanish in the limit.

In the following example, we will show that although the problem is inhomogeneous, noise terms will not appear. This confirms the fact that for the noise terms to appear between the first two components, then it is necessary for the exact solution to be part of u_0. This was examined in details in [124].

Example 3. Solve the inhomogeneous advection problem

$$u_t + \frac{1}{2}(u^2)_x = x, \ u(x,0) = 2. \tag{20}$$

Solution.

Operating with L_t^{-1} gives

$$u(x,t) = 2 + xt - \frac{1}{2}L_t^{-1}L_x(u^2). \tag{21}$$

The decomposition method admits the use of

$$u(x,t) = \sum_{n=0}^{\infty} u_n(x,t), \tag{22}$$

and

$$u^2(x,t) = \sum_{n=0}^{\infty} A_n, \tag{23}$$

into (21) to obtain

$$\sum_{n=0}^{\infty} u_n(x,t) = 2 + xt - \frac{1}{2}L_t^{-1}L_x\left(\sum_{n=0}^{\infty} A_n\right). \tag{24}$$

This gives the recursive relation

$$\begin{aligned}
u_0(x,t) &= 2 + xt, \\
u_{k+1}(x,t) &= -\frac{1}{2}L_t^{-1}L_x(A_k), \ k \geq 0,
\end{aligned} \tag{25}$$

that gives

$$\begin{aligned}
u_0(x,t) &= 2 + xt, \\
u_1(x,t) &= -\frac{1}{2}L_t^{-1}L_x(A_0) = -t^2 - \frac{1}{3}xt^3, \\
u_2(x,t) &= -\frac{1}{2}L_t^{-1}L_x(A_1) = \frac{5}{12}t^4 + \frac{2}{15}xt^5, \\
u_3(x,t) &= -\frac{1}{2}L_t^{-1}L_x(A_2) = -\frac{61}{60}t^6 - \frac{17}{315}xt^7.
\end{aligned} \tag{26}$$

Although the advection problem (20) is inhomogeneous, it is clear that the noise terms do not appear in u_0 and u_1. Consequently, the series solution is given by

$$\begin{aligned}
u(x,t) &= 2\left(1 - \frac{1}{2}t^2 + \frac{5}{24}t^4 - \frac{61}{120}t^6 + \cdots\right) \\
&+ x\left(t - \frac{1}{3}t^3 + \frac{2}{15}t^5 - \frac{17}{315}t^7 + \cdots\right),
\end{aligned} \tag{27}$$

and as a result, the exact solution is given by

$$u(x,t) = 2\operatorname{sech} t + x\tanh t. \tag{28}$$

Example 4. Solve the homogeneous nonlinear problem

$$u_t + u^2 u_x = 0, \; u(x,0) = 3x. \tag{29}$$

Solution.

Proceeding as before we find

$$u(x,t) = 3x - L_t^{-1}(u^2 u_x), \tag{30}$$

so that

$$\sum_{n=0}^{\infty} u_n(x,t) = 3x - L_t^{-1}\left(\sum_{n=0}^{\infty} A_n\right). \tag{31}$$

Adomian's method introduces the recursive relation

$$
\begin{aligned}
u_0(x,t) &= 3x, \\
u_{k+1}(x,t) &= -L_t^{-1}(A_k), \; k \geq 0,
\end{aligned}
\tag{32}
$$

which gives

$$
\begin{aligned}
u_0(x,t) &= 3x, \\
u_1(x,t) &= -L_t^{-1}(A_0) = -27x^2 t, \\
u_2(x,t) &= -L_t^{-1}(A_1) = 486x^3 t^2, \\
u_3(x,t) &= -L_t^{-1}(A_2) = -10935x^4 t^3,
\end{aligned}
\tag{33}
$$

and so on. Consequently, the series solution

$$u(x,t) = 3x - 27x^2 t + 486x^3 t^2 - 10935x^4 t^3 + \cdots. \tag{34}$$

Based on this, the solution can be expressed in the form

$$u(x,t) = \begin{cases} 3x & \text{for } t = 0 \\ \frac{1}{6t}(\sqrt{1 + 36xt} - 1) & \text{for } t > 0 \end{cases} \tag{35}$$

Exercises 9.2

Use the decomposition method and the noise terms phenomena where appropriate to solve the following nonlinear advection problems:

1. $u_t + uu_x = 1 - e^{-x}(t + e^{-x})$, $u(x,0) = e^{-x}$

2. $u_t + uu_x = 2t + x + t^3 + xt^2$, $u(x,0) = 0$

3. $u_t + uu_x = 2x^2t + 2xt^2 + 2x^3t^4$, $u(x,0) = 1$

4. $u_t + uu_x = 1 + t\cos x + \frac{1}{2}\sin 2x$, $u(x,0) = \sin x$

5. $u_t + uu_x = 0$, $u(x,0) = -x$

6. $u_t + uu_x = x$, $u(x,0) = -1$

7. $u_t + uu_x = 1 + x$, $u(x,0) = 0$

8. $u_t + uu_x - u = e^t$, $u(x,0) = 1 + x$

9. $u_t + uu_x = 0$, $u(x,0) = 4x$

10. $u_t + uu_x = 0$, $u(x,0) = x^2$

9.3 The Goursat Problem

In this section we will study the Goursat problem that arise in linear and nonlinear partial differential equations with mixed derivatives. Several numerical methods such as Runge-Kutta method, finite difference method, finite elements method, and geometric mean averaging of the functional values of $f(x, y, u, u_x, u_y)$ have been used to approach the problem. However, the linear and nonlinear Goursat models will be approached more effectively and rapidly by using the Adomian decomposition method.

The Goursat problem in its standard form is given by

$$u_{xy} = f(x, y, u, u_x, u_y),\ 0 \le x \le a,\ 0 \le y \le b, \tag{36}$$

$$u(x,0) = g(x),\ u(0,y) = h(y),\ g(0) = h(0) = u(0,0). \tag{37}$$

In an operator form, Eq. (36) can be rewritten as

$$L_x L_y u = f(x, y, u, u_x, u_y), \tag{38}$$

where

$$L_x = \frac{\partial}{\partial x},\ L_y = \frac{\partial}{\partial y}. \tag{39}$$

The inverse operators L_x^{-1} and L_y^{-1} can be defined as

$$L_x^{-1}(.) = \int_0^x (.)\,dx,\ L_y^{-1}(.) = \int_0^y (.)\,dy. \tag{40}$$

Because the Goursat problem (36) involves two distinct differential operators L_x and L_y, then the two inverse integral operator L_x^{-1} and L_y^{-1} will be used. Applying L_y^{-1} to both sides of (38) gives

$$L_x[L_y^{-1}L_y u(x,y)] = L_y^{-1}f(x, y, u, u_x, u_y). \tag{41}$$

It then follows that

$$L_x[u(x,y) - u(x,0)] = L_y^{-1} f(x, y, u, u_x, u_y), \tag{42}$$

or equivalently

$$L_x u(x, y) = L_x u(x, 0) + L_y^{-1} f(x, y, u, u_x, u_y). \tag{43}$$

Operating with L_x^{-1} on (43) yields

$$L_x^{-1} L_x u(x, y) = L_x^{-1} L_x u(x, 0) + L_x^{-1} L_y^{-1} f(x, y, u, u_x, u_y). \tag{44}$$

This gives

$$u(x, y) = u(x, 0) + u(0, y) - u(0, 0) + L_x^{-1} L_y^{-1} f(x, y, u, u_x, u_y), \tag{45}$$

or equivalently

$$u(x, y) = g(x) + h(y) - g(0) + L_x^{-1} L_y^{-1} f(x, y, u, u_x, u_y), \tag{46}$$

obtained upon using the conditions given in (36). Substituting

$$u(x, y) = \sum_{n=0}^{\infty} u_n(x, y), \tag{47}$$

into (46) leads to

$$\sum_{n=0}^{\infty} u_n(x, y) = g(x) + h(y) - g(0) + L_x^{-1} L_y^{-1} f(x, y, u, u_x, u_y). \tag{48}$$

Adomian's method admits the use of the recursive relation

$$\begin{aligned} u_0(x, y) &= \eta(x, y), \\ u_{k+1}(x, y) &= L_x^{-1} L_y^{-1} \sigma(u_k, u_{k_x}, u_{k_y}), \, k \geq 0, \end{aligned} \tag{49}$$

where
$$\eta(x, y) =$$

$$\begin{cases} g(x) + h(y) - g(0) & f = \sigma(u, u_x, u_y) \\ g(x) + h(y) - g(0) + L_x^{-1} L_y^{-1} \tau(x, y) & f = \tau(x, y) + \sigma(u, u_x, u_y) \end{cases} \tag{50}$$

In view of (49), the components $u_n(x, y), n \geq 0$ of the solution $u(x, y)$ can be computed. Consequently, the solution in a series form follows immediately. The resulting series solution may provide the exact solution. Otherwise, the n−term approximation ϕ_n can be used for numerical purposes. In the latter case, it can be shown that the distance between the exact solution and the n−term approximation decreases monotonically for all values of x and y as additional components evaluated.

In the following, four linear and nonlinear Goursat models will be discussed for illustrative purposes.

Example 1. Solve the following linear Goursat problem

$$u_{xy} = -x + u, \tag{51}$$

subject to the conditions

$$u(x, 0) = x + e^x, \ u(0, y) = e^y, \ u(0, 0) = 1. \tag{52}$$

Solution.

Following the previous discussion and using (45) we find

$$u(x, y) = x + e^x + e^y - 1 - \frac{1}{2}x^2 y + L_x^{-1} L_y^{-1} u(x, y), \tag{53}$$

and by using the series representation for $u(x, t)$ into (53) gives

$$\sum_{n=0}^{\infty} u_n(x, y) = x + e^x + e^y - 1 - \frac{1}{2}x^2 y + L_x^{-1} L_y^{-1} \left(\sum_{n=0}^{\infty} u_n(x, y) \right). \tag{54}$$

The recursive relation

$$\begin{aligned} u_0(x, y) &= x + e^x + e^y - 1 - \frac{1}{2}x^2 y, \\ u_{k+1}(x, y) &= L_x^{-1} L_y^{-1} u_k(x, y), \ k \geq 0, \end{aligned} \tag{55}$$

follows immediately. Consequently, the first three components of the solution $u(x, y)$ are given by

$$\begin{aligned} u_0(x, y) &= x + e^x + e^y - 1 - \frac{1}{2}x^2 y, \\ u_1(x, y) &= L_x^{-1} L_y^{-1} u_0(x, y), \\ &= \frac{1}{2}x^2 y + y(e^x - 1) + x(e^y - 1) - xy - \frac{1}{12}x^3 y^2, \\ u_2(x, y) &= L_x^{-1} L_y^{-1}, \\ &= \frac{1}{12}x^3 y^2 + \frac{1}{2}y^2(e^x - 1 - x) + \frac{1}{2}x^2(e^y - 1 - y) \\ &\quad - \frac{1}{4}x^2 y^2 - \frac{1}{144}y^4 x^3, \end{aligned} \tag{56}$$

This gives

$$
\begin{aligned}
u(x,y) &= x + e^x \left(1 + y + \tfrac{1}{2!}y^2 + \tfrac{1}{3!}y^3 + \cdots\right) \\
&\quad + e^y \left(1 + x + \tfrac{1}{2!}x^2 + \tfrac{1}{3!}x^3 + \cdots\right) \\
&\quad - \left(1 + x + y + xy + \tfrac{1}{2!}x^2 + \tfrac{1}{2!}y^2 + \tfrac{1}{3!}x^3 + \tfrac{1}{3!}y^3 + \tfrac{1}{2!}x^2y + \cdots\right),
\end{aligned}
\tag{57}
$$

or equivalently

$$
u(x,y) = x + e^x \left(1 + y + \tfrac{1}{2!}y^2 + \tfrac{1}{3!}y^3 + \cdots\right)
$$

$$
\begin{aligned}
&\quad + e^y \left(1 + x + \tfrac{1}{2!}x^2 + \tfrac{1}{3!}x^3 + \cdots\right) \\
&\quad - (1 + x + \tfrac{1}{2!}x^2 + \cdots)(1 + y + \tfrac{1}{2!}y^2 + \cdots).
\end{aligned}
\tag{58}
$$

Accordingly, the solution in a closed form is given by

$$
u(x,y) = x + e^{x+y},
\tag{59}
$$

obtained upon using the Taylor expansions for e^x and e^y.

Example 2. Solve the following linear Goursat problem

$$
u_{xy} = 4xy - x^2y^2 + u,
\tag{60}
$$

subject to the conditions

$$
u(x,0) = e^x, \; u(0,y) = e^y, \; u(0,0) = 1.
\tag{61}
$$

Solution.

Proceeding as before we find

$$
u(x,y) = x^2y^2 - \frac{1}{9}x^3y^3 - 1 + e^x + e^y + L_x^{-1}L_y^{-1}u(x,y).
\tag{62}
$$

This also gives

$$
\sum_{n=0}^{\infty} u_n(x,y) = x^2y^2 - \frac{1}{9}x^3y^3 - 1 + e^x + e^y + L_x^{-1}L_y^{-1}\left(\sum_{n=0}^{\infty} u_n(x,y)\right).
\tag{63}
$$

The decomposition method introduces the recursive relation

$$
\begin{aligned}
u_0(x,y) &= x^2y^2 - \tfrac{1}{9}x^3y^3 - 1 + e^x + e^y, \\
u_{k+1}(x,y) &= L_x^{-1}L_y^{-1}u_k(x,y), \; k \geq 0,
\end{aligned}
\tag{64}
$$

that leads to

$$
\begin{aligned}
u_0(x,y) &= x^2 y^2 - \tfrac{1}{9} x^3 y^3 - 1 + e^x + e^y, \\
u_1(x,y) &= L_x^{-1} L_y^{-1} u_0(x,y), \\
&= \tfrac{1}{9} x^3 y^3 - \tfrac{1}{144} x^4 y^4 - xy + y(e^x - 1) + x(e^y - 1), \\
u_2(x,y) &= L_x^{-1} L_y^{-1} u_1(x,y), \\
&= \tfrac{1}{144} x^4 y^4 - \tfrac{1}{3600} x^5 y^5 - \tfrac{1}{4} x^2 y^2 + \tfrac{1}{2} y^2 (e^x - 1 - x) \\
&\quad + \tfrac{1}{2} x^2 (e^y - 1 - y).
\end{aligned}
$$

(65)

In view of (65), the solution in a series form is given by

$$
\begin{aligned}
u(x,y) &= x^2 y^2 + e^x \left(1 + y + \tfrac{1}{2!} y^2 + \cdots\right) + e^y \left(1 + x + \tfrac{1}{2!} x^2 + \cdots\right) \\
&\quad - \left(1 + x + y + xy + \tfrac{1}{2!} x^2 + \tfrac{1}{2!} y^2 + \tfrac{1}{3!} x^3 + \tfrac{1}{3!} y^3 + \cdots\right),
\end{aligned}
$$

(66)

and in a closed form by

$$
u(x,y) = x^2 y^2 + e^{x+y}. \tag{67}
$$

Example 3. Solve the following nonlinear Goursat problem

$$
u_{xy} = e^{x+y} e^u, \tag{68}
$$

subject to the conditions

$$
u(x,0) = \ln 2 - 2\ln(1 + e^x), \quad u(0,y) = \ln 2 - 2\ln(1 + e^y). \tag{69}
$$

Solution.

Following the discussions presented above yields

$$
u(x,y) = 3\ln 2 - 2\ln(1 + e^x) - 2\ln(1 + e^y) + L_x^{-1} L_y^{-1} e^{x+y} e^u. \tag{70}
$$

Proceeding as before we obtain

$$
\begin{aligned}
\sum_{n=0}^{\infty} u_n(x,y) &= 3\ln 2 - 2\ln(1 + e^x) - 2\ln(1 + e^y) + e^y \\
&\quad + L_x^{-1} L_y^{-1} \left(\sum_{n=0}^{\infty} e^{x+y} A_n \right),
\end{aligned}
$$

(71)

where A_n are the Adomian polynomials for the nonlinear term e^u. The Adomian polynomials for the exponential nonlinearity e^u were calculated before and given by

$A_0 = e^{u_0}$,

$A_1 = u_1 e^{u_0}$,

$A_2 = (\frac{1}{2!} u_1^2 + u_2) e^{u_0}$,

$A_3 = (\frac{1}{3!} u_1^3 + u_1 u_2 + u_3) e^{u_0}$.

The decomposition method introduces the recursive relation

$$u_0(x, y) \quad = \quad 3 \ln 2 - 2 \ln(1 + e^x) - 2 \ln(1 + e^y), \tag{72}$$

$$u_{k+1}(x, y) \quad = \quad L_x^{-1} L_y^{-1} (e^{x+y} A_k), \; k \geq 0.$$

The first three components of the solution $u(x, y)$ are given by

$$u_0(x, y) \quad = \quad 3 \ln 2 - 2 \ln(1 + e^x) - 2 \ln(1 + e^y),$$

$$u_1(x, y) \quad = \quad L_x^{-1} L_y^{-1} e^{x+y} e^{u_0},$$

$$= \quad 8 L_x^{-1} L_y^{-1} \left[\frac{e^x}{(e^x + 1)^2} \times \frac{e^y}{(e^y + 1)^2} \right]$$

$$= \quad 2 \left[\frac{(e^x - 1)(e^y - 1)}{(e^x + 1)(e^y + 1)} \right],$$

$$u_2(x, y) \quad = \quad L_x^{-1} L_y^{-1} e^{x+y} u_1 e^{u_0}, \tag{73}$$

$$= \quad 16 L_x^{-1} L_y^{-1} \left[\frac{e^x(e^x - 1)}{(e^x + 1)^3} \times \frac{e^y(e^y - 1)}{(e^y + 1)^3} \right],$$

$$= \quad \left[\frac{(e^x - 1)(e^y - 1)}{(e^x + 1)(e^y + 1)} \right]^2,$$

$$u_3(x, y) \quad = \quad L_x^{-1} L_y^{-1} (\frac{1}{2!} u_1^2 + u_2) e^{u_0},$$

$$= \quad \frac{2}{3} \left[\frac{(e^x - 1)(e^y - 1)}{(e^x + 1)(e^y + 1)} \right]^3.$$

and so on. Note that the integrals involved above can be obtained by substituting $z = 1 + e^t, dz = e^t \, dt$. In view of (73), the solution in a series form is given by

$$u(x, y) \quad = \quad 3 \ln 2 - 2 \ln(1 + e^x) - 2 \ln(1 + e^y)$$

$$+ 2 \left(\sum_{n=1}^{\infty} \frac{K^n(x, y)}{n} \right), \tag{74}$$

where

$$K(x, y) = \frac{(e^x - 1)(e^y - 1)}{(e^x + 1)(e^y + 1)}. \tag{75}$$

Recall that the Taylor expansion for $\ln(1-t)$ is given by

$$\begin{aligned}
\ln(1-t) &= -(t + \tfrac{1}{2}t^2 + \tfrac{1}{3}t^3 + \cdots), \\
&= -\sum_{n=1}^{\infty} \frac{t^n}{n}.
\end{aligned} \tag{76}$$

This means that Eq. (74) becomes

$$\begin{aligned}
u(x,y) &= 3\ln 2 - 2\ln(1+e^x) - 2\ln(1+e^y) - 2\ln[1 - K(x,y)], \\
&= 3\ln 2 - 2\ln(1+e^x) - 2\ln(1+e^y) \\
&\quad -2\ln[1 - \frac{(e^x-1)(e^y-1)}{(e^x+1)(e^y+1)}], \\
&= 3\ln 2 - 2\ln(1+e^x) - 2\ln(1+e^y) \\
&\quad -2\ln 2\left(\frac{(e^x+e^y)}{(e^x+1)(e^y+1)}\right), \\
&= \ln 2 - 2\ln(e^x + e^y).
\end{aligned} \tag{77}$$

Example 4. Solve the following nonlinear Goursat problem

$$u_{xy} = \frac{2}{3}e^{3u}, \tag{78}$$

subject to the conditions

$$u(x,0) = \frac{1}{3}x - \frac{2}{3}\ln(1+e^x), \ u(0,y) = \frac{1}{3}y - \frac{2}{3}\ln(1+e^y). \tag{79}$$

Solution.

Proceeding as before we find

$$u(x,y) = \frac{x+y}{3} - \frac{2}{3}\ln(1+e^x) - \frac{2}{3}\ln(1+e^y) + \frac{2}{3}\ln 2 + \frac{2}{3}L_x^{-1}L_y^{-1}e^{3u}. \tag{80}$$

Substituting the series representation for the linear and the nonlinear terms into (80) we obtain

$$\begin{aligned}
\sum_{n=0}^{\infty} u_n(x,y) &= \frac{x+y}{3} - \frac{2}{3}\ln(1+e^x) - \frac{2}{3}\ln(1+e^y) + \frac{2}{3}\ln 2 \\
&\quad + \frac{2}{3}L_x^{-1}L_y^{-1}\left(\sum_{n=0}^{\infty} A_n\right),
\end{aligned} \tag{81}$$

where A_n are Adomian polynomials for the nonlinear term e^{3u} given by

$A_0 = e^{3u_0}$,

$A_1 = 3u_1 e^{3u_0}$,

$A_2 = (\frac{9}{2!}u_1^2 + 3u_2)e^{3u_0}$,

$A_3 = (\frac{27}{3!}u_1^3 + 9u_1 u_2 + 3u_3)e^{3u_0}$.

Following Adomian analysis, we set the recursive relation

$$u_0(x,y) = \frac{x+y}{3} + \frac{2}{3}\ln(1+e^x) + \frac{2}{3}\ln(1+e^y) + \frac{2}{3}\ln 2,$$

$$u_{k+1}(x,y) = \frac{2}{3}L_x^{-1}L_y^{-1}(A_k),\ k \geq 0, \tag{82}$$

so that

$$u_0(x,y) = \frac{x+y}{3} - \frac{2}{3}\ln(1+e^x) - \frac{2}{3}\ln(1+e^y) - \frac{2}{3}\ln 2,$$

$$u_1(x,y) = \frac{2}{3}L_x^{-1}L_y^{-1}e^{3u_0},$$

$$= L_x^{-1}L_y^{-1}\left[\frac{e^x}{(e^x+1)^2} \times \frac{e^y}{(1+e^y)^2}\right],$$

$$= \left[\frac{(e^x-1)(e^y-1)}{(e^x+1)(e^y+1)}\right], \tag{83}$$

$$u_2(x,y) = \frac{2}{3}L_x^{-1}L_y^{-1}3u_1 e^{3u_0},$$

$$= \frac{16}{3}L_x^{-1}L_y^{-1}\left[\frac{e^x(e^x-1)}{(e^x+1)^3} \times \frac{e^y(e^y-1)}{(e^y+1)^3}\right],$$

$$= \frac{1}{3}\left[\frac{(e^x-1)(e^y-1)}{(e^x+1)(e^y+1)}\right]^2,$$

and so on. Proceeding as before we obtain

$$u(x,y) = \frac{x+y}{3} - \frac{2}{3}\ln(1+e^x) - \frac{2}{3}\ln(1+e^y) + \frac{2}{3}\ln 2$$

$$+\frac{2}{3}\left(\sum_{n=1}^{\infty}\frac{K^n(x,y)}{n}\right), \tag{84}$$

where

$$K(x,y) = \frac{(e^x-1)(e^y-1)}{(e^x+1)(e^y+1)}. \tag{85}$$

Using the Taylor expansion for $\ln(1-t)$ gives

$$u(x,y) = \frac{x+y}{3} - \frac{2}{3}\ln(1+e^x) - \frac{2}{3}\ln(1+e^y) + \frac{2}{3}\ln 2$$

$$-\frac{2}{3}\ln[1-K(x,y)], \tag{86}$$

$$= \frac{x+y}{3} - \frac{2}{3}\ln(e^x+e^y).$$

Exercises 9.3

In Exercises 1 – 6, use Adomian decomposition method to solve the following linear Goursat problems:

1. $u_{xy} = -y + u$, $u(x,0) = e^x$, $u(0,y) = y + e^y$

2. $u_{xy} = 1 - xy + u$, $u(x,0) = e^x$, $u(0,y) = e^y$

3. $u_{xy} = x + y + u$, $u(x,0) = -x + e^x$, $u(0,y) = -y + e^y$

4. $u_{xy} = -x^2 + u$, $u(x,0) = x^2 + e^x$, $u(0,y) = e^y$

5. $u_{xy} = u + 2e^{x+y}$, $u(x,0) = xe^x$, $u(0,y) = ye^y$

6. $u_{xy} = u$, $u(x,0) = e^x$, $u(0,y) = e^y$

In Exercises 7 – 12, use Adomian decomposition method to solve the following nonlinear Goursat problems:

7. $u_{xy} = e^{2u}$, $u(x,0) = \frac{1}{2}x - \ln(1 + e^x)$, $u(0,y) = \frac{1}{2}y - \ln(1 + e^y)$

8. $u_{xy} = -e^{2u}$, $u(x,0) = \frac{1}{2}x - \ln(1 + e^x)$, $u(0,y) = -\frac{1}{2}y - \ln(1 + e^{-y})$

9. $u_{xy} = e^y e^{2u}$, $u(x,0) = \frac{1}{2}x - \ln(1 + e^x)$, $u(0,y) = -\ln(1 + e^y)$

10. $u_{xy} = e^x e^{2u}$, $u(x,0) = -\ln(1 + e^x)$, $u(0,y) = \frac{1}{2}y - \ln(1 + e^y)$

11. $u_{xy} = \frac{2}{5}e^y e^{5u}$, $u(x,0) = \frac{1}{5}x - \frac{2}{5}\ln(1 + e^x)$, $u(0,y) = -\frac{2}{5}\ln(1 + e^y)$

12. $u_{xy} = e^{x+y} e^{2u}$, $u(x,0) = -\ln(1 + e^x)$, $u(0,y) = -\ln(1 + e^y)$

9.4 The Klein-Gordon Equation

The Klein-Gordon equation is considered one of the most important mathematical models in quantum field theory. The equation appears in relativistic physics and is used to describe dispersive wave phenomena in general. In addition, it also appears in nonlinear optics and plasma physics. The Klein-Gordon equation arise in physics in linear and nonlinear forms.

The Klein-Gordon equation has been extensively studied by using traditional methods such as finite difference method, finite element method, or collocation method. Bäcklund transformations and the inverse scattering method were also applied to handle Klein-Gordon equation. The methods investigated the concepts of existence, uniqueness of the solution and the weak solution as well. The objectives of these studies were mostly focused on the determination of numerical solutions where a considerable volume of calculations is usually needed.

In this section, the Adomian decomposition method will be applied to obtain exact solutions if exist, and approximate to solutions for concrete problems.

9.4.1   Linear Klein-Gordon Equation

The linear Klein-Gordon equation in its standard form is given by

$$u_{tt}(x,t) - u_{xx}(x,t) + au(x,t) = h(x,t), \tag{87}$$

subject to the initial conditions

$$u(x,0) = f(x), \; u_t(x,0) = g(x), \tag{88}$$

where a is a constant and $h(x,t)$ is the source term. It is interesting to point here that if $a = 0$, Eq. (87) becomes the inhomogeneous wave equation that was introduced before. The linear Klein-Gordon equation is important in quantum mechanics. It is derived from the relativistic energy formula.

In an operator form, Eq. (87) can be rewritten as

$$L_t u(x,t) = u_{xx}(x,t) - au(x,t) + h(x,t), \tag{89}$$

where L_t is a second order differential operator and the inverse operator L_t^{-1} is a two-fold integral operator defined by

$$L_t^{-1}(.) = \int_0^t \int_0^t (.)dt\, dt. \tag{90}$$

Applying L_t^{-1} to both sides of (89) and using the initial conditions we find

$$u(x,t) = f(x) + tg(x) + L_t^{-1}(h(x,t)) + L_t^{-1}\left(u_{xx}(x,t) - au(x,t)\right). \tag{91}$$

Using the decomposition representation for $u(x,t)$ into both sides of (91) gives

$$
\begin{aligned}
\sum_{n=0}^{\infty} u_n(x,t) \;=\; & f(x) + tg(x) + L_t^{-1}(h(x,t)) + \\
& L_t^{-1}\left(\left(\sum_{n=0}^{\infty} u_n(x,t)\right)_{xx} - a\sum_{n=0}^{\infty} u_n(x,t)\right).
\end{aligned}
\tag{92}
$$

We can formally set the recursive relation

$$
\begin{aligned}
u_0(x,t) \;&=\; f(x) + tg(x) + L_t^{-1}(h(x,t)), \\
u_{k+1}(x,t) \;&=\; L_t^{-1}\left(u_{k_{xx}}(x,t) - au_k(x,t)\right), k \geq 0.
\end{aligned}
\tag{93}
$$

This completes the determination of the components of $u(x,t)$. The solution in a series form follows immediately. In many cases we can obtain inductively the exact solution. The algorithm discussed above will be explained through the following illustrative examples.

Example 1. Solve the following linear Klein-Gordon equation

$$u_{tt} - u_{xx} + u = 0, \quad u(x,0) = 0, \quad u_t(x,0) = x. \tag{94}$$

Solution.

Applying L_t^{-1} to both sides of (94) and using the decomposition series for $u(x,t)$ give

$$\sum_{n=0}^{\infty} u_n(x,t) = xt + L_t^{-1}\left(\left(\sum_{n=0}^{\infty} u_n(x,t)\right)_{xx} - \sum_{n=0}^{\infty} u_n(x,t)\right). \tag{95}$$

Close examination of (95) suggests that the recursive relation is

$$
\begin{aligned}
u_0(x,t) &= xt, \\
u_{k+1}(x,t) &= L_t^{-1}\left(u_{k_{xx}}(x,t) - u_k(x,t)\right), k \geq 0,
\end{aligned}
\tag{96}
$$

that in turn gives

$$
\begin{aligned}
u_0(x,t) &= xt, \\
u_1(x,t) &= L_t^{-1}(u_{0_{xx}}(x,t) - u_0(x,t)) = -\tfrac{1}{3!}xt^3, \\
u_2(x,t) &= L_t^{-1}(u_{1_{xx}}(x,t) - u_1(x,t)) = \tfrac{1}{5!}xt^5.
\end{aligned}
\tag{97}
$$

In view of (97) the series solution is given by

$$u(x,t) = x\left(t - \frac{1}{3!}t^3 + \frac{1}{5!}t^5 - \cdots\right), \tag{98}$$

and the exact solution is given by

$$u(x,t) = x \sin t. \tag{99}$$

Example 2. Solve the following linear inhomogeneous Klein-Gordon equation

$$u_{tt} - u_{xx} + u = 2\sin x, \quad u(x,0) = \sin x, \quad u_t(x,0) = 1. \tag{100}$$

Solution.

Proceeding as in Example 1 we find

$$\sum_{n=0}^{\infty} u_n(x,t) = \sin x + t + t^2 \sin x L_t^{-1}\left(\left(\sum_{n=0}^{\infty} u_n(x,t)\right)_{xx} - \sum_{n=0}^{\infty} u_n(x,t)\right). \tag{101}$$

Consequently, we set the relation

$$u_0(x,t) = \sin x + t + t^2 \sin x$$

$$u_{k+1}(x,t) = L_t^{-1}(u_{k_{xx}}(x,t) - u_k(x,t)), k \geq 0, \tag{102}$$

that gives

$$u_0(x,t) = \sin x + t + t^2 \sin x,$$

$$u_1(x,t) = L_t^{-1}(u_{0_{xx}}(x,t) - u_0(x,t)) = -t^2 \sin x - \tfrac{1}{6}t^4 \sin x - \tfrac{1}{3!}t^3, \tag{103}$$

$$u_2(x,t) = L_t^{-1}(u_{1_{xx}}(x,t) - u_1(x,t)) = \tfrac{1}{6}t^4 \sin x + \tfrac{1}{90}t^6 \sin x + \tfrac{1}{5!}t^5.$$

In view of (103), the series solution is given by

$$u(x,t) = \sin x + (t - \frac{1}{3!}t^3 + \frac{1}{5!}t^5 - \frac{1}{7}t^7 + \cdots), \tag{104}$$

where noise terms vanish in the limit. The solution in a closed form

$$u(x,t) = \sin x + \sin t, \tag{105}$$

follows immediately.

9.4.2   Nonlinear Klein-Gordon Equation

The nonlinear Klein-Gordon equation comes from quantum field theory and describes nonlinear wave interaction. The nonlinear Klein-Gordon equation in its standard form is given by

$$u_{tt}(x,t) - u_{xx}(x,t) + au(x,t) + F(u(x,t)) = h(x,t), \tag{106}$$

subject to the initial conditions

$$u(x,0) = f(x), \; u_t(x,0) = g(x), \tag{107}$$

where a is a constant, $h(x,t)$ is a source term and $F(u(x,t))$ is a nonlinear function of $u(x,t)$. The equation has been investigated using numerical methods such as finite difference method and the averaging techniques.

In a manner parallel to that used before, the decomposition method will be employed. The nonlinear term $F(u(x,t))$ will be equated to the infinite series of Adomian polynomials. Applying L_t^{-1} to both sides of (106) and using the initial conditions give

$$u(x,t) = f(x) + tg(x) + L_t^{-1}(h(x,t))$$

$$+ L_t^{-1}(u_{xx}(x,t) - au(x,t)) - L_t^{-1}(F(u(x,t))). \tag{108}$$

Using the decomposition series for the linear term $u(x,t)$, the infinite series of Adomian polynomials for the nonlinear term $F(u(x,t))$, and proceeding as before we obtain the recursive relation

$$
\begin{aligned}
u_0(x,t) &= f(x) + tg(x) + L_t^{-1}(h(x,t)), \\
u_{k+1}(x,t) &= L_t^{-1}(u_{k_{xx}}(x,t) - u_k(x,t)) - L_t^{-1}(A_k), k \geq 0,
\end{aligned}
\tag{109}
$$

that leads to

$$
\begin{aligned}
u_0(x,t) &= f(x) + tg(x) + L_t^{-1}(h(x,t)), \\
u_1(x,t) &= L_t^{-1}L_t^{-1}(u_{0_{xx}}(x,t) - u_0(x,t)) - L_t^{-1}(A_0), \\
u_2(x,t) &= L_t^{-1}L_t^{-1}(u_{1_{xx}}(x,t) - u_1(x,t)) - L_t^{-1}(A_1).
\end{aligned}
\tag{110}
$$

This completes the determination of the first few components of the solution.

Based on this determination, the solution in a series form is readily obtained. In many cases, a closed form solution can be obtained inductively.

The following examples will be used to illustrate the algorithm discussed above. The noise terms phenomenon and the modified decomposition method will be implemented in this illustration to accelerate the convergence.

Example 3. Solve the following nonlinear Klein-Gordon equation

$$
u_{tt} - u_{xx} + u^2 = x^2 t^2, \ u(x,0) = 0, \ u_t(x,0) = x.
\tag{111}
$$

Solution.

Following the discussion presented above we find

$$
\sum_{n=0}^{\infty} u_n(x,t) = xt + \frac{1}{12}x^2t^4 + L_t^{-1}\left(\left(\sum_{n=0}^{\infty} u_n(x,t)\right)_{xx}\right) - L_t^{-1}\left(\sum_{n=0}^{\infty} A_n\right).
\tag{112}
$$

We will approach the problem by using the noise terms phenomenon. Eq. (112) gives the recursive relation

$$
\begin{aligned}
u_0(x,t) &= xt + \frac{1}{12}x^2t^4, \\
u_{k+1}(x,t) &= L_t^{-1}u_{k_{xx}}(x,t) - L_t^{-1}A_k, k \geq 0,
\end{aligned}
\tag{113}
$$

that yields

$$
\begin{aligned}
u_0(x,t) &= xt + \frac{1}{12}x^2t^4, \\
u_1(x,t) &= L_t^{-1}u_{0_{xx}}(x,t) - L_t^{-1}A_0, \\
&= \frac{1}{180}t^6 - \frac{1}{12}x^2t^4 - \frac{1}{252}x^3t^7 + \frac{1}{12960}x^4t^{10}.
\end{aligned}
\tag{114}
$$

Canceling the noise term $\frac{1}{12}x^2t^4$ from the component u_0, and justifying that the remaining non-canceled term satisfies the equation, the exact solution

$$u(x,t) = xt, \tag{115}$$

is readily obtained.

In the following we will solve this example by using the modified decomposition method. As introduced before we split the terms assigned to the zeroth component $u_0(x,t)$ to the first two components $u_0(x,t)$ and $u_1(x,t)$. Thus the modified recursive relation can be rewritten in the scheme

$$
\begin{aligned}
u_0(x,t) &= xt, \\
u_1(x,t) &= \tfrac{1}{12}x^2t^4 + L_t^{-1}(u_{0_{xx}}(x,t)) - L_t^{-1}A_0, \\
u_{k+1}(x,t) &= L_t^{-1}(u_{k_{xx}}(x,t)) - L_t^{-1}A_k, \; k \geq 1.
\end{aligned}
\tag{116}
$$

This leads to

$$
\begin{aligned}
u_0(x,t) &= xt, \\
u_1(x,t) &= \tfrac{1}{12}x^2t^4 + L_t^{-1}(u_{0_{xx}}(x,t)) - L_t^{-1}A_0 = 0, \\
u_{k+1}(x,t) &= 0, \; k \geq 1.
\end{aligned}
\tag{117}
$$

Therefore, the exact solution is given by

$$u(x,t) = xt. \tag{118}$$

Example 4. Solve the following nonlinear Klein-Gordon equation

$$u_{tt} - u_{xx} + u^2 = 2x^2 - 2t^2 + x^4t^4, \; u(x,0) = u_t(x,0) = 0. \tag{119}$$

Solution.

The noise terms phenomenon will be used in this example. Proceeding as before gives

$$
\begin{aligned}
u_0(x,t) &= x^2t^2 - \tfrac{1}{6}t^4 + \tfrac{1}{30}x^4t^6, \\
u_{k+1}(x,t) &= L_t^{-1}(u_{k_{xx}}(x,t)) - L_t^{-1}A_k, \; k \geq 0.
\end{aligned}
\tag{120}
$$

Based on this relation the first two components are given by

$$
\begin{aligned}
u_0(x,t) &= x^2t^2 - \tfrac{1}{6}t^4 + \tfrac{1}{30}x^4t^6, \\
u_1(x,t) &= L_t^{-1}(u_{0_{xx}}(x,t)) - L_t^{-1}A_0 = \tfrac{1}{6}t^4 - \tfrac{1}{30}x^4t^6 + \cdots.
\end{aligned}
\tag{121}
$$

Canceling the noise terms in $u_0(x,t)$ that appear in $u_1(x,t)$ and justifying that the remaining term satisfies the equation leads to the exact solution

$$u(x,t) = x^2 t^2. \tag{122}$$

Next we formally show that the modified decomposition method accelerates the convergence of the solution and minimizes the size of calculations. The modified method introduces the relation

$$
\begin{aligned}
u_0(x,t) &= x^2 t^2, \\
u_1(x,t) &= -\tfrac{1}{6}t^4 + \tfrac{1}{30}x^4 t^6 + L_t^{-1}(u_{0_{xx}}(x,t) - A_0) = 0, \\
u_{k+1}(x,t) &= 0, \ k \geq 1.
\end{aligned}
\tag{123}
$$

This formally gives the exact solution

$$u(x,t) = x^2 t^2. \tag{124}$$

9.4.3   The Sine-Gordon Equation

The Sine-Gordon equation appeared first in differential geometry. This equation became the focus of a lot of research work because it appears in many physical phenomena such as the propagation of magnetic flux and the stability of fluid motions. The equation is considered an important nonlinear evolution equation that plays a major role in nonlinear physics.

The standard form of the Sine-Gordon equation is given by

$$u_{tt} - c^2 u_{xx} + \alpha \sin u = 0, \ u(x,0) = f(x), \ u_t(x,0) = g(x), \tag{125}$$

where c and α are constants. It is clear that this equation adds the nonlinear term $\sin u$ to the standard wave equation.

Several classical methods have been employed to handle the Sine-Gordon equation. The Bäcklund transformations, the similarity method, and the inverse scattering method are mostly used to investigate this equation.

However, the Sine-Gordon equation will be handled by using the Adomian decomposition method. Applying L_t^{-1} to (125) and using the initial conditions leads to

$$u(x,t) = f(x) + tg(x) + c^2 L_t^{-1}(u_{xx}(x,t)) - \alpha L_t^{-1}(\sin u(x,t)). \tag{126}$$

Noting that $\sin u$ is a nonlinear operator where the relevant Adomian polynomials have been derived before. Substituting the series decomposition for $u(x,t)$ and the infinite series of Adomian polynomials for $\sin u$ gives

$$\sum_{n=0}^{\infty} u_n(x,t) = f(x) + tg(x) + c^2 L_t^{-1}\left(\left(\sum_{n=0}^{\infty} u_n(x,t)\right)_{xx} - \alpha\left(\sum_{n=0}^{\infty} A_n\right)\right). \tag{127}$$

This gives the recursive relation

$$u_0(x,t) = f(x) + tg(x),$$

$$u_{k+1}(x,t) = c^2 L_t^{-1}(u_{k_{xx}}(x,t)) - \alpha L_t^{-1}(A_k), \ k \geq 0. \tag{128}$$

This will lead to the determination of the solution in a series form. This can be illustrated as follows.

Example 5. Solve the following Sine-Gordon equation

$$u_{tt} - u_{xx} = \sin u, \ u(x,0) = \frac{\pi}{2}, \ u_t(x,0) = 0. \tag{129}$$

Solution.

Using the recursive scheme (128) yields

$$u_0(x,t) = \tfrac{\pi}{2},$$

$$u_{k+1}(x,t) = L_t^{-1}(u_{k_{xx}}(x,t)) + L_t^{-1}(A_k), \ k \geq 0. \tag{130}$$

The first few Adomian polynomials for $\sin u$ are given by

$$A_0 = \sin u_0,$$

$$A_1 = u_1 \cos u_0,$$

$$A_2 = u_2 \cos u_0 - \tfrac{1}{2!} u_1^2 \sin u_0, \tag{131}$$

$$A_3 = u_3 \cos u_0 - u_2 u_1 \sin u_0 - \tfrac{1}{3!} u_1^3 \cos u_0.$$

Combining (130) and (131) leads to

$$u_0(x,t) = \tfrac{\pi}{2}, \ u_1(x,t) = \tfrac{1}{2}t^2, \ u_2(x,t) = 0,$$

$$u_3(x,t) = -\tfrac{1}{240}t^6, \ u_4(x,t) = 0, \ u_5(x,t) = \tfrac{1}{172800}t^{10}, \tag{132}$$

The series solution

$$u(x,t) = \frac{\pi}{2} + \frac{1}{2}t^2 - \frac{1}{240}t^6 + \frac{1}{172800}t^{10} + \cdots, \tag{133}$$

is readily obtained.

Example 6. Solve the following Sine-Gordon equation

$$u_{tt} - u_{xx} = \sin u, \ u(x,0) = \frac{\pi}{2}, \ u_t(x,0) = 1. \tag{134}$$

Solution.

Using the relation (128) gives

$$
\begin{aligned}
u_0(x,t) &= \tfrac{\pi}{2} + t, \\
u_{k+1}(x,t) &= L_t^{-1}(u_{k_{xx}}(x,t)) + L_t^{-1}(A_k), \ k \geq 0.
\end{aligned}
\tag{135}
$$

Using Adomian polynomials for $\sin u$ as shown above leads to the results

$$
\begin{aligned}
u_0(x,t) &= \tfrac{\pi}{2} + t, \\
u_1(x,t) &= 1 - \cos t, \\
u_2(x,t) &= \sin t - \tfrac{3}{4}t - \tfrac{1}{8}\sin 2t.
\end{aligned}
\tag{136}
$$

Summing these iterates yields

$$
u(x,t) = \frac{\pi}{2} + t + 1 - \cos t + \sin t - \frac{3}{4}t - \frac{1}{8}\sin 2t + \cdots,
\tag{137}
$$

so that the series solution

$$
u(x,t) = \frac{\pi}{2} + t + \frac{1}{2!}t^2 - \frac{1}{4!}t^4 + \cdots,
\tag{138}
$$

obtained upon using Taylor expansion for the trigonometric functions involved.

It is important to note that another form of the Sine-Gordon equation is sometimes used and given in the form

$$
u_{xt} = \sin u.
\tag{139}
$$

Recall that the initial value problem of (139) has been discussed before as a Goursat problem.

Exercises 9.4

In Exercises 1 – 5, use Adomian decomposition method to solve the linear equations:

1. $u_{tt} - u_{xx} - u = -\cos x \cos t$, $u(x,0) = \cos x$, $u_t(x,0) = 0$

2. $u_{tt} - u_{xx} - u = -\cos x \sin t$, $u(x,0) = 0$, $u_t(x,0) = \cos x$

3. $u_{tt} - u_{xx} - u = -\sin x \sin t$, $u(x,0) = 0$, $u_t(x,0) = \sin x$

4. $u_{tt} - u_{xx} - u = 0$, $u(x,0) = 0$, $u_t(x,0) = \sin x$

5. $u_{tt} - u_{xx} + u = 0$, $u(x,0) = 0$, $u_t(x,0) = \cosh x$

In Exercises 6 – 10, use the modified decomposition method to solve the nonlinear equations:

6. $u_{tt} - u_{xx} - u + u^2 = xt + x^2t^2$, $u(x,0) = 1$, $u_t(x,0) = x$

7. $u_{tt} - u_{xx} + u^2 = 1 + 2xt + x^2t^2$, $u(x,0) = 1$, $u_t(x,0) = x$

8. $u_{tt} - u_{xx} + u^2 = 6xt(x^2 + t^2) + x^6t^6$, $u(x,0) = 0$, $u_t(x,0) = 0$

9. $u_{tt} - u_{xx} + u^2 = (t^2 + x^2)^2$, $u(x,0) = x^2$, $u_t(x,0) = 0$

10. $u_{tt} - u_{xx} + u + u^2 = x^2\cos^2 t$, $u(x,0) = x$, $u_t(x,0) = 0$

In Exercises 11 – 15, find the ϕ_3 approximant of the solution of the following Sine-Gordon equations:

11. $u_{tt} - u_{xx} = \sin u$, $u(x,0) = \frac{\pi}{6}$, $u_t(x,0) = 0$

12. $u_{tt} - u_{xx} = \sin u$, $u(x,0) = \frac{\pi}{4}$, $u_t(x,0) = 0$

13. $u_{tt} - u_{xx} = \sin u$, $u(x,0) = 0$, $u_t(x,0) = 1$

14. $u_{tt} - u_{xx} = \sin u$, $u(x,0) = \pi$, $u_t(x,0) = 1$

15. $u_{tt} - u_{xx} = \sin u$, $u(x,0) = \frac{3\pi}{2}$, $u_t(x,0) = 1$

9.5   The Burgers' Equation

Burgers' equation is considered one of the fundamental model equations in fluid mechanics. The equation demonstrates the coupling between diffusion and convection processes.

The standard form of Burgers' equation is given by

$$u_t + uu_x = \nu u_{xx}, x \in R, t > 0, \tag{140}$$

where ν is a constant that defines the kinematic viscosity. If the viscosity $\nu = 0$, the equation is called *inviscid* Burgers' equation. The inviscid Burgers' equation governs gas dynamics. The inviscid Burgers' equation has been discussed before as a homogeneous case of the advection problem. The inviscid equation can be elegantly handled as discussed before in Section 9.2.

Nonlinear Burgers' equation is considered by most as a simple nonlinear partial differential equation incorporating both convection and diffusion in fluid dynamics. Burgers introduced this equation in [30] to capture some of the features of turbulent fluid in a channel caused by the interaction of the opposite effects of convection and diffusion. It is also used to describe the structure of shock waves, traffic flow, and acoustic transmission.

A great potential of research work has been invested on Burgers' equation. Several exact solutions have been derived by using distinct approaches. Appendix C contains many of these exact solutions. The *Carl-Hopf* transformation is the commonly used approach. The solution $u(x,t)$ was replaced by ψ_x where upon integrating the resulting equation we find

$$\psi_t + \frac{1}{2}\psi_x^2 = \nu\psi_{xx}. \tag{141}$$

Using the *Carl-Hopf* transformation

$$\psi = -2\nu \ln \phi, \tag{142}$$

so that

$$u(x,t) = \psi_x = -2\nu\frac{\phi_x}{\phi}, \tag{143}$$

transforms the nonlinear equation into the heat flow equation

$$\phi_t = \nu\phi_{xx}. \tag{144}$$

It is obvious that the difficult nonlinear Burgers' equation (140) has been converted to an easily solvable linear equation. This will lead to exact solutions, each solution depends on the given conditions.

Another technique for deriving solutions to Burgers' equation is the method of *symmetry reduction* in which solutions of the nonlinear Burgers' equation are found in terms of parabolic cylinder functions or Airy functions. The symmetry reduction method was applied in a modified way where the Burgers' equation was transformed to an ordinary differential equation.

However, it is the intention of this text to effectively apply the reliable Adomian decomposition method. We consider the Burgers' equation

$$u_t + uu_x = u_{xx}, \; u(x,0) = f(x). \tag{145}$$

Applying the inverse operator L_t^{-1} to (145) leads to

$$u(x,t) = f(x) + L_t^{-1}(u_{xx}) - L_t^{-1}(uu_x). \tag{146}$$

Using the decomposition series for the linear term $u(x,t)$ and the series of Adomian polynomials for the nonlinear term uu_x give

$$\sum_{n=0}^{\infty} u_n(x,t) = f(x) + L_t^{-1}\left(\left(\sum_{n=0}^{\infty}(u_n(x,t))\right)_{xx}\right) - L_t^{-1}\left(\sum_{n=0}^{\infty} A_n\right). \tag{147}$$

Identifying the zeroth component $u_0(x,t)$ by the term that arise from the initial condition and following the decomposition method, we obtain the recursive relation

$$\begin{aligned} u_0(x,t) &= f(x), \\ u_{k+1}(x,t) &= L_t^{-1}(u_{k_{xx}}) - L_t^{-1}A_k, k \geq 0. \end{aligned} \tag{148}$$

The Adomian polynomials for the nonlinear term uu_x have been derived in the form

$$
\begin{aligned}
A_0 &= u_{0_x} u_0, \\[4pt]
A_1 &= u_{0_x} u_1 + u_{1_x} u_0, \\[4pt]
A_2 &= u_{0_x} u_2 + u_{1_x} u_1 + u_{2_x} u_0, \\[4pt]
A_3 &= u_{0_x} u_3 + u_{1_x} u_2 + u_{2_x} u_1 + u_{3_x} u_0, \\[4pt]
A_4 &= u_{0_x} u_4 + u_{1_x} u_3 + u_{2_x} u_2 + u_{3_x} u_1 + u_{4_x} u_0.
\end{aligned}
\tag{149}
$$

In view of (148) and (149), the first few components can be identified by

$$
\begin{aligned}
u_0(x,t) &= f(x), \\[4pt]
u_1(x,t) &= L_t^{-1}(u_{0_{xx}}) - L_t^{-1} A_0, \\[4pt]
u_2(x,t) &= L_t^{-1}(u_{1_{xx}}) - L_t^{-1} A_1, \\[4pt]
u_3(x,t) &= L_t^{-1}(u_{2_{xx}}) - L_t^{-1} A_2.
\end{aligned}
\tag{150}
$$

Additional components can be elegantly computed to enhance the accuracy level. The solution in a series form follows immediately. However, the n-term approximant ϕ_n can be determined by

$$
\phi_n = \sum_{k=0}^{n-1} u_k(x,t).
\tag{151}
$$

In the following we list some of the derived exact solutions of Burgers' equation derived by many researchers:

$$
\begin{aligned}
u(x,t) &= 2\tan x, \; -2\cot x, \; -2\tanh x, \\[4pt]
u(x,t) &= \frac{x}{t}, \; \frac{x}{t} + \frac{2}{x+t} + \frac{x+t}{2t^2 - t}, \\[4pt]
u(x,t) &= \frac{-2e^{-t}\cos x}{1 + e^{-t}\sin x}, \\[4pt]
u(x,t) &= \frac{2e^{-t}\sin x}{1 + e^{-t}\cos x}.
\end{aligned}
\tag{152}
$$

A table of solutions of Burgers' equation can be found in Appendix C. The following examples will be used to illustrate the discussion carried out above.

Example 1. Solve the following Burgers' equation

$$
u_t + uu_x = u_{xx}, \; u(x,0) = x.
\tag{153}
$$

Solution.

Operating with L_t^{-1} and using (147) we find

$$\sum_{n=0}^{\infty} u_n(x,t) = x + L_t^{-1}\left(\left(\sum_{n=0}^{\infty} u_n(x,t)\right)_{xx}\right) - L_t^{-1}\left(\sum_{n=0}^{\infty} A_n\right). \tag{154}$$

This gives the recursive relation

$$\begin{aligned}
u_0(x,t) &= x, \\
u_{k+1}(x,t) &= L_t^{-1}\left(u_{k_{xx}}(x,t)\right) - L_t^{-1}(A_k), \, k \geq 0.
\end{aligned} \tag{155}$$

Using Adomian polynomials we obtain

$$\begin{aligned}
u_0(x,t) &= x, \\
u_1(x,t) &= L_t^{-1}\left(u_{0_{xx}}(x,t)\right) - L_t^{-1}(A_0) = -xt, \\
u_2(x,t) &= L_t^{-1}\left(u_{1_{xx}}(x,t)\right) - L_t^{-1}(A_1) = xt^2, \\
u_3(x,t) &= L_t^{-1}\left(u_{2_{xx}}(x,t)\right) - L_t^{-1}(A_2) = -xt^3.
\end{aligned} \tag{156}$$

Summing these iterates gives the series solution

$$u(x,t) = x(1 - t + t^2 - t^3 + \cdots). \tag{157}$$

Consequently, the exact solution is given by

$$u(x,t) = \frac{x}{1+t}. \tag{158}$$

Example 2. Solve the following Burgers' equation

$$u_t + uu_x = u_{xx}, \, u(x,0) = 1 - \frac{2}{x}, x > 0. \tag{159}$$

Solution.

Proceeding as before gives

$$\sum_{n=0}^{\infty} u_n(x,t) = 1 - \frac{2}{x} + L_t^{-1}\left(\left(\sum_{n=0}^{\infty} u_n(x,t)\right)_{xx}\right) - L_t^{-1}\left(\sum_{n=0}^{\infty} A_n\right). \tag{160}$$

Consequently, we set the recursive relation

$$\begin{aligned}
u_0(x,t) &= 1 - \frac{2}{x}, \\
u_{k+1}(x,t) &= L_t^{-1}\left(u_{k_{xx}}(x,t)\right) - L_t^{-1}(A_k), \, k \geq 0,
\end{aligned} \tag{161}$$

that gives

$$
\begin{aligned}
u_0(x,t) &= 1 - \tfrac{2}{x}, \\
u_1(x,t) &= L_t^{-1}\left(u_{0_{xx}}(x,t)\right) - L_t^{-1}\left(A_0\right) = L_t^{-1}\left(-\tfrac{2}{x^2}\right) = -\tfrac{2}{x^2}t, \\
u_2(x,t) &= L_t^{-1}\left(u_{1_{xx}}(x,t)\right) - L_t^{-1}\left(A_1\right) = L_t^{-1}\left(-\tfrac{4}{x^3}t\right) = -\tfrac{2}{x^3}t^2, \\
u_3(x,t) &= L_t^{-1}\left(u_{2_{xx}}(x,t)\right) - L_t^{-1}\left(A_2\right) = L_t^{-1}\left(-\tfrac{6}{x^4}t^2\right) = -\tfrac{2}{x^4}t^3.
\end{aligned}
\tag{162}
$$

The series solution

$$
u(x,t) = 1 - \frac{2}{x} - \frac{2}{x^2}t - \frac{2}{x^3}t^2 - \frac{2}{x^4}t^3 + \cdots,
\tag{163}
$$

is readily obtained. To determine the exact solution, Eq. (163) can be rewritten as

$$
u(x,t) = 1 - \frac{2}{x}\left(1 + \frac{t}{x} + \frac{t^2}{x^2} + \frac{t^3}{x^3} + \cdots\right) = 1 - \frac{2}{x}\left(\frac{1}{1 - \frac{t}{x}}\right) = 1 - \frac{2}{x - t}.
\tag{164}
$$

Example 3. Solve the following Burgers' equation

$$
u_t + u u_x = u_{xx}, \quad u(x,0) = 2\tan x.
\tag{165}
$$

Solution.

Following the analysis presented above gives

$$
\sum_{n=0}^{\infty} u_n(x,t) = 2\tan x + L_t^{-1}\left(\left(\sum_{n=0}^{\infty} u_n(x,t)\right)_{xx}\right) - L_t^{-1}\left(\sum_{n=0}^{\infty} A_n\right).
\tag{166}
$$

The recursive relation

$$
\begin{aligned}
u_0(x,t) &= 2\tan x, \\
u_{k+1}(x,t) &= L_t^{-1}\left(u_{k_{xx}}(x,t)\right) - L_t^{-1}\left(A_k\right), \, k \geq 0,
\end{aligned}
\tag{167}
$$

leads to the determination of the first few components:

$$
\begin{aligned}
u_0(x,t) &= 2\tan x, \\
u_1(x,t) &= L_t^{-1}\left(u_{0_{xx}}(x,t)\right) - L_t^{-1}\left(A_0\right) = 0, \\
u_{k+2}(x,t) &= 0, \, k \geq 0.
\end{aligned}
\tag{168}
$$

Thus, the exact solution is given by

$$
u(x,t) = 2\tan x.
\tag{169}
$$

Example 4. Solve the following Burgers' equation

$$u_t + uu_x = u_{xx}, \; u(0,t) = -\frac{2}{t}, \; u_x(0,t) = \frac{1}{t} + \frac{2}{t^2}. \tag{170}$$

Solution.

It is important to note that, unlike the initial value problems discussed in the previous examples, the boundary conditions are given in this example. Hence, it is appropriate in this case to solve in the x direction. For this reason we first rewrite (170) in an operator form by

$$L_x u(x,t) = u_t + uu_x, \tag{171}$$

where L_x is a second order differential operator and the inverse operator L_x^{-1} is a two-fold integral operator defined by

$$L_x^{-1}(.) = \int_0^x \int_0^x (.)dx \, dx. \tag{172}$$

Operating with L_x^{-1} on both sides of (171) gives

$$u(x,t) = -\frac{2}{t} + \left(\frac{1}{t} + \frac{2}{t^2}\right)x + L_x^{-1}(u_t) + L_x^{-1}(uu_x). \tag{173}$$

Substituting the linear term $u(x,t)$ by a series of components, and the nonlinear term uu_x by a series of Adomian polynomials, we obtain

$$\sum_{n=0}^{\infty} u_n(x,t) = -\frac{2}{t} + \left(\frac{1}{t} + \frac{2}{t^2}\right)x + L_x^{-1}\left(\left(\sum_{n=0}^{\infty} u_n(x,t)\right)_t\right) + L_x^{-1}\left(\sum_{n=0}^{\infty} A_n\right). \tag{174}$$

The recursive relation

$$\begin{aligned}
u_0(x,t) &= -\frac{2}{t} + \left(\frac{1}{t} + \frac{2}{t^2}\right)x, \\
u_{k+1}(x,t) &= L_x^{-1}\left(u_{k_t}(x,t)\right) + L_x^{-1}\left(A_k\right), k \geq 0,
\end{aligned} \tag{175}$$

gives

$$\begin{aligned}
u_0(x,t) &= -\frac{2}{t} + \left(\frac{1}{t} + \frac{2}{t^2}\right)x, \\
u_1(x,t) &= L_x^{-1}\left(u_{0_t}(x,t)\right) + L_x^{-1}\left(A_0\right) = -2\frac{x^2}{t^3} + \frac{2}{3}\frac{x^3}{t^4}, \\
u_2(x,t) &= L_x^{-1}\left(u_{1_t}(x,t)\right) + L_x^{-1}\left(A_1\right) = \frac{4}{3}\frac{x^3}{t^4} + \cdots.
\end{aligned} \tag{176}$$

Summing the resulting components, the series solution

$$u(x,t) = \frac{x}{t} - \frac{2}{t}\left(1 - \frac{x}{t} + \frac{x^2}{t^2} - \frac{x^3}{t^3} + \cdots\right), \tag{177}$$

is readily obtained. The exact solution

$$u(x,t) = \frac{x}{t} - \frac{2}{x+t},\qquad(178)$$

follows immediately.

Exercises 9.5

In Exercises 1 – 5, use Adomian decomposition method to solve the inviscid Burgers' equations:

1. $u_t + uu_x = 0,\ u(x,0) = x$

2. $u_t + uu_x = 0,\ u(x,0) = -x$

3. $u_t + uu_x = 0,\ u(x,0) = 2x$

4. $u_t + uu_x = 0,\ u(x,0) = -2x$

5. $u_t + uu_x = 0,\ u(x,0) = \dfrac{1}{1+x}$

In Exercises 6 – 10, use Adomian decomposition method to solve the following Burgers' equations:

6. $u_t + uu_x = u_{xx},\ u(x,0) = -x$

7. $u_t + uu_x = u_{xx},\ u(x,0) = 2x$

8. $u_t + uu_x = u_{xx},\ u(x,0) = 4\tan 2x$

9. $u_t + uu_x = u_{xx},\ u(0,t) = \frac{1}{2t-1},\ u_x(0,t) = \frac{2}{2t-1}$

10. $u_t + uu_x = u_{xx},\ u(0,t) = -\frac{2}{3t},\ u_x(0,t) = \frac{1}{t} + \frac{2}{9t^2}$

9.6　The Telegraph Equation

The standard form of the telegraph equation is given by

$$u_{xx} = au_{tt} + bu_t + cu,\qquad(179)$$

where $u = u(x,t)$, and a, b and c are constants related to the resistance, inductance, capacitance and conductance of the cable. Note that the telegraph equation is a linear partial differential equation. The telegraph equation arises in the propagation of electrical signals along a telegraph line. If we set $a = 0$ and $c = 0$, because of electrical properties of the cable, we then obtain

$$u_{xx} = bu_t,\qquad(180)$$

which is the standard linear heat equation discussed in Chapter 3. On the other hand, the electrical properties may lead to $b = 0$ and $c = 0$; hence we obtain

$$u_{xx} = au_{tt}, \tag{181}$$

which is the standard linear wave equation presented in Chapter 5.

We now proceed formally to apply the decomposition method in a parallel manner to the approach used for handling other physical models. Without loss of generality, consider the initial boundary value telegraph equation

$$u_{xx} = u_{tt} + u_t + u, \, 0 < x < L, \tag{182}$$

with boundary and initial conditions

$$\text{B.C} \quad u(0,t) = f(t), \, u_x(0,t) = g(t),$$
$$\text{I.C} \quad u(x,0) = h(x), \, u_t(x,0) = v(x). \tag{183}$$

In an operator form, Eq. (182) becomes

$$L_x u(x,t) = u_{tt} + u_t + u, \tag{184}$$

where L_x is a second order differential operator with respect to x. Consequently, the inverse operator L_x^{-1} is considered a two-fold integral operator so that

$$L_x^{-1} L_x u(x,t) = u(x,t) - u(0,t) - xu_x(0,t). \tag{185}$$

Operating with L_x^{-1} on both sides of (184), using the boundary conditions, and noting (185) we obtain

$$u(x,t) = f(t) + xg(t) + L_x^{-1}(u_{tt} + u_t + u). \tag{186}$$

It is normal to define the recursive relation by

$$u_0(x,t) \quad = \quad f(t) + xg(t),$$
$$u_{k+1}(x,t) \quad = \quad L_x^{-1}\left(u_{k_{tt}} + u_{k_t} + u_k\right), \, k \geq 0, \tag{187}$$

that in turn gives

$$u_0(x,t) \quad = \quad f(t) + xg(t),$$
$$u_1(x,t) \quad = \quad L_x^{-1}\left(u_{0_{tt}} + u_{0_t} + u_0\right),$$
$$u_2(x,t) \quad = \quad L_x^{-1}\left(u_{1_{tt}} + u_{1_t} + u_1\right),$$
$$u_3(x,t) \quad = \quad L_x^{-1}\left(u_{2_{tt}} + u_{2_t} + u_2\right). \tag{188}$$

Having determined the components of $u(x, t)$, the solution in a series form can thus be established upon summing these iterates. As indicated before, the resulting series may give the exact solution in a closed form.

The analysis presented above will be illustrated by discussing the following examples.

Example 1. Solve the following homogeneous telegraph equation:

$$u_{xx} = u_{tt} + u_t - u, \tag{189}$$

subject to the conditions

$$\text{B.C} \quad u(0, t) = e^{-2t}, \ u_x(0, t) = e^{-2t}, \tag{190}$$

$$\text{I.C} \quad u(x, 0) = e^x, \ u_t(x, 0) = -2e^x.$$

Solution.

Operating with L_x^{-1} on (189) and using the boundary conditions yields

$$u(x, t) = e^{-2t} + xe^{-2t} + L_x^{-1}(u_{tt} + u_t - u). \tag{191}$$

Following the discussions presented before gives

$$\sum_{n=0}^{\infty} u_n(x, t) = e^{-2t} + xe^{-2t} + L_x^{-1}\left(\left(\sum_{n=0}^{\infty} u_n\right)_{tt} + \left(\sum_{n=0}^{\infty} u_n\right)_t - \sum_{n=0}^{\infty} u_n\right). \tag{192}$$

The decomposition method suggests the relation

$$\begin{aligned} u_0(x, t) &= e^{-2t} + xe^{-2t} \\ u_{k+1}(x, t) &= L_x^{-1}\left(u_{k_{tt}} + u_{k_t} - u_k\right), k \geq 0, \end{aligned} \tag{193}$$

where the components of the solution $u(x, t)$ given by

$$\begin{aligned} u_0(x, t) &= e^{-2t} + xe^{-2t} \\ u_1(x, t) &= L_x^{-1}\left(u_{0_{tt}} + u_{0_t} - u_0\right) = \frac{1}{2!}x^2 e^{-2t} + \frac{1}{3!}x^3 e^{-2t}, \\ u_2(x, t) &= L_x^{-1}\left(u_{1_{tt}} + u_{1_t} - u_1\right) = \frac{1}{4!}x^4 e^{-2t} + \frac{1}{5!}x^5 e^{-2t}, \\ u_3(x, t) &= L_x^{-1}\left(u_{2_{tt}} + u_{2_t} - u_2\right) = \frac{1}{6!}x^6 e^{-2t} + \frac{1}{7!}x^7 e^{-2t}. \end{aligned} \tag{194}$$

follow immediately. In view of (194), the solution in a series form is given by

$$u(x, t) = e^{-2t}\left(1 + x + \frac{1}{2!}x^2 + \frac{1}{3!}x^3 + \frac{1}{4!}x^4 + \frac{1}{5!}x^5 + \cdots\right), \tag{195}$$

which gives the exact solution in the form

$$u(x,t) = e^{x-2t}. \tag{196}$$

Example 2. Solve the following homogeneous telegraph equation:

$$u_{xx} = u_{tt} + 4u_t + 4u, \tag{197}$$

subject to the conditions

$$\text{B.C} \quad u(0,t) = 1 + e^{-2t}, \; u_x(0,t) = 2,$$

$$\text{I.C} \quad u(x,0) = 1 + e^{2x}, \; u_t(x,0) = -2. \tag{198}$$

Solution.

Applying the two-fold integral operator L_x^{-1} on (197) gives

$$u(x,t) = 1 + e^{-2t} + 2x + L_x^{-1}(u_{tt} + 4u_t + 4u), \tag{199}$$

where using the decomposition series for $u(x,t)$ we obtain

$$\sum_{n=0}^{\infty} u_n(x,t) = 1 + e^{-2t} + 2x + L_x^{-1}\left(\left(\sum_{n=0}^{\infty} u_n\right)_{tt} + 4\left(\sum_{n=0}^{\infty} u_n\right)_t + 4\sum_{n=0}^{\infty} u_n\right). \tag{200}$$

A close observation of (200) suggests the recursive relation

$$u_0(x,t) = 1 + e^{-2t} + 2x$$

$$u_{k+1}(x,t) = L_x^{-1}(u_{k_{tt}} + 4u_{k_t} + 4u_k), \, k \geq 0. \tag{201}$$

In view of (201) we obtain

$$u_0(x,t) = 1 + e^{-2t} + 2x$$

$$u_1(x,t) = L_x^{-1}(u_{0_{tt}} + 4u_{0_t} + 4u_0) = 2x^2 + \tfrac{4}{3}x^3, \tag{202}$$

$$u_2(x,t) = L_x^{-1}(u_{1_{tt}} + 4u_{1_t} + 4u_1) = \tfrac{2}{3}x^4 + \tfrac{4}{15}x^5.$$

Other components can be computed in a similar manner. Consequently, the solution in a series form is given by

$$u(x,t) = e^{-2t} + \left(1 + 2x + \tfrac{1}{2!}(2x)^2 + \tfrac{1}{3!}(2x)^3 + \cdots\right), \tag{203}$$

so that the exact solution

$$u(x,t) = e^{2x} + e^{-2t},\tag{204}$$

is readily obtained.

Exercises 9.6

Use the decomposition method to solve the telegraph equations:

1. $u_{xx} = u_{tt} + u_t + u$
   $u(0,t) = e^{-t}, \ u_x(0,t) = e^{-t}$
   $u(x,0) = e^x, \ u_t(x,0) = -e^x$

2. $u_{xx} = \frac{1}{3}(u_{tt} + u_t + u)$
   $u(0,t) = e^t, \ u_x(0,t) = e^t$
   $u(x,0) = e^x, \ u_t(x,0) = e^x$

3. $u_{xx} = u_{tt} + 2u_t + u$
   $u(0,t) = 1 + e^{-t}, \ u_x(0,t) = 1$
   $u(x,0) = 1 + e^x, \ u_t(x,0) = -1$

4. $u_{xx} = u_{tt} + u_t + 4u$
   $u(0,t) = e^{-t}, \ u_x(0,t) = 2e^{-t}$
   $u(x,0) = e^{2x}, \ u_t(x,0) = -e^{2x}$

5. $u_{xx} = 2u_{tt} + 3u_t + u$
   $u(0,t) = 1 - e^{-t}, \ u_x(0,t) = 1$
   $u(x,0) = e^x - 1, \ u_t(x,0) = 1$

6. $u_{xx} = u_{tt} + 2u_t + u$
   $u(0,t) = e^{-t}, \ u_x(0,t) = 1$
   $u(x,0) = 1 + \sinh x, \ u_t(x,0) = -1$

7. $u_{xx} = u_{tt} + 2u_t + u$
   $u(0,t) = 1 - e^{-t}, \ u_x(0,t) = 0$
   $u(x,0) = \cosh x - 1, \ u_t(x,0) = 1$

8. $u_{xx} = \frac{4}{3}u_{tt} + \frac{1}{3}u_t + u$
   $u(0,t) = 1 + e^{-3t}, \ u_x(0,t) = 1$
   $u(x,0) = 1 + e^x, \ u_t(x,0) = -3$

9. $u_{xx} = u_{tt} + 4u_t + 4u$
   $u(0,t) = 1 + e^{-2t}, \ u_x(0,t) = 2$
   $u(x,0) = 1 + e^{2x}, \ u_t(x,0) = -2$

10. $u_{xx} = u_{tt} + 4u_t + 4u$
    $u(0,t) = e^{-2t}, \ u_x(0,t) = 2$
    $u(x,0) = 1 + \sinh 2x, \ u_t(x,0) = -2$

9.7 Schrodinger Equation

In this section, the linear and nonlinear Schrodinger equations will be investigated. It is well-known that this equation arises in the study of the time evolution of the wave function.

9.7.1 The Linear Schrodinger Equation

The initial value problem for the linear Schrodinger equation for a free particle with mass m is given by the following standard form

$$u_t = iu_{xx}, \ u(x,0) = f(x), \ x \in R, \ t > 0, \tag{205}$$

where $f(x)$ is continuous and square integrable. It is to be noted that Schrodinger equation (205) discusses the time evolution of a free particle. Moreover, the function $u(x,t)$ is complex, and Eq. (205) is a first order differential equation in t. The linear Schrodinger equation (205) is usually handled by using the spectral transform technique among other methods.

The Adomian decomposition method will be applied here to handle the linear and the nonlinear Schrodinger equations. To achieve this goal, we apply L_t^{-1} to both sides of (205) to obtain

$$u(x,t) = f(x) + iL_t^{-1}(u_{xx}), \tag{206}$$

and using the series representation for $u(x,t)$ yields

$$\sum_{n=0}^{\infty} u_n(x,t) = f(x) + iL_t^{-1}\left(\left(\sum_{n=0}^{\infty} u_n(x,t)\right)_{xx}\right). \tag{207}$$

Applying the decomposition method leads to the recursive scheme

$$\begin{aligned} u_0(x,t) &= f(x), \\ u_{k+1}(x,t) &= iL_t^{-1}(u_{k_{xx}}), k \geq 0. \end{aligned} \tag{208}$$

Using few iterations of (208) gives

$$\begin{aligned} u_0(x,t) &= f(x), \\ u_1(x,t) &= iL_t^{-1}(u_{0_{xx}}), \\ u_2(x,t) &= iL_t^{-1}(u_{1_{xx}}), \\ u_3(x,t) &= iL_t^{-1}(u_{2_{xx}}). \end{aligned} \tag{209}$$

Other components can be evaluated in a parallel manner. Having determined the first few components of $u(x,t)$, the solution in a series form is readily obtained.

The following examples will be used to illustrate the analysis discussed above.

Example 1. Solve the linear Schrodinger equation

$$u_t = iu_{xx}, \ u(x,0) = e^{ix} \tag{210}$$

Solution.

Following the discussions presented above we obtain

$$
\begin{aligned}
u_0(x,t) &= e^{ix}, \\
u_1(x,t) &= iL_t^{-1}\left(-e^{ix}\right) = -ite^{ix}, \\
u_2(x,t) &= L_t^{-1}\left(-te^{ix}\right) = -\tfrac{1}{2!}t^2 e^{ix}, \\
u_3(x,t) &= iL_t^{-1}\left(\tfrac{1}{2!}t^2 e^{ix}\right) = \tfrac{1}{3!}it^3 e^{ix}.
\end{aligned}
\tag{211}
$$

Summing these iterations yields the series solution

$$u(x,t) = e^{ix}\left(1 - (it) + \frac{1}{2!}(it)^2 - \frac{1}{3!}(it)^3 + \cdots\right), \tag{212}$$

that leads to the exact solution

$$u(x,t) = e^{i(x-t)}, \tag{213}$$

obtained upon using the Taylor expansion for e^{-it}.

Example 2. Solve the linear Schrodinger equation

$$
\begin{aligned}
u_t &= iu_{xx}, \\
u(x,0) &= \sinh x
\end{aligned}
\tag{214}
$$

Solution.

Proceeding as in Example 1, we obtain

$$
\begin{aligned}
u_0(x,t) &= \sinh x, \\
u_1(x,t) &= iL_t^{-1}\left(\sinh x\right) = it\sinh x, \\
u_2(x,t) &= -L_t^{-1}\left(t\sinh x\right) = -\tfrac{1}{2!}t^2 \sinh x, \\
u_3(x,t) &= -iL_t^{-1}\left(\tfrac{1}{2!}t^2 \sinh x\right) = -\tfrac{1}{3!}it^3 \sinh x.
\end{aligned}
\tag{215}
$$

Summing these components gives the series solution

$$u(x,t) = \sinh x \left(1 + (it) + \frac{1}{2!}(it)^2 + \frac{1}{3!}(it)^3 + \cdots \right), \tag{216}$$

and hence the exact solution is

$$u(x,t) = \sinh x \, e^{it}. \tag{217}$$

9.7.2 The Nonlinear Schrodinger Equation

We now turn to study the nonlinear Schrodinger equation (NLS) defined by its standard form

$$iu_t + u_{xx} + \gamma |u|^2 u = 0, \tag{218}$$

where γ is a constant and $u(x,t)$ is complex. The Schrodinger equation (218) generally exhibits solitary type solutions. A soliton, or solitary wave, is a wave where the speed of propagation is independent of the amplitude of the wave. Solitons usually occur in fluid mechanics.

The nonlinear Schrodinger equations that are commonly used are given by

$$iu_t + u_{xx} + 2|u|^2 u = 0, \tag{219}$$

and

$$iu_t + u_{xx} - 2|u|^2 u = 0. \tag{220}$$

Moreover, other forms of nonlinear Schrodinger equations are used as well depending on the constant γ. The inverse scattering method is usually used to handle the nonlinear Schrodinger equation where solitary type solutions were derived.

The nonlinear Schrodinger equation will be handled differently in this section by using the Adomian decomposition method. We start our analysis by considering the initial value problem

$$iu_t + u_{xx} + \gamma |u|^2 u = 0, \; u(x,0) = f(x). \tag{221}$$

In an operator form, Eq. (221) becomes

$$L_t u(x,t) = iu_{xx} + i\gamma |u|^2 u. \tag{222}$$

Applying L_t^{-1} to both sides of (222) gives

$$u(x,t) = f(x) + iL_t^{-1}u_{xx} + i\gamma L_t^{-1} F(u(x,t)), \tag{223}$$

where the nonlinear term $F(u(x,t))$ is given by

$$F(u(x,t)) = |u|^2 u. \tag{224}$$

Substituting the decomposition series for $u(x,t)$ and the series of Adomian polynomials for $F(u(x,t))$ into (223) to obtain

$$\sum_{n=0}^{\infty} u_n(x,t) = f(x) + iL_t^{-1}\left(\left(\sum_{n=0}^{\infty} u_n(x,t)\right)_{xx}\right) + i\gamma L_t^{-1}\left(\sum_{n=0}^{\infty} A_n\right). \qquad (225)$$

Adomian's analysis introduces the recursive relation

$$\begin{aligned}
u_0(x,t) &= f(x), \\
u_{k+1}(x,t) &= iL_t^{-1}(u_{k_{xx}}) + i\gamma L_t^{-1}(A_k), \ k \geq 0.
\end{aligned} \qquad (226)$$

Recall from complex analysis that

$$|u|^2 = u\bar{u}, \qquad (227)$$

where $\bar{u}$ is the conjugate of u. This means that (224) can be rewritten as

$$F(u) = u^2\bar{u}. \qquad (228)$$

In view of (228), and following the formal techniques used before to derive the Adomian polynomials, we can easily derive that $F(u)$ has the following polynomials representation

$$\begin{aligned}
A_0 &= u_0^2\bar{u}_0, \\
A_1 &= 2u_0u_1\bar{u}_0 + u_0^2\bar{u}_1, \\
A_2 &= 2u_0u_2\bar{u}_0 + u_1^2\bar{u}_0 + 2u_0u_1\bar{u}_1 + u_0^2\bar{u}_2, \\
A_3 &= 2u_0u_3\bar{u}_0 + 2u_1u_2\bar{u}_0 + 2u_0u_2\bar{u}_1 + u_1^2\bar{u}_1 + 2u_0u_1\bar{u}_2 + u_0^2\bar{u}_3.
\end{aligned} \qquad (229)$$

In conjunction with (226) and (229), we can easily determine the first few components by

$$\begin{aligned}
u_0(x,t) &= f(x), \\
u_1(x,t) &= iL_t^{-1}(u_{0_{xx}}) + i\gamma L_t^{-1}(A_0), \\
u_2(x,t) &= iL_t^{-1}(u_{1_{xx}}) + i\gamma L_t^{-1}(A_1), \\
u_3(x,t) &= iL_t^{-1}(u_{2_{xx}}) + i\gamma L_t^{-1}(A_2).
\end{aligned} \qquad (230)$$

Other components can be determined as well. This completes the determination of the series solution.

The analysis introduced above will be illustrated by discussing the following examples.

Example 3. Use the decomposition method to solve the following nonlinear Schrodinger equation.

$$iu_t + u_{xx} + 2|u|^2 u = 0, \; u(x,0) = e^{ix} \tag{231}$$

Solution.

Following the analysis presented above gives

$$\sum_{n=0}^{\infty} u_n(x,t) = e^{ix} + iL_t^{-1}\left(\left(\sum_{n=0}^{\infty} u_n(x,t)\right)_{xx}\right) + 2iL_t^{-1}\left(\sum_{n=0}^{\infty} A_n\right). \tag{232}$$

The decomposition method suggests the use of the recursive relation

$$
\begin{aligned}
u_0(x,t) &= f(x), \\
u_{k+1}(x,t) &= iL_t^{-1}\left(u_{k_{xx}}\right) + 2iL_t^{-1}\left(A_k\right), \; k \geq 0,
\end{aligned}
\tag{233}
$$

that in turn gives the first few components by

$$
\begin{aligned}
u_0(x,t) &= e^{ix}, \\
u_1(x,t) &= iL_t^{-1}\left(u_{0_{xx}}\right) + 2iL_t^{-1}\left(A_0\right) = ite^{ix}, \\
u_2(x,t) &= iL_t^{-1}\left(u_{1_{xx}}\right) + 2iL_t^{-1}\left(A_1\right) = -\tfrac{1}{2!}t^2 e^{ix}, \\
u_3(x,t) &= iL_t^{-1}\left(u_{2_{xx}}\right) + 2iL_t^{-1}\left(A_2\right) = -\tfrac{1}{3!}it^3 e^{ix}.
\end{aligned}
\tag{234}
$$

Accordingly, the series solution is given by

$$u(x,t) = e^{ix}\left(1 + it + \frac{1}{2!}(it)^2 + \frac{1}{3!}(it)^3 + \cdots\right), \tag{235}$$

that gives the exact solution by

$$u(x,t) = e^{i(x+t)}. \tag{236}$$

Example 4. Use the decomposition method to solve the following nonlinear Schrodinger equation.

$$iu_t + u_{xx} - 2|u|^2 u = 0, \; u(x,0) = e^{ix} \tag{237}$$

Solution.

Using the analysis of Example 3 yields

$$\sum_{n=0}^{\infty} u_n(x,t) = e^{ix} + iL_t^{-1}\left(\left(\sum_{n=0}^{\infty} u_n(x,t)\right)_{xx}\right) - 2i\left(\sum_{n=0}^{\infty} A_n\right). \tag{238}$$

This gives the recursive relation

$$u_0(x,t) = f(x),$$
$$u_{k+1}(x,t) = iL_t^{-1}(u_{k_{xx}}) - 2iL_t^{-1}(A_k), \; k \geq 0. \tag{239}$$

Using the Adomian polynomials A_n that were derived before, the first few components are given by

$$u_0(x,t) = e^{ix},$$
$$u_1(x,t) = iL_t^{-1}(u_{0_{xx}}) - 2iL_t^{-1}(A_0) = -3ite^{ix},$$
$$u_2(x,t) = iL_t^{-1}(u_{1_{xx}}) + 2iL_t^{-1}(A_1) = \tfrac{1}{2!}(3it)^2 e^{ix}, \tag{240}$$
$$u_3(x,t) = iL_t^{-1}(u_{2_{xx}}) + 2iL_t^{-1}(A_2) = -\tfrac{1}{3!}(3it)^3 e^{ix}.$$

In view of (240), the series solution is given by

$$u(x,t) = e^{ix}\left(1 - (3it) + \frac{1}{2!}(3it)^2 - \frac{1}{3!}(3it)^3 + \cdots\right). \tag{241}$$

The exact solution is therefore given by

$$u(x,t) = e^{i(x-3t)}. \tag{242}$$

It is important to note that the solution (242) is a solitary type solution. It can be written in the form $u(x,t) = f(x - ct)$ where $c = 3$ is the speed of wave propagation.

Exercises 9.7

In Exercises 1 – 5, use Adomian decomposition method to solve the following linear Schrodinger equations:

1. $u_t = iu_{xx}, \; u(x,0) = e^{2ix}$

2. $u_t = iu_{xx}, \; u(x,0) = \sin x$

3. $u_t = iu_{xx}, \; u(x,0) = \cosh x$

4. $u_t = iu_{xx}, \; u(x,0) = 1 + \cos 3x$

5. $u_t = iu_{xx}, \; u(x,0) = \sin 2x$

In Exercises 6 – 10, use Adomian decomposition method to solve the following nonlinear Schrodinger equations NLS:

6. $iu_t + u_{xx} + |u|^2 u = 0, \; u(x,0) = e^{2ix}$

7. $iu_t + u_{xx} + 2|u|^2 u = 0$, $u(x,0) = e^{-ix}$

8. $iu_t + u_{xx} + 2|u|^2 u = 0$, $u(x,0) = e^{2ix}$

9. $iu_t + u_{xx} - 2|u|^2 u = 0$, $u(x,0) = e^{2ix}$

10. $iu_t + u_{xx} + |u|^2 u = 0$, $u(0,t) = e^{8it}$, $u_x(0,t) = 3ie^{8it}$

9.8 Korteweg-deVries Equation

The Korteweg-deVries (KdV) equation in its simplest form is given by

$$u_t + auu_x + u_{xxx} = 0. \tag{243}$$

The KdV equation arises in the study of shallow water waves. In particular, the KdV equation is used to describe long waves traveling in canals. It is formally proved that this equation has solitary waves as solutions, hence it can have any number of solitons. The KdV equation has received a lot of attention and has been extensively studied. Several numerical and analytical techniques were employed to study the solitary waves that result from this equation. The solitary waves of this equation will be presented in Chapter 11.

In this section, the decomposition method will be used to handle the KdV equation. We first consider the initial value problem

$$u_t + auu_x + bu_{xxx} = 0, \; u(x,0) = f(x), \tag{244}$$

where a and b are constants. In an operator form, the KdV equation becomes

$$L_t u = -bu_{xxx} - auu_x. \tag{245}$$

Applying L_t^{-1} on both sides of (245) yields

$$u(x,t) = f(x) - bL_t^{-1} u_{xxx} - aL_t^{-1} F(u(x,t)), \tag{246}$$

where the nonlinear term $F(u(x,t))$ is

$$F(u(x,t)) = uu_x. \tag{247}$$

Using the decomposition identification for the linear and nonlinear terms yields

$$\sum_{n=0}^{\infty} u_n(x,t) = f(x) - bL_t^{-1} \left(\sum_{n=0}^{\infty} u_n(x,t) \right)_{xxx} - aL_t^{-1} \left(\sum_{n=0}^{\infty} A_n \right). \tag{248}$$

The typical approach of Adomian's method is the introduction of the recursive relation

$$u_0(x,t) = f(x),$$
$$u_{k+1}(x,t) = -bL_t^{-1}(u_{k_{xxx}}) - aL_t^{-1}(A_k), k \geq 0. \tag{249}$$

The components $u_n, n \geq 0$ can be elegantly calculated by

$$
\begin{aligned}
u_0(x,t) &= f(x), \\
u_1(x,t) &= -bL_t^{-1}\left(u_{0_{xxx}}\right) - aL_t^{-1}(A_0), \\
u_2(x,t) &= -bL_t^{-1}\left(u_{1_{xxx}}\right) - aL_t^{-1}(A_1), \\
u_3(x,t) &= -bL_t^{-1}\left(u_{2_{xxx}}\right) - aL_t^{-1}(A_2),
\end{aligned}
\tag{250}
$$

where Adomian polynomials A_n for the nonlinearity uu_x were derived before and used in advection and Burgers' problems. Summing the computed components (250) gives the solution in a series form. The discussion presented above will be illustrated as follows.

Example 1. Solve the following homogeneous KdV equation:

$$
u_t - 6uu_x + u_{xxx} = 0,\ u(x,0) = 6x.
\tag{251}
$$

Solution.

Proceeding as before we find

$$
\sum_{n=0}^{\infty} u_n(x,t) = 6x - L_t^{-1}\left(\left(\sum_{n=0}^{\infty} u_n(x,t)\right)_{xxx}\right) + 6L_t^{-1}\left(\sum_{n=0}^{\infty} A_n\right).
\tag{252}
$$

A close observation of (252) admits the recursive relation

$$
\begin{aligned}
u_0(x,t) &= 6x, \\
u_{k+1}(x,t) &= -L_t^{-1}\left(u_{k_{xxx}}\right) + 6L_t^{-1}\left(A_k\right),\ k \geq 0,
\end{aligned}
\tag{253}
$$

that gives the first few components by

$$
\begin{aligned}
u_0(x,t) &= 6x, \\
u_1(x,t) &= -L_t^{-1}\left(u_{0_{xxx}}\right) + 6L_t^{-1}(A_0) = 6^3 xt, \\
u_2(x,t) &= -L_t^{-1}\left(u_{1_{xxx}}\right) + L_t^{-1}(A_1) = 6^5 xt^2, \\
u_3(x,t) &= -L_t^{-1}\left(u_{2_{xxx}}\right) + 6L_t^{-1}(A_2) = 6^7 xt^3.
\end{aligned}
\tag{254}
$$

In view of (254), the solution in a series form is given by

$$
u(x,t) = 6x(1 + (36t) + (36t)^2 + (36t)^3 + \cdots),
\tag{255}
$$

and in a closed form by

$$u(x,t) = \frac{6x}{1 - 36t}, \ |36t| < 1. \tag{256}$$

Example 2. Solve the following KdV equation:

$$u_t - 6uu_x + u_{xxx} = 0, \ u(x,0) = \frac{1}{6}(x - 1). \tag{257}$$

Solution.

Proceeding as in Example 1 gives

$$\sum_{n=0}^{\infty} u_n(x,t) = \frac{1}{6}(x-1) - L_t^{-1}\left(\left(\sum_{n=0}^{\infty} u_n(x,t)\right)_{xxx}\right) + 6L_t^{-1}\left(\sum_{n=0}^{\infty} A_n\right). \tag{258}$$

This gives the relation

$$\begin{aligned}
u_0(x,t) &= \tfrac{1}{6}(x-1), \\
u_{k+1}(x,t) &= -L_t^{-1}(u_{k_{xx}}) + 6L_t^{-1}(A_k), k \geq 0,
\end{aligned} \tag{259}$$

and as a result we find

$$\begin{aligned}
u_0(x,t) &= \tfrac{1}{6}(x-1), \\
u_1(x,t) &= -L_t^{-1}(u_{0_{xxx}}) + 6L_t^{-1}(A_0) = \tfrac{1}{6}(x-1)t, \\
u_2(x,t) &= -L_t^{-1}(u_{1_{xxx}}) + L_t^{-1}(A_1) = \tfrac{1}{6}(x-1)t^2, \\
u_3(x,t) &= -L_t^{-1}(u_{2_{xxx}}) + 6L_t^{-1}(A_2) = \tfrac{1}{6}(x-1)t^3.
\end{aligned} \tag{260}$$

The solution in a series form is therefore given by

$$u(x,t) = \frac{1}{6}(x - 1)\left(1 + t + t^2 + t^3 + \cdots\right), \tag{261}$$

and in a closed form by

$$u(x,t) = \frac{1}{6}\left(\frac{x-1}{1-t}\right), \ |t| < 1. \tag{262}$$

Exercises 9.8

1. Show that $u = -\dfrac{2}{x^2}$ is a solution of $u_t + 6uu_x + u_{xxx} = 0$

2. Show that $u = \dfrac{1}{6}\dfrac{x-3}{t-3}$ is a solution of $u_t + 6uu_x + u_{xxx} = 0$

3. Show that $u = \dfrac{2}{(x-2)^2}$ is a solution of $u_t - 6uu_x + u_{xxx} = 0$

4. Show that $u = \dfrac{6x(x^3-24t)}{(x^3+12t)^2}$ is a solution of $u_t - 6uu_x + u_{xxx} = 0$

5. Show that $u = \dfrac{4(x-6t)^2-3}{4(x-6t)^2+1}$ is a solution of $u_t + 6u^2u_x + u_{xxx} = 0$

Use Adomian decomposition method to solve the following KdV equations:

6. $u_t + 6uu_x + u_{xxx} = 0,\ u(x,0) = x$

7. $u_t - 6uu_x + u_{xxx} = 0,\ u(x,0) = \frac{2}{x^2}$

8. $u_t - 6uu_x + u_{xxx} = 0,\ u(x,0) = \frac{1}{12}(x-2)$

9. $u_t - 6uu_x + u_{xxx} = 0,\ u(x,0) = \frac{2}{(x-3)^2}$

10. $u_t - 6uu_x + u_{xxx} = 0,\ u(x,0) = \frac{1}{18}(x-4)$

9.9 Fourth Order Parabolic Equation

We close this chapter by discussing the fourth order parabolic linear partial differential equation with constant and variable coefficients.

9.9.1 Equations with Constant Coefficients

In what follows we study the fourth order parabolic linear partial differential equation with constant coefficients of the form

$$\frac{\partial^2 u}{\partial t^2} + \frac{\partial^4 u}{\partial x^4} = f(x,t),\ 0 \le x \le 1, t > 0, \tag{263}$$

with initial conditions

$$u(x,0) = g(x),\ \ u_t(x,0) = h(x). \tag{264}$$

It is worth mentioning that the fourth order parabolic equation (263) governs the transverse vibrations of a homogeneous beam. In addition, the equation (263), subject to specific initial and boundary conditions, was handled numerically by the finite difference method and by the alternating group explicit method. It is our main goal in this chapter to employ the Adomian decomposition method to physical applications.

In an operator form, Eq. (263) can be rewritten as

$$L_t u(x,t) = f(x,t) - L_x u(x,t), \tag{265}$$

where L_t is a second order partial derivative with respect to t, and L_x is a fourth order partial derivative with respect to x. Operating with the two-fold integral operator L_t^{-1} and using the decomposition series for $u(x,t)$ give

$$\sum_{n=0}^{\infty} u_n(x,t) = g(x) + th(x) + L_t^{-1}f(x,t) - L_t^{-1}L_x\left(\sum_{n=0}^{\infty} u_n(x,t)\right). \qquad (266)$$

It follows that

$$\begin{aligned}
u_0(x,t) &= g(x) + th(x) + L_t^{-1}f(x,t),\\
u_{k+1}(x,t) &= -L_t^{-1}L_x\left(\sum_{n=0}^{\infty} u_n(x,t)\right).
\end{aligned} \qquad (267)$$

Using few iterations we obtain

$$\begin{aligned}
u_0(x,t) &= g(x) + th(x) + L_t^{-1}f(x,t),\\
u_1(x,t) &= -L_t^{-1}L_x\left(u_0\right),\\
u_2(x,t) &= -L_t^{-1}L_x\left(u_1\right),\\
u_3(x,t) &= -L_t^{-1}L_x\left(u_2\right).
\end{aligned} \qquad (268)$$

The series solution follows immediately upon summing the components obtained in (268).

It is important to point out that the homogeneous and the inhomogeneous cases will be illustrated by discussing the following examples. For the inhomogeneous case, the noise terms will play a major role in accelerating the convergence of the solution.

Example 1. Solve the following homogeneous fourth order equation:

$$\frac{\partial^2 u}{\partial t^2} + \frac{\partial^4 u}{\partial x^4} = 0, \qquad (269)$$

with initial conditions

$$u(x,0) = \cos x, \; u_t(x,0) = -\sin x. \qquad (270)$$

Solution.

Operating with the two-fold integral operator L_t^{-1}, and representing $u(x,t)$ by the decomposition series of components we obtain

$$\sum_{n=0}^{\infty} u_n(x,t) = \cos x - t\sin x - L_t^{-1}L_x\left(\sum_{n=0}^{\infty} u_n(x,t)\right). \qquad (271)$$

The recursive scheme

$$
\begin{aligned}
u_0(x,t) &= \cos x - t \sin x, \\
u_{k+1}(x,t) &= -L_t^{-1} L_x \left(\sum_{n=0}^{\infty} u_k(x,t) \right), \ k \geq 0,
\end{aligned}
\tag{272}
$$

follows immediately. Using few iterations we obtain

$$
\begin{aligned}
u_0(x,t) &= \cos x - t \sin x, \\
u_1(x,t) &= -L_t^{-1} L_x \left(u_0(x,t) \right) = -\tfrac{1}{2!} t^2 \cos x + \tfrac{1}{3!} t^3 \sin x, \\
u_2(x,t) &= -L_t^{-1} L_x \left(u_1(x,t) \right) = \tfrac{1}{4!} t^4 \cos x - \tfrac{1}{5!} t^5 \sin x.
\end{aligned}
\tag{273}
$$

Therefore, the series solution is given by

$$
u(x,t) = \cos x \left(1 - \frac{1}{2!} t^2 + \frac{1}{4!} t^4 - \cdots \right) - \sin x \left(t - \frac{1}{3!} t^3 + \frac{1}{5!} t^5 - \cdots \right).
\tag{274}
$$

Consequently, the exact solution is given by

$$
u(x,t) = \cos x \cos t - \sin x \sin t = \cos(x+t).
\tag{275}
$$

Example 2. Solve the following inhomogeneous fourth order equation:

$$
\frac{\partial^2 u}{\partial t^2} + \frac{\partial^4 u}{\partial x^4} = (\pi^4 - 1) \sin \pi x \sin t,
\tag{276}
$$

with initial conditions
$$
u(x,0) = 0, \ u_t(x,0) = \sin \pi x.
\tag{277}
$$

Solution.

Proceeding as in Example 1 we obtain

$$
\sum_{n=0}^{\infty} u_n(x,t) = t \sin \pi x + (\pi^4 - 1) \sin \pi x (t - \sin t) - L_t^{-1} L_x \left(\sum_{n=0}^{\infty} u_n(x,t) \right).
\tag{278}
$$

This gives the relation

$$
\begin{aligned}
u_0(x,t) &= t \sin \pi x + (\pi^4 - 1) \sin \pi x (t - \sin t), \\
u_{k+1}(x,t) &= -L_t^{-1} L_x \left(\sum_{n=0}^{\infty} u_k(x,t) \right), \ k \geq 0.
\end{aligned}
\tag{279}
$$

To use the noise terms phenomenon we determine the first two components, hence we find

$$
\begin{aligned}
u_0(x,t) &= \sin \pi x \sin t + \pi^4 \sin \pi x (t - \sin t), \\
u_1(x,t) &= -L_t^{-1} L_x \left(u_0(x,t) \right), \\
&= -\pi^4 \sin \pi x (t - \sin t) + \pi^8 \sin \pi x (\tfrac{1}{3!} t^3 + \sin t - t).
\end{aligned}
\tag{280}
$$

A close examination of the first two components shows the appearance of the noise term $\pi^4 \sin \pi x (t - \sin t)$ with opposite signs in $u_0(x,t)$ and $u_1(x,t)$. By canceling this term from $u_0(x,t)$ and checking that the remaining term justifies the equation give the exact solution

$$
u(x,t) = \sin \pi x \sin t. \tag{281}
$$

9.9.2 Equations with Variable Coefficients

In what follows we investigate the variable coefficient fourth-order parabolic partial differential equation of the form

$$
\frac{\partial^2 u}{\partial t^2} + \mu(x) \frac{\partial^4 u}{\partial x^4} = 0, \ \mu(x) > 0, \ a < x < b, t > 0, \tag{282}
$$

where $\mu(x) > 0$ is the ratio of flexural rigidity of the beam to its mass per unit length.

The initial conditions associated with (282) are of the form

$$
\begin{aligned}
u(x,0) &= f(x), \ a \le x \le b, \\
u_t(x,0) &= g(x), \ a \le x \le b,
\end{aligned}
\tag{283}
$$

and the boundary conditions are given by

$$
\begin{aligned}
u(a,t) &= h(t), \quad u(b,t) = r(t), \ t > 0, \\
u_{xx}(a,t) &= s(t), \quad u_{xx}(b,t) = q(t), \ t > 0,
\end{aligned}
\tag{284}
$$

where the functions $f(x), g(x), h(t), r(t), s(t)$ and $q(t)$ are continuous functions.

In an operator form, Eq. (282) becomes

$$
L_t u = -\mu(x) \frac{\partial^4 u}{\partial x^4}, \ \mu(x) > 0. \tag{285}
$$

Operating with the two fold integral operator and using the initial conditions yields

$$
u(x,t) = f(x) + t g(x) - L_t^{-1} \left(\mu(x) \frac{\partial^4 u}{\partial x^4} \right). \tag{286}
$$

Using the series representation of $u(x,t)$ leads to

$$\sum_{n=0}^{\infty} u_n(x,t) = f(x) + tg(x) - L^{-1}\left(\mu(x)\frac{\partial^4}{\partial x^4}\left(\sum_{n=0}^{\infty} u_n(x,t)\right)\right). \tag{287}$$

This gives the recurrence relation

$$
\begin{aligned}
u_0(x,t) &= f(x) + tg(x), \\
u_{k+1}(x,t) &= L^{-1}\left(\mu(x)\frac{\partial^4 u_k}{\partial x^4}\right), \ k \geq 0,
\end{aligned} \tag{288}
$$

so that

$$
\begin{aligned}
u_0(x,t) &= f(x) + tg(x), \\
u_1(x,t) &= -L^{-1}\left(\mu(x)\frac{\partial^4 u_0}{\partial x^4}\right), \\
u_2(x,t) &= -L^{-1}\left(\mu(x)\frac{\partial^4 u_1}{\partial x^4}\right), \\
u_3(x,t) &= -L^{-1}\left(\mu(x)\frac{\partial^4 u_2}{\partial x^4}\right).
\end{aligned} \tag{289}
$$

In view of (289), the series solution follows immediately.

Example 3. Solve the fourth order parabolic equation

$$\frac{\partial^2 u}{\partial t^2} + \left(\frac{1}{x} + \frac{x^4}{120}\right)\frac{\partial^4 u}{\partial x^4} = 0, \ \frac{1}{2} < x < 1, t > 0, \tag{290}$$

subject to the initial conditions

$$u(x,0) = 0, \ u_t(x,0) = 1 + \frac{x^5}{120}, \ \frac{1}{2} < x < 1, \tag{291}$$

and the boundary conditions

$$
\begin{aligned}
u(\tfrac{1}{2},t) &= \left(1 + \frac{(1/2)^5}{120}\right)\sin t, \quad u(1,t) = \left(\frac{121}{120}\right)\sin t, \ t > 0, \\
\frac{\partial^2 u}{\partial x^2}(\tfrac{1}{2},t) &= \tfrac{1}{6}(\tfrac{1}{2})^3\sin t, \qquad \frac{\partial^2 u}{\partial x^2}(1,t) = \tfrac{1}{6}\sin t, \ t > 0,
\end{aligned} \tag{292}
$$

Solution.

Adomian's analysis gives the recurrence relation

$$
\begin{aligned}
u_0(x,t) &= \left(1 + \frac{x^5}{120}\right)t, \\
u_{k+1}(x,t) &= -L^{-1}\left(\left(\frac{1}{x} + \frac{x^4}{120}\right)\frac{\partial^4 u_k}{\partial x^4}\right), \ k \geq 0,
\end{aligned} \tag{293}
$$

that gives

$$
\begin{aligned}
u_0(x,t) &= \left(1 + \tfrac{x^5}{120}\right)t, \\
u_1(x,t) &= -\left(1 + \tfrac{x^5}{120}\right)\tfrac{t^3}{3!}, \\
u_2(x,t) &= \left(1 + \tfrac{x^5}{120}\right)\tfrac{t^5}{5!}.
\end{aligned}
\tag{294}
$$

The solution in a series form is

$$
u(x,t) = \left(1 + \frac{x^5}{120}\right)\left(t - \frac{t^3}{3!} + \frac{t^5}{5!} - \cdots\right),
\tag{295}
$$

and in a closed form by

$$
u(x,t) = \left(1 + \frac{x^5}{120}\right)\sin t.
\tag{296}
$$

We close our analysis by discussing the following nonhomogeneous equation. The decomposition method will be combined with the effect of the noise terms phenomenon. This will facilitate the convergence of the solution.

Example 4. We finally consider the nonhomogeneous parabolic equation

$$
\frac{\partial^2 u}{\partial t^2} + (1+x)\frac{\partial^4 u}{\partial x^4} = (x^4 + x^3 - \frac{6}{7!}x^7)\cos t, \ 0 < x < 1, t > 0,
\tag{297}
$$

subject to the initial conditions

$$
u(x,0) = \frac{6}{7!}x^7, \ u_t(x,0) = 0, \ 0 < x < 1,
\tag{298}
$$

and the boundary conditions

$$
\begin{aligned}
u(0,t) &= 0, \quad u(1,t) = \tfrac{6}{7!}\cos t, \ t > 0, \\
\tfrac{\partial^2 u}{\partial x^2}(0,t) &= 0, \quad \tfrac{\partial^2 u}{\partial x^2}(1,t) = \tfrac{1}{20}\cos t, \ t > 0,
\end{aligned}
\tag{299}
$$

Solution.

Following our discussions above, we obtain

$$
\begin{aligned}
u_0(x,t) &= \tfrac{6}{7!}x^7\cos t + (x^4 + x^3)(1 - \cos t), \\
u_{k+1}(x,t) &= -L^{-1}\left((1+x)\tfrac{\partial^4 u_k}{\partial x^4}\right), \ k \geq 0,
\end{aligned}
\tag{300}
$$

that gives

$$
\begin{aligned}
u_0(x,t) &= \tfrac{6}{7!}x^7\cos t + (x^3 + x^4)(1 - \cos t), \\
u_1(x,t) &= -(x^4 + x^3)(1 - \cos t) - 24(1 + x)(\tfrac{t^2}{2!} - 1 + \cos t), \\
u_2(x,t) &= 24(1 + x)(\tfrac{t^2}{2!} - 1 + \cos t),
\end{aligned}
\tag{301}
$$

and so on. It is clear that the noise terms $\pm(x^4 + x^3)(1 - \cos t)$ appear $u_0(x, t)$ and $u_1(x, t)$. Canceling this term from $u_0(x, t)$ and justifying that the remaining non-canceled term of $u_0(x, t)$ justifies the equation gives the solution by

$$u(x, t) = \frac{6}{7!}x^7 \cos t, \tag{302}$$

that satisfies the boundary conditions.

Exercises 9.9

In Exercises 1 – 5, use the Adomian decomposition method to solve the following homogeneous fourth order equations:

1. $u_{tt} + u_{xxxx} = 0$, $u(x, 0) = \sin x$, $u_t(x, 0) = \cos x$

2. $u_{tt} + u_{xxxx} = 0$, $u(x, 0) = \sin x$, $u_t(x, 0) = 0$

3. $u_{tt} + u_{xxxx} = 0$, $u(x, 0) = \cos x$, $u_t(x, 0) = 0$

4. $u_{tt} + u_{xxxx} = 0$, $u(x, 0) = 1 + \cos x$, $u_t(x, 0) = \sin x$

5. $u_{tt} + u_{xxxx} = 0$, $u(x, 0) = 2$, $u_t(x, 0) = \sin x$

In Exercises 6 – 10, use the Adomian decomposition method to solve the following inhomogeneous fourth order equations:

6. $u_{tt} + u_{xxxx} = 15 \sin 2x \cos t$, $u(x, 0) = \sin 2x$, $u_t(x, 0) = 0$

7. $u_{tt} + u_{xxxx} = 2e^{x+t}$, $u(x, 0) = e^x$, $u_t(x, 0) = e^x$

8. $u_{tt} + u_{xxxx} = 12 \sin 2x \sin 2t$, $u(x, 0) = 0$, $u_t(x, 0) = 2 \sin 2x$

9. $u_{tt} + u_{xxxx} = 2e^{x-t}$, $u(x, 0) = e^x$, $u_t(x, 0) = -e^x$

10. $u_{tt} + u_{xxxx} = -3 \sin(x + 2t)$, $u(x, 0) = \sin x$, $u_t(x, 0) = 2 \sin x$

In Exercises 11 – 15, use Adomian decomposition method to solve the following fourth order equations with variable coefficients, $0 < x < 1$:

11. $u_{tt} + (\frac{x}{\sin x} - 1)u_{xxxx} = 0$, $u(x, 0) = -u_t(x, 0) = x - \sin x$

12. $u_{tt} + \frac{x^4}{360}u_{xxxx} = 0$, $u(x, 0) = 0$, $u_t(x, 0) = \frac{x^6}{720}$

13. $u_{tt} + (\frac{x}{\cos x} - 1)u_{xxxx} = 0$, $u(x, 0) = -u_t(x, 0) = x - \cos x$

14. $u_{tt} + (1 + \frac{x^4}{360}x^4)u_{xxxx} = \frac{5}{2}x^2 \sin t$, $u(x, 0) = 0$, $u_t(x, 0) = \frac{5}{6!}x^6$

15. $u_{tt} + (1 + \frac{3!}{7!})u_{xxxx} = x^3(\sin t + \cos t)$, $u(x, 0) = u_t(x, 0) = \frac{6}{7!}x^7$

Chapter 10

Numerical Applications and Padé Approximants

10.1 Introduction

In this chapter we will apply the Adomian decomposition method to handle linear and nonlinear differential equations numerically. Because the decomposition method provides a rapidly convergent series and faster than existing numerical techniques, it is therefore considered an efficient, reliable and easy to use from a computational viewpoint. It is to be noted that few terms are usually needed to supply a reliable result much closer to the exact value. The overall error can be significantly decreased by computing additional terms of the decomposition series.

The common numerical techniques that are usually used are the *finite differences* method, *finite element* method, and Galerkin method. The finite differences method handles the differential equation by replacing the derivatives in the equation with difference quotients. The finite elements method reduces any partial differential equation to a system of ordinary differential equations. However, the Galerkin method approximates the solution of a differential equation by a finite linear combinations of basic functions with specific properties. In addition, other techniques are used such as Crank-Nicolson method, perturbation methods and collocation method.

Recently, several useful comparative discussions have been conducted between the decomposition method and other numerical approaches. The studies have formally proved that the decomposition method is faster and more efficient to use in numerical applications as well as in analytical approaches. A useful comparative study conducted by Bellomo and Monaco [19] between Adomian's method and the perturbation techniques for nonlinear random differential equations. The conclusion made in [19] is that Adomian's method provides a reliable method and it reduces the size of calculations if compared with the perturbation techniques. The obtained results were confirmed by employing numerical examples. Rach [95] performed comparisons of Adomian's method with Picard's method. The study emphasized the reality that Adomian's method is different in structure than Picard's method. Moreover, it was shown that Adomian's method gives approximations of a high degree of accuracy. In [128], a comparison between the decomposition method and Taylor series was conducted to emphasize the fact that the decomposition method gives a rapid convergent series with a minimal volume of computational work. Other studies were conducted to compare the decomposition method with other existing numerical techniques to confirm the efficiency of the method.

As will be seen later, the n-term approximant ϕ_n

$$\phi_n = \sum_{n=0}^{n-1} u_n, \tag{1}$$

offers a very good approximation for quite low values of n. The accuracy level can be significantly enhanced by computing additional components of the solution.

It is the purpose of this chapter to employ the decomposition method for handling differential equations numerically. Moreover, comparison between decomposition method and other existing techniques will be carried out for specific cases.

The decomposition method has been outlined before in previous chapters and has been extensively employed in the text. The best way to describe the use of the decomposition method for numerical studies is to work on several examples, ordinary and partial differential equations. Our approach will begin first with ordinary differential equations. The partial differential equations will be investigated as well.

10.2   Ordinary Differential Equations

10.2.1   Perturbation problems

It is useful to consider a comparative study between the decomposition method and the perturbation technique. Two illustrative perturbation problems will be examined.

Example 1. Consider the Duffing equation

$$\frac{d^2y}{dt^2} + y + \epsilon y^3 = 0, \, y(0) = 1, y^{'}(0) = 0 \tag{2}$$

Solution.

The comparative study will be carried out by applying the two methods separately.

The Decomposition Method

Applying the two-fold integral operator L_t^{-1} to both sides of (2) gives

$$y(t) = 1 - L_t^{-1} y - \epsilon L_t^{-1} y^3. \tag{3}$$

Using the series representation for y and y^3 into (3) yields

$$\sum_{n=0}^{\infty} y_n = 1 - L_t^{-1}\left(\sum_{n=0}^{\infty} y_n\right) - \epsilon L_t^{-1}\left(\sum_{n=0}^{\infty} A_n\right), \tag{4}$$

where A_n are Adomian polynomials for y^3. The decomposition method suggests the recursive relation

$$
\begin{aligned}
y_0(t) &= 1, \\
y_{k+1} &= -L_t^{-1}(y_k) - \epsilon L_t^{-1}(A_k), \, k \geq 0.
\end{aligned}
\tag{5}
$$

The components $y_n(t)$ can be elegantly determined by

$y_0(t) = 1,$
$y_1(t) = -L_t^{-1}(y_0) - \epsilon L_t^{-1}(A_0) = -\frac{1}{2!}(1 + \epsilon)t^2,$
$y_2(t) = -L_t^{-1}(y_1) - \epsilon L_t^{-1}(A_1) = \frac{1}{4!}(1 + 4\epsilon + 3\epsilon^2)t^4,$
$y_3(t) = -L_t^{-1}(y_2) - \epsilon L_t^{-1}(A_2) = -\frac{1}{6!}(1 + 25\epsilon + 51\epsilon^2 + 27\epsilon^3)t^6.$

Consequently, the ϕ_4 approximant is given by

$$
\begin{aligned}
\phi_4 &= \sum_{n=0}^{3} y_n(t), \\
&= \left(1 - \frac{1}{2!}t^2 + \frac{1}{4!}t^4 - \frac{1}{6!}t^6\right) - \epsilon\left(\frac{1}{2!}t^2 - \frac{1}{3!}t^4 + \frac{25}{6!}t^6\right) + O(\epsilon^2).
\end{aligned}
\tag{6}
$$

The Perturbation Method

We next approach the problem (2) by applying the perturbation technique. To obtain a perturbative solution for (2), we first represent $y(t)$ by a power series in ϵ as

$$y(t) = \sum_{n=0}^{\infty} \epsilon^n y_n(t). \tag{7}$$

This means that the initial condition can be reduced to a set of initial conditions defined by

$$y_0(0) = 1, \, y_0'(0) = 0, \, y_k(0) = y_k'(0) = 0, \, k \geq 1. \tag{8}$$

Substituting (7) into (2) and equating coefficients of like powers of ϵ gives the differential equations

$$
\begin{aligned}
y_0'' + y_0 &= 0, \, y_0(0) = 1, \, y_0'(0) = 0, \\
y_1'' + y_1 &= -y_0^3, \, y_1(0) = 0, \, y_1'(0) = 0 \\
y_2'' + y_2 &= -3y_0^2 y_1, \, y_2(0) = 0, \, y_2'(0) = 0.
\end{aligned}
\tag{9}
$$

Solving the first homogeneous equation and using the result in solving the inhomogeneous equation gives

$$y_0 = \cos t,$$

$$y_1 = -\frac{1}{32}\cos t - \frac{3}{8}t\sin t + \frac{1}{32}\cos 3t.$$

(10)

In view of (10), the first-order perturbative solution to the Duffing equation is given by

$$y(t) = \cos t + \epsilon\left(-\frac{1}{32}\cos t - \frac{3}{8}t\sin t + \frac{1}{32}\cos 3t\right) + O(\epsilon^2).$$

(11)

A close examination of the decomposition method and the perturbation method clearly shows that the decomposition method can calculate the components elegantly where we integrated simple terms of the form t^n. However, using the perturbation technique requires solving homogeneous and inhomogeneous differential equations with trigonometric functions and trigonometric identities, such as $\cos 3t$ in this example. This shows that the perturbation technique suffers from the cumbersome work especially if a higher order solution is sought. On the other hand, we can easily evaluate additional components of the decomposition series as much as we like.

Example 2. Solve the equation

$$y' = y^2 \sin(\epsilon t), \quad y(0) = 1$$

(12)

Solution.

It is to be noted that the nonlinear equation (12) can be solved by using the separation of variables method where the analytic solution is

$$y = \frac{\epsilon}{(\epsilon - 1) + \cos \epsilon t}.$$

(13)

We next carry out the comparison between the decomposition method and the perturbation technique.

The Decomposition Method

In an operator form, Eq. (12) can be rewritten as

$$L_t y(t) = y^2 \sin(\epsilon t).$$

(14)

Applying the inverse operator L_t^{-1} to both sides of (14) yields

$$y(t) = 1 + L_t^{-1} y^2 \sin(\epsilon t).$$

(15)

Using the decomposition assumptions for $y(t)$ and for the nonlinear term y^2 gives

$$\sum_{n=0}^{\infty} y_n(t) = 1 + L_t^{-1}\left(\sin(\epsilon t)\sum_{n=0}^{\infty} A_n\right). \tag{16}$$

This suggests the recursive relation

$$
\begin{aligned}
y_0(t) &= 1, \\
y_{k+1} &= L_t^{-1}\left(\sin(\epsilon t)\sum_{n=0}^{\infty} A_k\right), \ k \geq 0.
\end{aligned} \tag{17}
$$

This gives the first four components by

$$
\begin{aligned}
y_0(t) &= 1, \\
y_1(t) &= L_t^{-1}\left(\sin(\epsilon t)\, A_0\right) = \tfrac{1}{\epsilon}(1 - \cos(\epsilon t)), \\
y_2(t) &= L_t^{-1}\left(\sin(\epsilon t)\, A_1\right) = \tfrac{1}{\epsilon^2}\left(1.5 - 2\cos(\epsilon t) + 0.5\cos(2\epsilon t)\right), \\
y_3(t) &= L_t^{-1}\left(\sin(\epsilon t)\, A_1\right) = \tfrac{1}{\epsilon^3}\left(1 - 3\cos(\epsilon t) + 3\cos^2(\epsilon t) - \cos^3(\epsilon t)\right), \\
y_4(t) &= L_t^{-1}\left(\sin(\epsilon t)\, A_2\right), \\
&= \tfrac{1}{\epsilon^4}\left(1 - 4\cos(\epsilon t) + 6\cos^2(\epsilon t) - \tfrac{8}{3}\cos^3(\epsilon t) + \tfrac{5}{6}\cos^4(\epsilon t)\right).
\end{aligned} \tag{18}
$$

This gives the ϕ_5 approximant by

$$
\begin{aligned}
\phi_5 &= \sum_{n=0}^{4} y_n, \\
&= 1 + \tfrac{1}{\epsilon}(1 - \cos(\epsilon t)) + \tfrac{1}{\epsilon^2}\left(1.5 - 2\cos(\epsilon t) + 0.5\cos(2\epsilon t)\right) \\
&\quad + \tfrac{1}{\epsilon^3}\left(1 - 3\cos(\epsilon t) + 3\cos^2(\epsilon t) - \cos^3(\epsilon t)\right) \\
&\quad + \tfrac{1}{\epsilon^4}\left(1 - 4\cos(\epsilon t) + 6\cos^2(\epsilon t) - \tfrac{8}{3}\cos^3(\epsilon t) + \tfrac{5}{6}\cos^4(\epsilon t)\right).
\end{aligned} \tag{19}
$$

The Perturbation Method

To obtain a perturbative solution for (12), we first represent $y(t)$ by a power series in ϵ as

$$y(t) = \sum_{n=0}^{\infty} \epsilon^n y_n(t). \tag{20}$$

This means that the initial condition can be reduced to a set of initial conditions defined by

$$y_0(0) = 1, \ y_0'(0) = 0, \ y_k(0) = y_k'(0) = 0, \ k \geq 1. \tag{21}$$

Substituting (20) into (12) gives

$$y_0' + \epsilon y_1' + \epsilon^2 y_2' + \epsilon^3 y_3' + \epsilon^4 y_4' + \cdots =$$

$$\left(y_0^2 + 2\epsilon y_0 y_1 + \epsilon^2 (2y_0 y_2 + y_1^2)\right) \times \left(\epsilon t - \frac{1}{3!}\epsilon^3 t^3\right). \tag{22}$$

Equating coefficients of like powers of ϵ in (22) leads to the set of differential equations

$$
\begin{aligned}
y_0' &= 0, \; y_0(0) = 1, \\
y_1' &= ty_0^2, \; y_1(0) = 0, \\
y_2' &= 2ty_0 y_1, \; y_2(0) = 0, \\
y_3' &= t(2y_0 y_2 + y_1^2) - \tfrac{1}{3!}t^3 y_0^2, \; y_3(0) = 0, \\
y_4' &= t(y_0 y_3 + 2y_1 y_2 + y_0 y_3) - \tfrac{2}{3!}t^3 y_0 y_1, \; y_4(0) = 0.
\end{aligned}
\tag{23}
$$

Solving the resulting equations and using the relevant initial conditions give

$$y_0(t) = 1, \; y_1(t) = \frac{1}{2}t^2, \; y_2(t) = \frac{1}{4}t^4, \; y_3(t) = \frac{1}{8}t^6 - \frac{1}{24}t^4, \; y_4(t) = \frac{1}{16}t^8 - \frac{5}{72}t^6. \tag{24}$$

The perturbative solution is therefore given by

$$y(t) = 1 + \frac{1}{2}\epsilon t^2 + \frac{1}{4}\epsilon^2 t^4 + \epsilon^3 (\frac{1}{8}t^6 - \frac{1}{24}t^4) + \epsilon^4 \left(\frac{1}{16}t^8 - \frac{5}{72}t^6\right) + O(\epsilon^5). \tag{25}$$

Table 1

Performance of the perturbation and the decomposition methods

x	u_{analytic}	$u_{\text{perturbation}}$	$u_{\text{decomposition}}$
0.8	1.03304	1.03304	1.03304
1.5	1.12649	1.12702	1.12647
2.0	1.24896	1.25658	1.24857
3.0	1.80713	2.02780	1.77501

Table 1 above shows the performance of the perturbation and the decomposition methods by considering $\epsilon = 0.1$. Comparing the performance of the decomposition method and the perturbation method in this example clearly shows that the decomposition method encountered the difficulties that arise from $\sin \epsilon t$, where components of the decomposition series have been computed directly. However, using the perturbation technique, the Taylor expansion of $\sin \epsilon t$ has been used to control the powers of ϵ in both sides of (22).

In addition, it is to be noted that approximating $\sin(\epsilon t)$ in using the perturbation technique by the first two terms of the Taylor expansion will affect the approximation numerically. However, the decomposition method has been applied directly without using any expansion. More importantly from the point of view of numerical purposes, the calculations in Table 1 show improvements of the decomposition method over perturbation method.

10.2.2 Nonperturbed Problems

In this section, nonperturbed ordinary differential equations will be handled from numerical viewpoint. A comparison will be carried here between the decomposition method and the Taylor series method. To achieve our goal, we study the following examples:

Example 3. Consider the first order ordinary differential equation

$$y' + y = \frac{1}{1 + x^2}, \, y(0) = 0 \tag{26}$$

Solution.

It is to be noted that the solution of the first order linear equation (26) cannot be found in a closed form [27]. We next carry out the comparison between the decomposition method and the Taylor series method.

The Decomposition Method

In an operator form, Eq. (26) can be rewritten as

$$L_x y = \frac{1}{1 + x^2} - y, \, y(0) = 0. \tag{27}$$

Applying the inverse operator L_x^{-1} to both sides of (27) yields

$$y(x) = \arctan x - L_x^{-1} y(x). \tag{28}$$

Substituting the series expression for $y(x)$ carries (28) into

$$\sum_{n=0}^{\infty} y_n(x) = \arctan x - L_x^{-1} \left(\sum_{n=0}^{\infty} y_n(x) \right). \tag{29}$$

This suggests the recursive relation

$$\begin{aligned} y_0(x) &= \arctan x, \\ y_{k+1}(x) &= -L_x^{-1} y_k(x), \, k \geq 0. \end{aligned} \tag{30}$$

This gives the first three components by

$$
\begin{aligned}
y_0(x) &= \arctan x, \\
y_1(x) &= -L_x^{-1}(y_0(x)) = -x\arctan x + \tfrac{1}{2}\ln(1+x^2), \\
y_2(x) &= -L_x^{-1}(y_1(x)) = \tfrac{1}{2}\left((x^2-1)\arctan x + x - x\ln(1+x^2)\right),
\end{aligned}
\tag{31}
$$

where tables of integrals in Appendix A are used. This gives the ϕ_3 approximant by

$$
\begin{aligned}
\phi_3 &= \sum_{n=0}^{2} y_n(x), \\
&= \arctan x - x\arctan x + \tfrac{1}{2}\ln(1+x^2) \\
&\quad + \tfrac{1}{2}\left[(x^2-1)\arctan x + x - x\ln(1+x^2)\right].
\end{aligned}
\tag{32}
$$

The Taylor Series Method

This equation can be handled by using the integrating factor μ given by

$$
\mu = e^x.
\tag{33}
$$

The solution of Eq. (26) is therefore given by the expression

$$
y(x) = \int_0^x \frac{e^t}{1+t^2}\, dt.
\tag{34}
$$

It is clear that a closed form solution is not obtainable in this problem [27]. Accordingly, we use the Taylor series method where we introduce the solution in the form of an infinite series

$$
y(x) = \sum_{n=0}^{\infty} a_n x^n.
\tag{35}
$$

Our goal is now to determine the coefficients $a_n, n \geq 0$. Substituting (35) into (26) gives

$$
\left(\sum_{n=0}^{\infty} a_n x^n\right)_x + \sum_{n=0}^{\infty} a_n x^n = \sum_{n=0}^{\infty} (-1)^n x^{2n}.
\tag{36}
$$

Note that the given condition $y(0) = 0$ gives $a_0 = 0$. The other coefficients $a_n, n \geq 1$ can be determined by equating the coefficients of like powers of x where we find

$$
\begin{aligned}
a_1 &= 1, \quad a_2 = -\tfrac{1}{2!}, \quad a_3 = -\tfrac{1}{3!} \\
a_4 &= \tfrac{1}{4!}, \quad a_5 = \tfrac{23}{5!}, \quad a_6 = -\tfrac{23}{6!}.
\end{aligned}
\tag{37}
$$

In view of (37), the solution in a series form is given by

$$y(x) = x - \frac{1}{2!}x^2 - \frac{1}{3!}x^3 + \frac{1}{4!}x^4 + \frac{23}{5!}x^5 - \frac{23}{6!}x^6 + \cdots. \tag{38}$$

Example 4. Solve the second order linear equation

$$y'' + 2xy' = 0, \; y(0) = 0, \; y'(0) = \frac{2}{\sqrt{\pi}} \tag{39}$$

Solution.

It is to be noted that the exact solution of this equation is given by

$$y(x) = \text{erf}(x), \tag{40}$$

where the error function $\text{erf}(x)$ is defined by

$$\text{erf}(x) = \frac{2}{\sqrt{\pi}} \int_0^x e^{-u^2} \, du, \tag{41}$$

and its complementary function $\text{erfc}(x)$ is defined by

$$\text{erfc}(x) = \frac{2}{\sqrt{\pi}} \int_x^\infty e^{-u^2} \, du, \tag{42}$$

such that

$$\text{erf}(x) + \text{erfc}(x) = 1. \tag{43}$$

The Decomposition Method

Applying the two-fold integral operator L_x^{-1} on (39) gives

$$y(x) = \frac{2}{\sqrt{\pi}}x - 2L_x^{-1}xy'(x). \tag{44}$$

Proceeding as in the previous examples, the first four components of the solution $y(x)$ can be determined as follows:

$$y_0(x) = \frac{2}{\sqrt{\pi}}x,$$

$$y_1(x) = -2L_x^{-1}(xy_0'(x)) = -\frac{2}{3\sqrt{\pi}}x^3,$$

$$y_2(x) = -2L_x^{-1}(xy_1'(x)) = \frac{1}{5\sqrt{\pi}}x^5,$$

$$y_3(x) = -2L_x^{-1}(xy_2'(x)) = -\frac{1}{21\sqrt{\pi}}x^7,$$

$$y_4(x) = -2L_x^{-1}(xy_3'(x)) = \frac{1}{108\sqrt{\pi}}x^9.$$

Combining the obtained components gives the ϕ_4 approximant

$$
\begin{aligned}
\phi_5 &= y_0 + y_1 + y_2 + y_3 + y_4, \\
&= \frac{2}{\sqrt{\pi}} \left[x - \frac{x^3}{3} + \frac{x^5}{5 \cdot 2!} - \frac{x^7}{7 \cdot 3!} + \frac{x^9}{9 \cdot 4!} \right].
\end{aligned}
\tag{45}
$$

The Taylor Series Method

To determine the series solution, several derivatives of $y(x)$ will be evaluated in this problem. To achieve this goal, we simply differentiate Eq. (39) successively [144] to obtain

$$
\begin{aligned}
y''(x) &= -2xy'(x), \\
y'''(x) &= -2xy''(x) - 2y'(x), \\
y^{(iv)}(x) &= -2xy'''(x) - 4y''(x), \\
y^{(v)}(x) &= -2xy^{(iv)}(x) - 6y'''(x), \\
y^{(vi)}(x) &= -2xy^{(v)}(x) - 8y^{(iv)}(x), \\
y^{(vii)}(x) &= -2xy^{(vi)}(x) - 10y^{(v)}(x).
\end{aligned}
\tag{46}
$$

Substituting the initial conditions

$$
\begin{aligned}
y(0) &= 0, \\
y'(0) &= \frac{2}{\sqrt{\pi}},
\end{aligned}
\tag{47}
$$

into (46) gives

$$
\begin{aligned}
y''(0) = 0, \quad y'''(0) &= -\frac{4}{\sqrt{\pi}}, \quad y^{(iv)}(0) = 0 \\
y^{(v)}(0) = \frac{24}{\sqrt{\pi}}, \quad y^{(vi)}(0) &= 0, \quad y^{(vii)}(0) = -\frac{240}{\sqrt{\pi}}.
\end{aligned}
\tag{48}
$$

Recall that the Taylor expansion of $y(x)$ is given by

$$
y(x) = \sum_{n=0}^{\infty} \frac{1}{n!} y^{(n)}(x).
\tag{49}
$$

Combining (48) and (49) gives the Taylor series solution

$$
y(x) = \frac{2}{\sqrt{\pi}} \left[x - \frac{x^3}{3} + \frac{x^5}{5 \cdot 2!} - \frac{x^7}{7 \cdot 3!} + \frac{x^9}{9 \cdot 4!} \right].
\tag{50}
$$

It is clear that the Taylor solution provided the same approximant obtained above by using the decomposition method. However, the Taylor method required more work than the decomposition method.

Table 2 below shows that by adding more terms in the decomposition series, we can easily enhance the accuracy level.

Table 2

Performance of various approximants of the decomposition series

x	$\mathrm{erf}(x)$	ϕ_3	ϕ_4	ϕ_5
0.2	0.22270	0.22270	0.22270	0.22270
0.4	0.42839	0.42844	0.42830	0.42839
0.6	0.60386	0.60456	0.60380	0.60386
0.8	0.74210	0.74710	0.74147	0.74225
1.0	0.84270	0.86509	0.83822	0.84410

Exercises 10.2

In Exercises 1 – 5, use Adomian decomposition method to find the perturbative approximation for each equation.

1. Find the ϕ_3 approximant for the equation:
   $y'' = (\epsilon x - 1)y, \; y(0) = 1, y'(0) = 0$

2. Find the ϕ_3 approximant for the equation:
   $y' = (\epsilon x + 1)y, \; y(0) = 1$

3. Find the ϕ_4 approximant for the linear damping oscillator equation:
   $u'' + 2\epsilon u' + u = 0, \; u(0) = 1, \; u'(0) = 0$

4. Find the ϕ_4 approximant for the Van der Pol equation:
   $u'' + \epsilon(u^2 - 1)u' + u = 0, \; u(0) = 1, \; u'(0) = 0$

5. Find the ϕ_3 approximant for the nonlinear equation:
   $y' = y^2 \cos(\epsilon x), \; y(0) = 1$

In Exercises 6 – 10, use Adomian decomposition method to find the ϕ_4 approximant for each equation:

6. $y'' + 2xy' = 0, \; y(0) = 0, \; y'(0) = 1$

7. $y'' + 2xy' = 0, \; y(0) = 0, \; y'(0) = \dfrac{4}{\sqrt{\pi}}$

8. $y'' - 2xy' = 0$, $y(0) = 0$, $y'(0) = 1$

9. $y'' = 12x^2 - y^2$, $y(0) = 0$, $y'(0) = 0$

10. $y'' = 2 + y' + y^2$, $y(0) = 0$, $y'(0) = 0$

10.3 Partial Differential Equations

In this section, the decomposition method will be applied to partial differential equations to study the numerical approximations of the solutions. The basic outlines of the method are well known from previous chapters.

Example 1. Find the ϕ_5 approximation of the solution of the heat equation

$$
\begin{aligned}
u_t &= u_{xx}, \ 0 < x < \pi, \ t > 0 \\
u(0,t) &= 0, \ u(\pi,t) = 0 \\
u(x,0) &= \sin x
\end{aligned}
\tag{51}
$$

Solution.

Recall that the heat equation (51) has been solved in Chapter 3 by using the decomposition and the separation of variables methods where we can easily show that the exact solution is given by

$$u(x,t) = e^{-t}\sin x. \tag{52}$$

To determine the ϕ_5 approximation, we apply the inverse operator L_t^{-1} to both sides of (51) to obtain

$$u(x,t) = \sin x + L_t^{-1}(u_{xx}). \tag{53}$$

Substituting $u(x,t) = \sum_{n=0}^{\infty} u_n(x,t)$ and using the resulting recursive relation, we can determine the first five components recurrently by

$$
\begin{aligned}
u_0(x,t) &= \sin x, \quad u_1(x,t) = -t\sin x, \quad u_2(x,t) = \tfrac{1}{2!}t^2\sin x, \\
u_3(x,t) &= -\tfrac{1}{3!}t^3\sin x, \quad u_4(x,t) = \tfrac{1}{4!}t^4\sin x.
\end{aligned}
\tag{54}
$$

Consequently, the ϕ_5 approximation is given by

$$\phi_5 = \sin x \left(1 - t + \frac{1}{2!}t^2 - \frac{1}{3!}t^3 + \frac{1}{4!}t^4\right). \tag{55}$$

Table 3 below shows the error that results from using the approximation (55) for $t = .5$ and for appropriate values of x, where $error = |\text{exact value} - \phi_5|$.

Table 3
Comparison of exact value and ϕ_5

x	exact value	ϕ_5	error
0.5	0.290786	0.290901	1.15E-4
1.0	0.510378	0.510580	2.02E-4
1.5	0.605011	0.605251	2.4E-4
2.0	0.551517	0.551735	2.18E-4
2.5	0.362992	0.363135	1.43E-4
3.0	0.085594	0.085628	3.44E-5

The table clearly shows that errors can be significantly decreased by evaluating additional components.

Example 2. Find the ϕ_5 approximation of the solution of the wave equation

$$u_{tt} = u_{xx},\ 0 < x < \pi,\ t > 0$$
$$u(0,t) = 0,\ u(\pi,t) = \pi \tag{56}$$
$$u(x,0) = x,\ u_t(x,0) = \sin x$$

Solution.

The wave equation (56) has been investigated in Chapter 5 by using the decomposition and the separation of variables methods where we can easily show that the exact solution is given by

$$u(x,t) = x + \sin x \sin t. \tag{57}$$

To determine the ϕ_5 approximation, we apply the two-fold inverse operator L_t^{-1} to both sides of (56) to obtain

$$u(x,t) = x + t\sin x + L_t^{-1}(u_{xx}). \tag{58}$$

Substituting $u(x,t) = \sum_{n=0}^{\infty} u_n(x,t)$ and proceeding as before, we can determine the first five components recurrently by

$$u_0(x,t) = x + t\sin x, \quad u_1(x,t) = -\frac{1}{3!}t^3\sin x, \quad u_2(x,t) = \frac{1}{5!}t^5\sin x,$$
$$u_3(x,t) = -\frac{1}{7!}t^7\sin x, \quad u_4(x,t) = \frac{1}{9!}t^9\sin x. \tag{59}$$

Consequently, the ϕ_5 approximation is given by

$$\phi_5 = \sin x \left(1 - t + \frac{1}{2!}t^2 - \frac{1}{3!}t^3 + \frac{1}{4!}t^4 \right). \tag{60}$$

Example 3. Find the ϕ_5 approximation of the solution of the nonlinear partial differential equation

$$u_t + u^2 u_x = 0, \ u(x,0) = x, \ x \in R, t > 0 \tag{61}$$

Solution.

Operating with L_t^{-1} on (61) gives

$$u(x, t) = x + L_t^{-1}(u^2 u_x). \tag{62}$$

Substituting $u(x, t) = \sum_{n=0}^{\infty} u_n(x, t)$, representing the nonlinear term $u^2 u_x$ by a series of Adomian polynomials, and proceeding as before, we find

$$
\begin{aligned}
u_0(x, t) &= x, \\
u_1(x, t) &= -L_t^{-1}(A_0) = -x^2 t, \\
u_2(x, t) &= -L_t^{-1}(A_1) = 2x^3 t^2, \\
u_3(x, t) &= -L_t^{-1}(A_2) = -5x^4 t^3, \\
u_4(x, t) &= -L_t^{-1}(A_3) = 14x^5 t^4.
\end{aligned} \tag{63}
$$

Consequently, the ϕ_5 approximation is given by

$$\phi_5 = x \left(1 - xt + 2x^2 t^2 - 5x^3 t^3 + 14x^4 t^4 \right). \tag{64}$$

Table 4 below shows the errors obtained if the approximation (64) is used.

Table 4
Errors of using ϕ_5 for several values of x and t

$t \backslash x$	0.1	0.2	0.3	0.4	0.5
0.1	1.01E-9	2.30E-8	2.82E-7	1.53E-6	5.67E-6
0.2	1.1E-8	7.65E-7	8.25E-6	4.40E-5	1.60E-4
0.3	9.4E-8	5.50E-6	5.81E-5	3.04E-4	1.09E-3
0.4	3.83E-7	2.20E-5	2.28E-4	1.17E-3	4.15E-3
0.5	1.14E-6	6.40E-5	6.53E-4	3.32E-3	1.16E-2

It is to be noted that the exact solution is given by

$$u(x,t) = \begin{cases} x & t = 0 \\ \frac{1}{2t}(\sqrt{1 + 4xt} - 1) & t > 0, 1 + 4xt > 0 \end{cases} \tag{65}$$

Exercises 10.3

In Exercises 1 – 5, use Adomian decomposition method to find the ϕ_4 approximation for each equation.

1. $u_t = u_{xx}$, $u(x,0) = \cos x$, $u(0,t) = e^{-t}$, $u(\pi,t) = 0$

2. $u_{tt} = u_{xx}$, $u(x,0) = \cos x$, $u(0,t) = \cos t$, $u(\pi,t) = -\cos t$

3. $u_t + \frac{1}{36}xu_{xx}^2 = x^3$, $u(x,0) = 0$

4. $u_t + u^2u_x = 0$, $u(x,0) = 4x$

5. $u_t + uu_x = x$, $u(x,0) = 2$

In Exercises 6 – 10, use Adomian decomposition method to find the ϕ_3 approximation for each equation.

6. $u_t + uu_x^2 = 0$, $u(x,0) = x$

7. $u_t + uu_x = x^2$, $u(x,0) = 0$

8. $u_t + u_x + u^2 = 0$, $u(x,0) = x$

9. $u_t = iu_{xx}$, $u(x,0) = \cosh x$

10. $u_t + u_{xx} = t\sinh x$, $u(x,0) = 1$

10.4   The Padé Approximants

In this section, the powerful Padé approximants will be investigated. Our main concern will be directed on two ways. First, we will discuss the construction of Padé approximants for functions and polynomials. Next, we will explore the implementation of Padé approximants in boundary value problems where the domain is unbounded.

Polynomials are frequently used to approximate power series. However, polynomials tend to exhibit oscillations that may produce an approximation error bounds. In addition, polynomials can never blow up in a finite plane; and this makes the singularities not apparent. To overcome these difficulties, the Taylor series is best manipulated by Padé approximants for numerical approximations.

Padé approximant represents a function by the ratio of two polynomials [17,18]. The coefficients of the polynomials in the numerator and in the denominator are determined by using the coefficients in the Taylor expansion of the function. Padé rational approximations are widely used in numerical analysis and fluid mechanics, because they are more efficient than polynomials.

To explore the need of Padé approximants, we consider the function

$$f(x) = \sqrt{\frac{1 + 3x}{1 + x}}. \tag{66}$$

The Taylor series of $f(x)$ in (66) is given by

$$f(x) = 1 + x - \frac{3}{2}x^2 + \frac{5}{2}x^3 - \frac{37}{8}x^4 + \frac{75}{8}x^5 - \frac{327}{16}x^6 + \frac{753}{16}x^7 + O(x^8). \tag{67}$$

The Taylor series (67) is often used to approximate $f(x)$ for values of x within the radius of convergence. However, if the polynomial obtained from using a finite number of the Taylor series (67) is to be evaluated for large positive values of x, such as $x = \infty$, the series or any truncated number of terms of (67) will definitely fail to provide a converging expression. Padé introduced a powerful tool that should be combined with power series for calculations work. This is highly needed especially for boundary value problems where, for specific cases, the domain of validity is unbounded. Using power series, isolated from other concepts, is not always useful because the radius of convergence of the series may not contain the two boundaries [28].

Padé approximant, symbolized by [m/n], is a rational function defined by

$$[m/n] = \frac{a_0 + a_1 x + a_2 x^2 + \cdots + a_m x^m}{1 + b_1 x + b_2 x^2 + \cdots + b_n x^n}, \tag{68}$$

where we considered $b_0 = 1$, and numerator and denominator have no common factors. If we selected $m = n$, then the approximants [n/n] are called diagonal approximants.

Notice that in (68) there are $m + 1$ independent numerator coefficients and n independent denominator coefficients, making altogether $m + n + 1$ unknowns [26]. This suggests that [m/n] Padé approximant fits the power series [26] of $f(x)$ through the orders $1, x, x^2, \cdots x^{m+n}$.

In addition, the Padé approximant will converge on the entire real axis if the function $f(x)$ has no singularities. It was discussed by many that the diagonal Padé approximants, where $m = n$, are more accurate and efficient. Based on this, our study will be focused only on diagonal approximants.

In the following we will introduce the simple and the straightforward method to construct Padé approximants. Suppose that $f(x)$ has a Taylor series given by

$$f(x) = \sum_{k=0}^{\infty} c_k x^k. \tag{69}$$

Assuming that $f(x)$ can be manipulated by the diagonal Padé approximant defined in (68), where $m = n$. This admits the use of

$$\frac{a_0 + a_1 x + a_2 x^2 + \cdots + a_n x^n}{1 + b_1 x + b_2 x^2 + \cdots + b_n x^n} = c_0 + c_1 x + c_2 x^2 + \cdots + c_{2n} x^{2n}. \tag{70}$$

By using cross multiplication in (70) we find

$$
\begin{aligned}
a_0 + a_1 x + a_2 x^2 + \cdots + a_n x^n &= c_0 + (c_1 + b_1 c_0)x + (c_2 + b_1 c_1 + b_2 c_0)x^2 \\
&\quad + (c_3 + b_1 c_2 + b_2 c_1 + b_3 c_0)x^3 + \cdots.
\end{aligned}
\tag{71}
$$

Equating powers of x leads to

$$\text{coefficient of } x^0: \quad a_0 \;=\; c_0,$$

$$\text{coefficient of } x^1: \quad a_1 \;=\; c_1 + b_1 c_0,$$

$$\text{coefficient of } x^2: \quad a_2 \;=\; c_2 + b_1 c_1 + b_2 c_0,$$

$$\text{coefficient of } x^3: \quad a_3 \;=\; c_3 + b_1 c_2 + b_2 c_1 + b_3 c_0,$$

$$\vdots$$

$$\text{coefficient of } x^n: \quad a_n \;=\; c_n + \sum_{k=1}^{n} b_k c_{n-k}.$$

Notice that coefficients of $x^{n+1}, x^{n+2}, \cdots x^{2n}$ should be equated to zero. This completes the determination of the constants of the polynomials in the numerator and in the denominator. The simple procedure outlined above will be illustrated by discussing the following examples.

Example 1. Find the Padé approximants [2/2] and [3/3] for the function

$$f(x) = \sqrt{\frac{1 + 3x}{1 + x}} \tag{72}$$

Solution.

The Taylor series for $f(x)$ of (72) is given by

$$f(x) = 1 + x - \frac{3}{2}x^2 + \frac{5}{2}x^3 - \frac{37}{8}x^4 + \frac{75}{8}x^5 - \frac{327}{16}x^6 + \frac{753}{16}x^7 + O(x^8). \tag{73}$$

The [2/2] approximant is defined by

$$[2/2] = \frac{a_0 + a_1 x + a_2 x^2}{1 + b_1 x + b_2 x^2}. \tag{74}$$

To determine the five coefficients of the two polynomials, the [2/2] approximant must fit the Taylor series of $f(x)$ in (73) through the orders of $1, x, \cdots, x^4$, hence we set

$$\frac{a_0 + a_1 x + a_2 x^2}{1 + b_1 x + b_2 x^2} = 1 + x - \frac{3}{2}x^2 + \frac{5}{2}x^3 - \frac{37}{8}x^4 + \cdots. \tag{75}$$

Cross multiplying yields

$$1 + (b_1 + 1)x + (b_1 + b_2 - \tfrac{3}{2})x^2 + (-\tfrac{3}{2}b_1 + b_2 + \tfrac{5}{2})x^3$$

$$+(\frac{5}{2}b_1 - \frac{3}{2}b_2 - \frac{37}{8})x^4 = a_0 + a_1 x + a_2 x^2. \tag{76}$$

Equating powers of x leads to

coefficients of x^4: $\frac{5}{2}b_1 - \frac{3}{2}b_2 - \frac{37}{8}$ $= 0,$

coefficients of x^3: $-\frac{3}{2}b_1 + b_2 + \frac{5}{2}$ $= 0,$

coefficients of x^2: $b_1 + b_2 - \frac{3}{2}$ $= a_2,$

coefficients of x^1: $b_1 + 1$ $= a_1,$

coefficients of x^0: 1 $= a_0.$

The solution of this system of equations is

$$a_0 = 1, \quad a_1 = \tfrac{9}{2}, \quad a_2 = \tfrac{19}{4},$$

$$b_1 = \tfrac{7}{2}, \quad b_2 = \tfrac{11}{4}. \tag{77}$$

Consequently, the [2/2] Padé approximant is

$$[2/2] = \frac{1 + \frac{9}{2}x + \frac{19}{4}x^2}{1 + \frac{7}{2}x + \frac{11}{4}x^2}. \tag{78}$$

Two conclusions can be made here. First, we note that the Taylor series for the [2/2] approximant is given by

$$[2/2]_{\text{Taylor}} = 1 + x - \frac{3}{2}x^2 + \frac{5}{3}x^3 - \frac{37}{8}x^4 + \frac{149}{16}x^5 - \frac{159}{8}x^6 + O(x^7). \tag{79}$$

A close examination of the Taylor series of the approximant [2/2] given by (79) and the Taylor series of $f(x)$ given by (73), one can easily conclude that the two series are consistent up to x^4 of each. This is normal because in order to determine the five coefficients a_0, a_1, a_2, b_1, b_2, it was necessary to use the terms of orders $1, x, \cdots, x^4$ in Taylor series (73) of $f(x)$. Second, it was difficult to use Taylor series (73) when x is large, say $x = \infty$. However, the limit of Padé approximant (78) as $x \to \infty$ is $\frac{a_2}{b_2}$. In other words, as $x \to \infty$ we obtain

$$\sqrt{3} = \frac{19}{11} = 1.72727. \tag{80}$$

To determine the Padé approximant [3/3], we first set

$$[3/3] = \frac{a_0 + a_1 x + a_2 x^2 + a_3 x^3}{1 + b_1 x + b_2 x^2 + b_3 x^3}. \tag{81}$$

To determine the seven coefficients of the two polynomials, we use the terms through orders $1, x, \cdots, x^6$ in Taylor series of $f(x)$ in (73), hence we set

$$\frac{a_0 + a_1 x + a_2 x^2 + a_3 x^3}{1 + b_1 x + b_2 x^2 + b_3 x^3} = 1 + x - \frac{3}{2} x^2 + \frac{5}{2} x^3 - \frac{37}{8} x^4$$
$$+ \frac{75}{8} x^5 - \frac{327}{16} x^6 + \cdots. \tag{82}$$

Cross multiplying, equating coefficients of like powers of x and solving the resulting system of equations lead to

$$a_0 = 1, \quad a_1 = \frac{13}{2}, \quad a_2 = \frac{27}{2}, \quad a_3 = \frac{81}{8}$$
$$b_1 = \frac{11}{2}, \quad b_2 = \frac{19}{2}, \quad b_3 = \frac{41}{8}. \tag{83}$$

This gives

$$[3/3] = \frac{1 + \frac{13}{2} x + \frac{27}{2} x^2 + \frac{71}{8} x^3}{1 + \frac{11}{2} x + \frac{19}{2} x^2 + \frac{41}{8} x^3}. \tag{84}$$

It is to be noted that the Taylor series of (84) and the Taylor series (73) are consistent up to term of order x^6. In addition, the limit of Padé approximant [3/3] as $x \to \infty$ is $\frac{a_3}{b_3}$. In other words, as $x \to \infty$ we find

$$\sqrt{3} = \frac{a_3}{b_3} = \frac{71}{41} = 1.731707, \tag{85}$$

a better approximation to $\sqrt{3}$ compared to that obtained from [2/2].

Example 2.
    Establish the Padé approximants [2/2] and [3/3] for

$$f(x) = e^{-x} \tag{86}$$

Solution.

The Taylor expansion for the exponential function is

$$e^{-x} = 1 - x + \frac{x^2}{2!} - \frac{x^3}{3!} + \frac{x^4}{4!} - \frac{x^5}{5!} - \frac{x^6}{6!} + O(x^7). \tag{87}$$

The [2/2] approximant is defined by

$$[2/2] = \frac{a_0 + a_1 x + a_2 x^2}{1 + b_1 x + b_2 x^2}. \tag{88}$$

To determine the five coefficients of the two polynomials in the numerator and the denominator, we use the Taylor series of $f(x)$ in (87) as discussed before, hence we set

$$\frac{a_0 + a_1 x + a_2 x^2}{1 + b_1 x + b_2 x^2} = 1 - x + \frac{x^2}{2!} - \frac{x^3}{3!} + \frac{x^4}{4!} + \cdots. \tag{89}$$

Cross multiplying yields

$$1 + (b_1 - 1)x + (-b_1 + b_2 + \tfrac{1}{2})x^2 + (\tfrac{1}{2}b_1 - b_2 - \tfrac{1}{6})x^3$$
$$+ (-\tfrac{1}{6}b_1 + \tfrac{1}{2}b_2 + \tfrac{1}{24})x^4 = a_0 + a_1 x + a_2 x^2. \tag{90}$$

Equating powers of x leads to

coefficient of x^4: $-\tfrac{1}{6}b_1 + \tfrac{1}{2}b_2 + \tfrac{1}{24}$ $= 0,$

coefficient of x^3: $\tfrac{1}{2}b_1 - b_2 - \tfrac{1}{6}$ $= 0,$

coefficient of x^2: $-b_1 + b_2 + \tfrac{1}{2}$ $= a_2,$

coefficient of x^1: $b_1 - 1$ $= a_1,$

coefficient of x^0: 1 $= a_0.$

This system of equations gives

$$a_0 = 1, \quad a_1 = -\tfrac{1}{2}, \quad a_1 = \tfrac{1}{12},$$
$$b_1 = \tfrac{1}{2}, \quad b_2 = \tfrac{1}{12}, \tag{91}$$

so that the Padé approximant is

$$[2/2] = \frac{1 - \tfrac{1}{2}x + \tfrac{1}{12}x^2}{1 + \tfrac{1}{2}x + \tfrac{1}{12}x^2}. \tag{92}$$

In a similar way, we can derive the Padé approximant $[3/3]$ by

$$[3/3] = \frac{1 - \tfrac{1}{2}x + \tfrac{1}{10}x^2 - \tfrac{1}{120}x^3}{1 + \tfrac{1}{2}x + \tfrac{1}{10}x^2 + \tfrac{1}{120}x^3}. \tag{93}$$

Note that as $x \to \infty$, Padé approximants fluctuate between -1 and 1 as can be easily seen from (92) and (93). In fact, $e^{-x} \to 0$ as $x \to \infty$.

   Table 5 below shows the numerical approximations for e^{-x} for several values of x. We can easily observe that the Padé approximant $[3/3]$ provides better approximation than the Taylor expansion. In general, the Padé approximants give small error near $x = 0$, but the error increases as $|x|$ increases.

Table 5
Numerical approximation for e^{-x} by Taylor series and Padé approximants

x	e^{-x}	[3/3]	Taylor
0.0	1	1	1
0.2	0.818731	0.818731	0.818731
0.4	0.670320	0.670320	0.670320
0.6	0.548812	0.548811	0.548817
0.8	0.449329	0.449328	0.449367
1.0	0.367879	0.367876	0.368056

Example 3. Establish the Padé approximants [2/2] and [4/4] for

$$f(x) = \cos x \tag{94}$$

Solution.

The Taylor expansion for the exponential function is

$$\cos x = 1 - \frac{x^2}{2!} + \frac{x^4}{4!} - \frac{x^6}{6!} + \frac{x^8}{8!} + O(x^{10}). \tag{95}$$

It is useful to note that $\cos x$ and its Padé approximants are even functions. To minimize the size of calculations, we substitute

$$z = x^2, \tag{96}$$

into (95) to obtain

$$\cos z^{\frac{1}{2}} = 1 - \frac{z}{2!} + \frac{z^2}{4!} - \frac{z^3}{6!} + \frac{z^4}{8!} + O(z^5). \tag{97}$$

The [2/2] approximant in this case is defined by

$$[2/2] = \frac{a_0 + a_1 z}{1 + b_1 z}. \tag{98}$$

To determine the three coefficients of the two polynomials, we proceed as before to find

$$a_0 = 1, \quad a_1 = -\tfrac{5}{12}, \tag{99}$$

$$b_1 = \tfrac{1}{12},$$

so that the Padé approximant is

$$[2/2] = \frac{1 - \frac{5}{12}z}{1 + \frac{1}{12}z},$$
(100)

or equivalently

$$[2/2] = \frac{1 - \frac{5}{12}x^2}{1 + \frac{1}{12}x^2},$$
(101)

To determine the Padé approximant $[4/4]$, we set

$$[4/4] = \frac{a_0 + a_1 z + a_2 z^2}{1 + b_1 z + b_2 z^2}.$$
(102)

Proceeding as before gives

$$[4/4] = \frac{1 - \frac{115}{252}z + \frac{313}{15120}z^2}{1 + \frac{11}{252}z + \frac{13}{15120}z^2},$$
(103)

or by

$$[4/4] = \frac{1 - \frac{115}{252}x^2 + \frac{313}{15120}x^4}{1 + \frac{11}{252}x^2 + \frac{13}{15120}x^4}.$$
(104)

Example 4. Establish the Padé approximants $[3/3]$ and $[4/4]$ for

$$f(x) = \frac{\ln(1 + x)}{x}$$
(105)

Solution.

The Taylor expansion for function in (105) is

$$\frac{\ln(1 + x)}{x} = 1 - \frac{x}{2} + \frac{x^2}{3} - \frac{x^3}{4} + \frac{x^4}{5} - \frac{x^5}{6} + \frac{x^6}{7} - \frac{x^7}{8} + \frac{x^8}{9} + O(x^{10}).$$
(106)

To establish $[3/3]$ approximant, we set

$$[3/3] = \frac{a_0 + a_1 x + a_2 x^2 + a_3 x^3}{1 + b_1 x + b_2 x^2 + b_3 x^3}.$$
(107)

To determine the unknowns, we proceed as before and therefore we set

$$\frac{a_0 + a_1 x + a_2 x^2 + a_3 x^3}{1 + b_1 x + b_2 x^2 + b_3 x^3} = 1 - \frac{x}{2} + \frac{x^2}{3} - \frac{x^3}{4} + \frac{x^4}{5} - \frac{x^5}{6} + \frac{x^6}{7}.$$
(108)

Cross multiplying and proceeding as before we find

$$a_0 = 1, \quad a_1 = \tfrac{17}{14}, \quad a_2 = \tfrac{1}{3}, \quad a_3 = \tfrac{1}{140},$$
$$b_1 = \tfrac{1}{7}, \quad b_2 = \tfrac{6}{7}, \quad b_3 = \tfrac{4}{35},$$
(109)

so that the Padé approximant is

$$[3/3] = \frac{420 + 510x + 140x^2 + 3x^3}{420 + 720x + 360x^2 + 48x^3}. \tag{110}$$

To determine the Padé approximant [4/4], we set

$$[4/4] = \frac{a_0 + a_1x + a_2x^2 + a_3x^3 + a_4x^4}{1 + b_1x + b_2x^2 + b_3x^3 + b_4x^4}. \tag{111}$$

Proceeding as before we obtain

$$[4/4] = \frac{3780 + 6510x + 3360x^2 + 505x^3 + 6x^4}{3780 + 8400x + 6300x^2 + 1800x^3x^3 + 150x^4}. \tag{112}$$

We close this section by pointing out that many symbolic computer languages, such as Maple and Mathematica have a built-in function that finds Padé approximants when a Taylor series is used. In Appendix D, we list Padé tables for several well-known functions.

Exercises 10.4

1. (a) Establish the Padé approximants [2/2] and [3/3] for

$$f(x) = \sqrt{\frac{1 + 5x}{1 + x}}$$

   (b) Use the result in part (a) to approximate $\sqrt{5}$

2. (a) Establish the Padé approximants [2/2] and [3/3] for

$$f(x) = \sqrt{\frac{1 + 13x}{1 + x}}$$

   (b) Use the result in part (a) to approximate $\sqrt{13}$

3. (a) Establish the Padé approximants [2/2] and [3/3] for

$$f(x) = \sin x$$

   (b) Use the result in part (a) to approximate $\sin 1$

4. Establish the Padé approximants [2/2] and [3/3] for

$$f(x) = e^x$$

5. Establish the Padé approximants [2/2] and [3/3] for

$$f(x) = \frac{\ln(1 - x)}{x}$$

6. Establish the Padé approximants [2/2] and [3/3] for

$$f(x) = \frac{\tan x}{x}$$

7. Establish the Padé approximants [3/3] and [4/4] for

$$f(x) = \tanh^{-1} x$$

8. Establish the Padé approximants [3/3] and [4/4] for

$$f(x) = \sinh x$$

9. Establish the Padé approximants [3/3] and [4/4] for

$$f(x) = \frac{1}{1+x}$$

10. Establish the Padé approximants [3/3] and [4/4] for

$$f(x) = \frac{\arctan x}{x}$$

11. Establish the Padé approximants [3/3] and [4/4] for

$$f(x) = e^{\sin x}$$

12. Establish the Padé approximants [3/3] and [4/4] for

$$f(x) = e^{\tan x}$$

10.5   Padé approximants and Boundary Value Problems

In the previous section we have discussed the Padé approximants which have the advantage of manipulating the polynomial approximation into a rational functions of polynomials. By this manipulation we gain more information about the mathematical behavior of the solution. In addition, we have studied that power series are not useful for large values of x, say $x = \infty$. Boyd [28] and others have formally shown that power series in isolation are not useful to handle boundary value problems. This can be attributed to the possibility that the radius of convergence may not be sufficiently large to contain the boundaries of the domain. Based on this, it is essential to combine the series solution, obtained by the decomposition method or any series solution method, with the Padé approximants to provide an effective tool to handle boundary value problems on an infinite or semi-infinite domains. Recall that the Padé approximants can be easily evaluated by using built-in function in manipulation languages such as Maple or Mathematica.

In this section, the boundary value problems on an infinite or semi-infinite intervals will be investigated. Our approach stems mainly from the combination of the decomposition method and the diagonal approximants. The decomposition method is well addressed in the text and can be assumed known.

In what follows, we outline the basic steps to be followed for handling the boundary value problems on an unbounded domain of validity. In the first step, we use the decomposition method or the modified decomposition method to derive a series solution. In the second step, we form the diagonal Padé approximants [n/n], because it is the most accurate and efficient approximation. Recall that

$$[n/n] = \frac{a_0 + a_1 x + a_2 x^2 + \cdots + a_n x^n}{1 + b_1 x + b_2 x^2 + \cdots + b_n x^n}. \tag{113}$$

In the last step, the most effective use of the diagonal approximant is that it can be used to evaluate the limit as $x \to \infty$. In this case

$$\lim_{x \to \infty} [n/n] = \frac{a_n}{b_n}. \tag{114}$$

However, if the boundary condition at $x = \infty$ is given by

$$u(\infty) = 0, \tag{115}$$

it follows immediately that

$$a_n(\alpha) = 0, \tag{116}$$

where α is a parameter.

It is interesting to note that solving the resulting polynomial (116) frequently leads to a set of roots. It is normal to discard complex roots and other roots that do not satisfy physical properties [28]. To better approximate the root α in (116), several Padé approximants should be established where the obtained roots converge to the accurate approximation of α.

To give a clear overview of the steps introduced above, three physical and population growth models, described by ordinary differential equations, will be discussed.

Model I: The Blasius equation.

We first consider the Blasius equation

$$y''' + \frac{1}{2} y y'' = 0, \tag{117}$$

subject to the conditions

$$y(0) = 0, \; y'(0) = 1, \; y''(0) = \alpha, \; \alpha > 0, \tag{118}$$

where α can be determined by using

$$\lim_{x \to -\infty} y' = 0. \tag{119}$$

Following Adomian's method we find

$$
\begin{aligned}
y(x) = {}& x + \frac{1}{2}\alpha x^2 - \frac{1}{48}\alpha x^4 - \frac{1}{240}\alpha^2 x^5 + \frac{1}{960}\alpha x^6 \\
& + \frac{11}{20160}\alpha^2 x^7 + (\frac{11}{161280}\alpha^3 - \frac{1}{21504}\alpha)x^8 \\
& - \frac{43}{967680}\alpha^2 x^9 + (-\frac{5}{387072}\alpha^3 + \frac{1}{552960}\alpha)x^{10} \\
& + (-\frac{5}{4257792}\alpha^4 + \frac{587}{212889600}\alpha^2)x^{11} + O(x^{12}).
\end{aligned}
\tag{120}
$$

To determine the constant α, we should use the condition

$$y'(x) = 0, \text{ as } x \to -\infty. \tag{121}$$

It is clear that this condition cannot be applied directly to the series of $y'(x)$, where

$$
\begin{aligned}
y'(x) = {}& 1 + \alpha x - \tfrac{1}{12}\alpha x^3 - \tfrac{1}{48}\alpha^2 x^4 + \tfrac{1}{160}\alpha x^5 + \tfrac{11}{2880}\alpha^2 x^6 \\
& + (\tfrac{11}{20160}\alpha^3 - \tfrac{1}{2688}\alpha)x^7 - \tfrac{43}{107520}\alpha^2 x^8 \\
& + (-\tfrac{25}{193536}\alpha^3 + \tfrac{1}{55296}\alpha)x^9 \\
& + (-\tfrac{5}{387072}\alpha^4 + \tfrac{587}{19353600}\alpha^2)x^{10} + O(x^{11}).
\end{aligned}
\tag{122}
$$

As stated above, the constant α can be evaluated by establishing the Padé approximants to $y'(x)$ in (122). Using computer tools, we list the first two diagonal approximants by

$$[2/2] = \frac{12 + 9\alpha x + (1 - 3\alpha^2)x^2}{12 - 3\alpha x + x^2}, \tag{123}$$

and

$$[3/3] = \frac{(300\alpha^2 - 40) + (300\alpha^3 - 70\alpha)x - 3x^2 + (\frac{45}{4}\alpha^3 - 3\alpha)x^3}{(300\alpha^2 - 40) - 30\alpha x + (-3 + 30\alpha^2)x^2 + (\frac{25}{4}\alpha^3 - \frac{10}{3}\alpha)x^3}, \tag{124}$$

where the approximants [4/4] and [5/5] are computed but not listed. The condition (121) means that we should set the coefficient of x of highest power in the numerator polynomial to 0. In view of this, we obtain the following equations:

$$
\begin{aligned}
-3\alpha^2 + 1 &= 0, \\
45\alpha^3 - 12\alpha &= 0, \\
\tfrac{135}{112}\alpha^6 - \tfrac{189}{64}\alpha^4 + \tfrac{51}{28}\alpha^2 - \tfrac{169}{560} &= 0, \\
-\tfrac{12555}{56}\alpha^9 + \tfrac{308295}{224}\alpha^7 - \tfrac{226701}{112}\alpha^5 + \tfrac{300603}{448}\alpha^3 - \tfrac{113681}{1680}\alpha &= 0,
\end{aligned}
\tag{125}
$$

obtained from [2/2], [3/3], [4/4], and [5/5] respectively. Solving these equations independently, we list the results of the roots of α in Table 6. As stated before, complex and negative roots may be obtained for α, where $\alpha > 0$.

Table 6
Results of α by using several Padé approximants

Padé approximants	roots
[2/2]	0.577350, -0.577350
[3/3]	0, 0.516398, -0.516398
[4/4]	0.522703, -0.522703,
[5/5]	complex roots

Discarding the complex roots, and noting that $\alpha > 0$, indicates that the approximations of the roots of α converge to

$$\alpha = 0.522703. \tag{126}$$

Model II: Volterra's Population Model.

Volterra introduced a model for population growth of a species in a closed system. The model is characterized by the nonlinear integro-differential equation

$$\kappa \frac{du}{dx} = u - u^2 - u \int_0^t u(x)\, dx, \quad u(0) = .15, \tag{127}$$

where κ is a prescribed parameter. To study the mathematical behavior of the scaled population of identical individuals $u(t)$, we first set

$$y(t) = \int_0^t u(x)\, dx, \tag{128}$$

so that

$$y'(t) = u(t), \quad y''(t) = u'(t). \tag{129}$$

Substituting (128) and (129) into (127), and using $\kappa = 0.25$ for numerical purposes we find

$$y'' = 4y' - 4(y')^2 - 4yy', \quad y(0) = 0, y'(0) = 0.15. \tag{130}$$

It is interesting to point out that the population growth model (127) and the related nonlinear differential equation (130) have been investigated by using many analytic and numerical techniques such as phase-plane and Runge-Kutta methods.

However, the decomposition method will be implemented here. Applying the inverse operator L_t^{-1} to both sides of (130) and using the initial conditions we obtain

$$y(t) = 0.15t + L_t^{-1}(4y' - 4(y')^2 - 4yy'). \qquad (131)$$

Following Adomian decomposition method and proceeding as before, the series solution

$$
\begin{aligned}
u(t) \quad = \quad & 0.15 + .51t + 0.669t^2 + 0.1246t^3 - 0.85901t^4 - 1.186918t^5 \\
& +0.109061t^6 + 1.562226t^7 + 1.865912t^8 \qquad (132) \\
& -0.162759t^9 - 2.791820t^{10} + O(t^{11}),
\end{aligned}
$$

is readily obtained upon using $u(t) = y'(t)$.

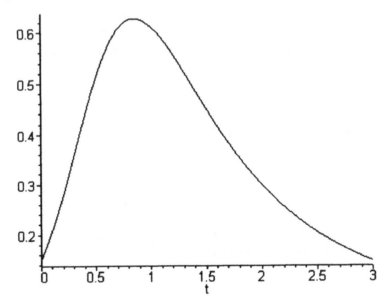

Fig. 1. The Padé approximant [4/4] shows the rapid growth followed by a slow exponential decay.

It is interesting to point out here that the focus of studies performed on this population model was on the phenomenon of the rapid growth of $u(t)$ to a certain peak along the logistic curve followed by the exponential decay as $t \to \infty$. As indicated before, the series solution (132) is not useful in isolation of other concepts. We cannot conduct the analysis to study the behavior of the solution $u(t)$ at $t = \infty$ by using the series (132). Consequently, the series (132) should be manipulated to construct several Padé approximants where the performance of the approximants show superiority over series solutions. Using computer tools we obtain the following approximants

$$[4/4] = \frac{.15 + .253224t + .169713t^2 + .051551t^3 + .005221t^4}{1 - 1.711841t + 2.491682t^2 - 1.323902t^3 + .571875t^4}, \qquad (133)$$

and

$$[5/5] =$$

$$\frac{.15 - .126408t - .468575t^2 - .371927t^3 - .119542t^4 - .010943t^5}{1 - 4.242723t + 6.841424t^2 - 7.64848t^3 + 3.946155t^4 - 1.44476t^5}, \qquad (134)$$

Figure 1 above shows the behavior of $u(t)$ and explore the rapid growth that will reach a peak followed by a slow exponential decay. This behavior cannot be obtained if we graph the converted polynomial of the series solution (132).

Model III: Thomas-Fermi Model.

The Thomas-Fermi model plays a major role in mathematical physics. This problem was developed to model the effective nuclear charge in heavy atoms. The model was introduced to investigate the potentials and charge densities of atoms having numerous electrons. The Thomas-Fermi model is characterized by the non-linear equation

$$y'' = \frac{y^{\frac{3}{2}}}{x^{\frac{1}{2}}}, \qquad (135)$$

subject to the boundary conditions

$$y(0) = 1, \ \lim_{x \to \infty} y(x) = 0. \qquad (136)$$

A considerable amount of research work has been invested in this important model. The focus of study was on obtaining an approximate solution to (135) and to determine a highly accurate value for the initial slope of the potential $y'(0)$. The importance of the initial slope $y'(0)$ is that it plays a major role in determining the energy of a neutral atom.

To avoid the cumbersome work that will arise from the radical power in $y^{\frac{3}{2}}$, the modified decomposition method will be implemented. The method will facilitate our approach and will reduce the size of calculations.

Applying L_x^{-1} to both sides of (135) gives

$$y(x) = 1 + Bx + L_x^{-1}\left(x^{-\frac{1}{2}}y^{\frac{3}{2}}\right). \qquad (137)$$

Using the decomposition assumption for $y(x)$ and $y^{\frac{3}{2}}$ yields

$$\sum_{n=0}^{\infty} y_n = 1 + Bx + L_x^{-1}\left(x^{-\frac{1}{2}}\sum_{n=0}^{\infty} A_n\right). \qquad (138)$$

The modified decomposition method introduces the use of the recursive relation of the form

$$y_0(x) = 1,$$

$$y_1(x) = Bx + L_x^{-1}\left(x^{-\frac{1}{2}}A_0\right),$$ $\qquad(139)$

$$y_{k+1}(x) = L_x^{-1}\left(x^{-\frac{1}{2}}A_k\right), \ k \geq 1.$$

This gives the first few components

$$y_0(x) = 1,$$

$$y_1(x) = Bx + L_x^{-1}\left(x^{-\frac{1}{2}}A_0\right) = Bx + \frac{4}{3}x^{\frac{3}{2}}$$

$$y_2(x) = L_x^{-1}\left(x^{-\frac{1}{2}}A_1\right) = \frac{2}{5}Bx^{\frac{5}{2}} + \frac{1}{3}x^3,$$ $\qquad(140)$

$$y_3(x) = L_x^{-1}\left(x^{-\frac{1}{2}}A_2\right) = \frac{3}{70}B^2x^{\frac{7}{2}} + \frac{2}{15}Bx^4 + \frac{2}{27}x^{\frac{9}{2}}.$$

In view of (140), the series solution is given by

$$y(x) = 1 + Bx + \frac{4}{3}x^{\frac{3}{2}} + \frac{2}{5}Bx^{\frac{5}{2}} + \frac{1}{3}x^3 + \frac{3}{70}B^2x^{\frac{7}{2}} + \frac{2}{15}Bx^4 + \frac{2}{27}x^{\frac{9}{2}} + \cdots. \quad(141)$$

To achieve our goal of studying the mathematical behavior of the potential $y(x)$ and to determine the initial slope of the potential $y'(0)$, Padé approximants of different degrees should be established. To form Padé approximants it is useful to set

$$x^{\frac{1}{2}} = t,$$ $\qquad(142)$

into (141) to obtain

$$y(t) = 1 + Bt^2 + \frac{4}{3}t^3 + \frac{2}{5}Bt^5 + \frac{1}{3}t^6 + \frac{3}{70}B^2t^7 + \frac{2}{15}Bt^8 + \cdots. \quad(143)$$

Using any manipulation language such as Maple or Mathematica we find

$$[2/2] = \frac{9B^2 - 12Bt + (9B^3 + 16)t^2}{9B^2 - 12Bt + 16t^2},$$ $\qquad(144)$

$$[4/4] = \frac{G(t)}{H(t)},$$

where

$$G(t) = 27B^4 + 97B + \left(\frac{140}{9} + \frac{33}{10}B^3\right)t + \left(\frac{675}{28}B^5 + \frac{437}{5}B^2\right)t^2$$ $\qquad(145)$

$$+ \left(\frac{453}{14}B^4 + \frac{1070}{9}B\right)t^3 + \left(-\frac{81}{28}B^6 - \frac{1096}{175}B^3 + \frac{455}{27}\right)t^4,$$

and

$$H(t) = 27B^4 + 97B + \left(\tfrac{140}{9} + \tfrac{33}{10}B^3\right)t - \left(\tfrac{81}{28}B^5 + \tfrac{48}{5}B^2\right)t^2$$
$$- \left(\tfrac{243}{35}B^4 + 26B\right)t^3 - \left(\tfrac{186}{175}B^3 + \tfrac{35}{9}\right)t^4. \tag{146}$$

Other Padé approximants are also computed. To determine the initial slope $B = y'(0)$, we use the boundary condition at $t = \infty$ given by

$$\lim_{t\to\infty} y(t) = 0, \tag{147}$$

in the established Padé approximants. This means that we should equate the coefficient of x of highest power in the numerator of each approximant by zero. The resulting values of the initial slope $B = y'(0)$ are tabulated in Table 7 as shown below.

Table 7

Approximations of the initial slope $y'(0)$

Padé approximants	Initial slope B
[2/2]	-1.211414
[4/4]	-1.550526
[7/7]	-1.586021

A better approximation has been obtained by evaluating more components of $y(x)$ and higher degree Padé approximants. Note that the accurate numerical solution of B is given by

$$B = y'(0) = -1.588071. \tag{148}$$

Exercises 10.5

Use the Adomian decomposition method and the Padé approximants to study the following models:

1. $y'' = 10y' - 10(y')^2 - 10yy'$, $y(0) = 0$, $y'(0) = .2$

2. $y'' = 5y' - 5(y')^2 - 5yy'$, $y(0) = 0$, $y'(0) = .1$

3. Find the constant α in the Flierl-Petiashvili equation:

$$u_{rr} + \tfrac{1}{r}u_r - u - u^2 = 0, \ u(0) = \alpha, \ u'(0) = 0, \ \lim_{r\to\infty} u(r) = 0$$

Chapter 11

Solitons and Compactons

11.1 Introduction

In 1834, John Scott Russell was the first to observe the solitary waves. He observed a large protrusion of water slowly traveling on the Edinburgh-Glasgow canal without change in shape. The bulge of water, that he observed and called "great wave of translation", was traveling along the channel of water for a long period of time while still retaining its shape. The remarkable discovery motivated Russell to conduct physical laboratory experiments to emphasize his observance and to study these solitary waves. He empirically derived the relation

$$c^2 = g(h + a), \tag{1}$$

that determines the speed c of the solitary wave, where a is the maximum amplitude above the water surface, h is the finite depth and g is the acceleration of gravity. The solitary waves are therfore called gravity waves.

The discovery of solitary waves inspired scientists to conduct a huge size of research work to study this concept. Two Dutchmen Korteweg and deVries derived a nonlinear partial differential equation, well known by the KdV equation, to model the height of the surface of shallow water in the presence of solitary waves. The KdV equation also describes the propagation of plasma waves in a dispersive medium. The KdV equation was introduced before in Chapter 8 where it was handled in a traditional way. The KdV equation in its simplest form is given by

$$u_t + auu_x + u_{xxx} = 0, \tag{2}$$

where it indicates that dispersion and nonlinearity might occur. The solitary wave solutions are assumed to be of the form

$$u(x, t) = f(x - ct), \tag{3}$$

387

where c is the speed of the wave propagation, and $f(z), f'(z), f''(z) \to 0$ as $z \to \pm\infty$, $z = x - ct$.

In 1965, Zabusky and Kruskal [143] investigated the interaction of solitary waves and the recurrence of initial states. They discovered that solitary waves undergo nonlinear interaction following the KdV equation. Further, the waves emerge from this interaction retaining its shape and amplitude. The remarkable discovery, that solitary waves retain its identities and that its character resembles particle like behavior, motivated Zabuski and Kruskal [143] to call these solitary waves *solitons*. Zabuski and Kruskal [143] marked the birth of the soliton, a name intended to signify particle like quantities. The interaction of two solitons emphasized the reality of the preservation of shapes and speeds and of the steady pulse like character of solitons.

A great deal of research work has been invested in recent years for the study of the soliton concept. Hirota [62] constructed the N- soliton solutions of the evolution equation by reducing it to the bilinear form. The bilinear formalism established by Hirota [62] was a very helpful tool in the study of the nonlinear equations. Nimmo and Freeman [90] introduced an alternative formulation of the N-soliton solutions in terms of some function of the Wronskian determinant of N functions.

The soliton concept appeared in the context of nonlinear lattices before it became a reality in many branches of science. Active research works have emerged worldwide in a diverse branches of scientific fields to study the soliton concept. It is now well known that solitons appear as a result of a balance between weak nonlinearity and dispersion. The soliton concept has attracted a huge size of studies due to its significant role in various scientific fields such as fluid dynamics, astrophysics and plasma physics.

For more details about the historical development of soliton and its physical structure, see the references [23,29,32,43,48,62,73,90,100,140,143] and the references therein.

Recently, in 1993, Rosenau and Hyman [100] discovered a class of solitary waves with compact support that are termed compactons. Compactons are defined by solitary waves with the remarkable soliton property that after colliding with other compactons, they reemerge with the same coherent shape [100]. These particle like waves exhibit elastic collision that are similar to the soliton interaction associated with completely integrable PDEs supporting an infinite number of conservation laws.

It was found in [100] that when the wave dispersion is purely nonlinear, some novel features may be observed and the most remarkable one is the existence of the so called compactons. The definitions given so far for compactons are:

(i) compactons are solitons with finite wavelength;

(ii) compactons are solitary waves with compact support;

(iii) compactons are solitons free of exponential tails;

(iv) compactons are solitons characterized by the absence of infinite wings;

(v) compactons are robust soliton like solutions.

Two important features of compactons structure are observed, namely:

(i) unlike the standard KdV soliton where $f(z) \to 0$ as $z \to \infty$, the compacton is characterized by the absence of the exponential tails or wings, where $f(z)$ does not tend to 0 as $z \to \infty$;

(ii) unlike the standard KdV soliton where width narrows as the amplitude increases, the width of the compacton is independent of the amplitude.

The role of nonlinear dispersion in the formation of patterns in liquid drops was investigated by Rosenau and Hyman [100]. The study in [100] was carried out by considering a genuinely nonlinear dispersive equation $K(n,n)$, a special type of the KdV equation, that was subjected to experimental and analytical studies. The remarkable discovery by Rosenau and Hyman [100] is that solitary waves may compactify under the influence of nonlinear dispersion which is capable of causing deep qualitative changes in the nature of genuinely nonlinear phenomena. It was shown that certain solutions of the $K(n,n)$ equation characterized by the absence of infinite wings can be constructed, and termed compactons. The derived results are new and of substantial interest.

The genuinely nonlinear dispersive $K(n,n)$ equations, a family of nonlinear KdV like equations is of the form

$$u_t + a(u^n)_x + (u^n)_{xxx} = 0, a > 0, n > 1, \tag{4}$$

which supports compact solitary traveling structures for $a > 0$. The existence and stability of the compact entities was examined in [51].

It is important to note that Eq. (4) with $+a$ is called the focusing branch. However, the equation

$$u_t - a(u^n)_x + (u^n)_{xxx} = 0, a > 0, n > 1, \tag{5}$$

is called the defocusing branch that was studied in [100] and in [137]. The studies revealed that Eq. (5) supports solutions with solitary patterns having cusps or infinite slopes. Further, it was shown that while compactons are the essence of the focusing branch $(+a)$, spikes, peaks and cusps are the hallmark of the defocusing branch $(-a)$. This in turn means that the focusing branch (4) and the defocusing branch (5) represent two different models, each leading to a different physical structure. The remarkable discovery of compactons has led, in turn, to an intense study over the last few years. The study of compactons may give insight into many scientific processes [78] such as the super deformed nuclei, preformation of cluster in hydrodynamic models, the fission of liquid drops and inertial fusion.

For more details about compactons, see [47,51,72,78,79,100-103,137] and the references therein.

11.2 Solitons

In this section we will study the solitary wave solutions of some of the well known nonlinear equations that exhibit solitons. It is interesting to point out that there is no precise definition of a soliton. However, a soliton can be defined as a solution of

a nonlinear partial differential equation that exhibits the following properties:
(i) the solution should demonstrate a wave of permanent form;
(ii) the solution is localized, which means that the solution either decays exponentially to zero such as the solitons provided by the KdV equation, or converges to a constant at infinity such as the solitons given by the Sine-Gordon equation;
(iii) the soliton interacts with other solitons preserving its character.

One basic expression of a solitary wave solution is of the form

$$u(x,t) = f(x - ct), \tag{6}$$

where c is the speed of wave propagation. For $c > 0$, the wave moves in the positive x direction, whereas the wave moves in the negative x direction for $c < 0$. More importantly, as will be seen later, the solutions of nonlinear equations may be a sech^2, sech, or $\arctan(e^{\alpha(x-ct)})$ function. Different methods were developed to obtain solitons. The inverse scattering transform method and the bilinear formalism were developed and implemented in [2] and [62] respectively. However, in this section we will use the direct substitution of the standard formula (6) and solve the obtained ordinary differential equation or by using Adomian decomposition method if initial condition is given.

In what follows, some of the well known nonlinear equations will be studied.

11.2.1 The KdV Equation

The nonlinear dispersive equation formulated by Korteweg and de Vries (KdV) in its simplest form is given by

$$u_t - 6uu_x + u_{xxx} = 0, \, x \in R, \tag{7}$$

with $u = u(x,t)$ is a differentiable function. We shall assume that the solution $u(x,t)$, along with its derivatives, tends to zero as $\mid x \mid \to \infty$.

Several different approaches, such as Bäcklund transformation, a bilinear form, and a Lax pair have been used independently by which soliton and multi-soliton solutions for nonlinear evolution equations are obtained.

As mentioned before, solitary wave solution can be written as

$$u(x,t) = f(x - ct), \tag{8}$$

where c is the soliton speed. Using (8) into (7) gives

$$-cf' - 6ff' + f''' = 0, z = x - ct. \tag{9}$$

Integrating (9) gives

$$-cf - 3f^2 + f'' = 0. \tag{10}$$

Multiplying (10) by $2f'$ and integrating the resulting equation we find

$$(f')^2 = cf^2 + 2f^3, \tag{11}$$

an ordinary differential equation with explicit solution

$$f(z) = -\frac{c}{2}\operatorname{sech}^2 \frac{\sqrt{c}}{2} z. \tag{12}$$

Combining (12) and (8) gives

$$u(x,t) = -\frac{c}{2}\operatorname{sech}^2 \frac{\sqrt{c}}{2}(x - ct). \tag{13}$$

It is obvious that $f(x,t)$ in (12), along with its derivatives, tends to zero as $|x| \to \infty$
In the following, Adomian decomposition method will be implemented to obtain a solitary wave solution for the KdV equation

$$\begin{aligned}
u_t - 6uu_x + u_{xxx} &= 0,\ x \in R,\\
u(x,0) &= -2\frac{k^2 e^{kx}}{(1+e^{kx})^2},
\end{aligned} \tag{14}$$

where $c = k^2$. Applying the inverse operator L_t^{-1} on both sides of (14) and using the decomposition series for $u(x,t)$ yields

$$\sum_{n=0}^{\infty} u_n(x,t) = -2\frac{k^2 e^{kx}}{(1+e^{kx})^2} + L^{-1}\left(6\left(\sum_{n=0}^{\infty} A_n\right) - \left(\sum_{n=0}^{\infty} u_n\right)_{xxx}\right). \tag{15}$$

Proceeding as before, Adomian decomposition method gives the recurrence relation

$$\begin{aligned}
u_0(x,t) &= -2\frac{k^2 e^{kx}}{(1+e^{kx})^2},\\
u_{k+1}(x,t) &= L^{-1}\left(6 A_k - u_{k_{xxx}}\right),\ k \geq 0,
\end{aligned} \tag{16}$$

that in turn gives

$$\begin{aligned}
u_0(x,t) &= -2\frac{k^2 e^{kx}}{(1+e^{kx})^2},\\
u_1(x,t) &= L^{-1}\left(6A_0 - u_{0_{xxx}}\right) = -2\frac{k^5 e^{kx}(e^{kx}-1)}{(1+e^{kx})^3} t,\\
u_2(x,t) &= L^{-1}\left(6 A_1 - u_{1_{xxx}}\right) = -\frac{k^8 e^{kx}(e^{2kx}-4e^{kx}+1)}{(1+e^{kx})^4} t^2.
\end{aligned} \tag{17}$$

In view of (17), the solution in a series form is given by

$$u(x,t) = -2\frac{k^2 e^{kx}}{(1+e^{kx})^2} - 2\frac{k^5 e^{kx}(e^{kx}-1)}{(1+e^{kx})^3} t - \frac{k^8 e^{kx}(e^{2kx}-4e^{kx}+1)}{(1+e^{kx})^4} t^2 + \cdots, \tag{18}$$

so that the exact solution

$$u(x,t) = u(x,t) = -\frac{c}{2}\operatorname{sech}^2 \frac{\sqrt{c}}{2}(x - ct), \tag{19}$$

is readily obtained.

It is worth noting that another form of the KdV equation given by

$$u_t + 6uu_x + u_{xxx} = 0 \tag{20}$$

can be proved to have the solitary wave solution

$$u(x,t) = u(x,t) = \frac{c}{2} \operatorname{sech}^2 \frac{\sqrt{c}}{2}(x - ct), \tag{21}$$

11.2.2   The Modified KdV Equation

We next consider the modified KdV (mKdV) equation of the form

$$
\begin{aligned}
u_t + 6u^2 u_x + u_{xxx} &= 0, \\
u(x,0) &= g(x).
\end{aligned}
\tag{22}
$$

We shall assume that the solution $u(x,t)$, along with its derivatives, tends to zero as $\mid x \mid \to \infty$.

Following the discussions made above, we look for a traveling wave solution in the form

$$u(x,t) = f(x - ct), \tag{23}$$

where c is the soliton speed, $z = x - ct$, $f(z)$, $f'(z)$ and $f''(z)$ tend to 0 as $\mid x \mid \to \infty$.

Substituting (23) into (22) gives

$$-cf' + 6f^2 f' + f''' = 0. \tag{24}$$

Integrating (24) gives

$$-cf + 2f^3 + f'' = A, \tag{25}$$

or equivalently

$$f'' = cf - 2f^3, \tag{26}$$

where we assumed that the constant of integration is zero. The exact solution of Eq. (22)

$$f(z) = \pm\sqrt{c}\operatorname{sech}\sqrt{c}\,z, \tag{27}$$

that gives

$$u(x,t) = \pm\sqrt{c}\operatorname{sech}\sqrt{c}\,(x - ct). \tag{28}$$

The intermediate calculations between (26) and (27) are left as an exercise.

It is interesting to point out that the KdV equation has the soliton solution in terms of sech^2 function, whereas the solution of the mKdV equation is in terms of sech function.

We next consider a generalized KdV equation of the form

$$u_t + (n+1)(n+2)u^n u_x + u_{xxx} = 0, \ x \in R, \ n = 1, 2, \cdots$$
$$u(x,0) = g(x),$$

(29)

Substituting (23) into (29) yields a differential equation for $f(z)$

$$-cf' + (n+1)(n+2)f^n f' + f''' = 0.$$

(30)

Integrating (30) gives

$$-cf + (n+2)f^{n+1} + f'' = 0,$$

(31)

so that

$$f''(z) = cf - (n+2)f^{n+1},$$

(32)

with exact solution

$$f(z) = \left(\frac{1}{2}c\,\text{sech}^2(\tfrac{1}{2}n\sqrt{c}\,z)\right)^{\frac{1}{n}}.$$

(33)

Combining (33) and (23) gives

$$u(x,t) = \left(\frac{1}{2}c\,\text{sech}^2(\tfrac{1}{2}n\sqrt{c}\,(x-ct))\right)^{\frac{1}{n}}.$$

(34)

In the following, the decomposition method will be applied for a specific modified KdV equation defined by

$$u_t + 6u^2 u_x + u_{xxx} = 0,$$
$$u(x,0) = 2\frac{ke^{kx}}{(1+e^{2kx})}.$$

(35)

with $u = u(x,t)$ is a sufficiently-often differentiable function.

Operating with L^{-1} yields

$$\sum_{n=0}^{\infty} u_n(x,t) = 2\frac{ke^{kx}}{(1+e^{2kx})} - L^{-1}\left(6\left(\sum_{n=0}^{\infty} A_n\right) + \left(\sum_{n=0}^{\infty} u_n\right)_{xxx}\right).$$

(36)

Adomian's method admits the use recurrence relation

$$u_0(x,t) = 2\frac{ke^{kx}}{(1+e^{2kx})},$$
$$u_{k+1}(x,t) = -L^{-1}\left(6A_k + u_{k_{xxx}}\right), k \geq 0,$$

(37)

that in turn gives

$$u_0(x,t) = 2\frac{ke^{kx}}{(1+e^{2kx})},$$

$$u_1(x,t) = -L^{-1}\left(6A_0 + u_{0_{xxx}}\right) = -2\frac{k^4 e^{kx}(1-e^{2kx})}{(1+e^{2kx})^2}t, \tag{38}$$

$$u_2(x,t) = -L^{-1}\left(6\,A_1 + u_{1_{xxx}}\right) = \frac{k^7 e^{kx}(1-6e^{2kx}+e^{4kx})}{(1+e^{2kx})^3}t^2.$$

The solution in a series form is given by

$$u(x,t) = 2\frac{ke^{kx}}{(1+e^{2kx})} - 2\frac{k^4 e^{kx}(1-e^{2kx})}{(1+e^{2kx})^2}t + \frac{k^7 e^{kx}(1-6e^{2kx}+e^{4kx})}{(1+e^{2kx})^3}t^2 + \cdots, \tag{39}$$

so that the exact solution

$$u(x,t) = u(x,t) = \pm\sqrt{c}\,\mathrm{sech}\sqrt{c}\,(x-ct), \tag{40}$$

is readily obtained noting that $c = k^2$.

11.2.3   The Sine-Gordon Equation

The Sine-Gordon equation is given by

$$u_{tt} - u_{xx} + \sin u = 0. \tag{41}$$

The sine-Gordon equation arises in the study of superconductor transmission lines, crystals, geometry of surfaces, laser pulses, pendula motions, and in the propagation of magnetic flux.

To determine solitary wave solutions of Eq. (41) we let

$$u(x,t) = f(x - ct), \tag{42}$$

that carries the Sine-Gordon equation into

$$(c^2 - 1)f'' f' + (\sin f)f' = 0, \tag{43}$$

obtained after multiplying it by f', noting that $z = x - ct$. Integrating (43) gives the first order equation

$$\frac{1}{2}(c^2 - 1)(f')^2 - \cos f = C, \tag{44}$$

where C is a constant of integration. It is appropriate to select $C = -1$ so that f approaches zero as z approaches infinity. This means that Eq. (44) becomes

$$(f')^2 = \frac{4}{1 - c^2}\sin^2(\frac{f}{2}),\ |c| < 1. \tag{45}$$

We can easily prove that one solution of Eq. (45) is given by

$$f(z) = 4\arctan\left[\exp\left(-\frac{z}{\sqrt{1-c^2}}\right)\right],\tag{46}$$

so that the solitary wave solution is

$$u(x,t) = 4\arctan\left[\exp\left(-\frac{x-ct}{\sqrt{1-c^2}}\right)\right].\tag{47}$$

Recall that the solitary wave solutions of the KdV and the modified KdV equations are given by sech^2 and sech function. The solution obtained in (47) shows that the solitary wave solution in terms of $\arctan(e^{\alpha z})$. Moreover, we can easily observe that $u(x,t) \to 0$ as $x \to \infty$, and $u(x,t) \to 2\pi$ as $x \to -\infty$. A solution for which $u(x,t)$ increases by 2π is called a *kink*, and one which decreases by 2π an *antikink*. Another solution of the Sine-Gordon equation can be derived in the form

$$u(x,t) = 4\arctan\left[\frac{\sinh(\frac{ct}{\sqrt{1-c^2}})}{c\cosh(\frac{x}{\sqrt{1-c^2}})}\right].\tag{48}$$

11.2.4   The Boussinesq Equation

A well known model of nonlinear dispersive waves was proposed by Boussinesq in the form

$$u_{tt} = u_{xx} + 3(u^2)_{xx} + u_{xxxx}, \; a \leq x \leq b.\tag{49}$$

The Boussinesq equation (1) describes motions of long waves in shallow water under gravity and in a one-dimensional nonlinear lattice. Using $u = f(z), z = (x-ct)$ into (49) gives

$$(c^2 - 1)f'' = 3(f^2)'' + f^{(iv)},\tag{50}$$

where integrating twice yields

$$(c^2 - 1)f - 3f^2 = f''.\tag{51}$$

Multiplying both sides of (51) by $2f'$ and integrating gives

$$u(x,t) = \frac{c}{2}\mathrm{sech}^2\left[\frac{\sqrt{c}}{2}x + \frac{\sqrt{c}}{2}\sqrt{1+c}\,t\right].\tag{52}$$

The intermediate calculations between (51) and (52) are left an exercise.

In what follows we will use the modified decomposition method to determine the solitary wave solutions of a specific form of the Boussinesq equation, defined by

$$\begin{aligned} u_{tt} &= u_{xx} + 3(u^2)_{xx} + u_{xxxx}, \; -80 \leq x \leq 80, \\ u(x,0) &= 2\frac{ak^2e^{kx}}{(1+ae^{kx})^2}, \; u_t(x,0) = -2\frac{ak^3\sqrt{1+k^2}e^{kx}(ae^{kx}x-1)}{(1+ae^{kx})^3}. \end{aligned}\tag{53}$$

Applying the inverse operator L_t^{-1} yields

$$\sum_{n=0}^{\infty} u_n(x,t) = 2\frac{ak^2 e^{kx}}{(1+ae^{kx})^2} - 2\frac{ak^3\sqrt{1+k^2}e^{kx}(ae^{kx}x-1)}{(1+ae^{kx})^3} t$$
$$+ L_t^{-1}\left(3\left(\sum_{n=0}^{\infty} A_n\right) + \left(\sum_{n=0}^{\infty} u_n\right)_{xx} + \left(\sum_{n=0}^{\infty} u_n\right)_{xxxx}\right). \tag{54}$$

The modified decomposition method gives the recurrence relation

$$u_0(x,t) = 2\frac{ak^2 e^{kx}}{(1+ae^{kx})^2},$$

$$u_1(x,t) = -2\frac{ak^3\sqrt{1+k^2}e^{kx}(ae^{kx}x-1)}{(1+ae^{kx})^3} t + L_t^{-1}\left(3A_0 + u_{0_{xx}} + u_{0_{xxxx}}\right), \tag{55}$$

$$u_{k+1}(x,t) = L_t^{-1}\left(3A_k + u_{k_{xx}} + u_{k_{xxxx}}\right), k \geq 1.$$

Consequently, we obtain

$$u_0(x,t) = 2\frac{ak^2 e^{kx}}{(1+ae^{kx})^2},$$

$$u_1(x,t) = -2\frac{ak^3\sqrt{1+k^2}e^{kx}(ae^{kx}x-1)}{(1+ae^{kx})^3} t + L_t^{-1}\left(3A_0 + u_{0_{xx}} + u_{0_{xxxx}}\right), \tag{56}$$

$$= -2\frac{ak^3\sqrt{1+k^2}e^{kx}(ae^{kx}x-1)}{(1+ae^{kx})^3} t + \frac{ak^4 e^{kx}(1+k^2)(a^2 e^{2kx}-4ae^{kx}+1)}{(1+ae^{kx})^4} t^2,$$

where u_2 and u_3 are determined but not listed. In view of (56), the solution in a series form is given by

$$u(x,t) = 2\frac{ak^2 e^{kx}}{(1+ae^{kx})^2} - 2ak^3\sqrt{1+k^2}e^{kx}\frac{(ae^{kx}x-1)}{(1+ae^{kx})^3} t$$
$$+ ak^4(1+k^2)e^{kx}\frac{(a^2 e^{2kx}-4ae^{kx}+1)}{(1+ae^{kx})^4} t^2 + \cdots, \tag{57}$$

and consequently, we find that the exact solution is

$$u(x,t) = 2\frac{ak^2 e^{kx+k\sqrt{1+k^2}\,t}}{(1+ae^{kx+k\sqrt{1+k^2}\,t})^2}, \tag{58}$$

or equivalently

$$u(x,t) = \frac{ak^2}{2}\mathrm{sech}^2\left[\frac{kx}{2} + \frac{k}{2}\sqrt{1+k^2}\,t\right], \tag{59}$$

where $c = k^2$.

11.2.5 The Kadomtsev-Petviashvili Equation

In 1970, Kadomtsev and Petviashvili generalized the KdV equation to two space variables and formulated the well-known Kadomtsev-Petviashvili equation to provide an explanation of the general weakly dispersive waves. The KP equation is of the form

$$(u_t - 6uu_x + u_{xxx})_x + 3u_{yy} = 0. \tag{60}$$

Adomian's method will be used to solve the specific KP equation

$$u_{xt} - 6u_x^2 - 6uu_{xx} + u_{xxxx} + 3u_{yy} = 0, \tag{61}$$

with the initial condition

$$u(x, y, 0) = \frac{-8e^{2x+2y}}{(1 + e^{2x+2y})^2}, \tag{62}$$

and the boundary conditions are zero at the boundary. Operating with the inverse operator L_{xt}^{-1}

$$L_{xt}^{-1}(.) = \int_{-80}^{x} \int_{0}^{t} (.)dt \, dx, \tag{63}$$

gives the relation

$$\sum_{n=0}^{\infty} u_n(x, y, t) = \frac{-8e^{2x+2y}}{(1 + e^{2x+2y})^2}$$
$$+ L_{xt}^{-1} \left(6 \left(\sum_{n=0}^{\infty} A_n \right) + 6 \left(\sum_{n=0}^{\infty} B_n \right) - \left(\sum_{n=0}^{\infty} u_n \right)_{xxxx} - 3 \left(\sum_{n=0}^{\infty} u_n \right)_{yy} \right), \tag{64}$$

where A_n and B_n are Adomian polynomials for u_x^2 and uu_{xx} respectively. This means that the first few components are derived as follows:

$$
\begin{aligned}
u_0(x, y, t) &= -\frac{8e^{2x+2y}}{(1+e^{2x+2y})^2}, \\
u_1(x, y, t) &= L_{xt}^{-1} \left(6A_0 + 6B_0 - (u_0)_{xxxx} - 3(u_0)_{yy} \right), \\
&= -112 \left(\frac{(-1+e^{2x+2y})e^{2x+2y}}{(1+e^{2x+2y})^3} + 112 \frac{(-1+e^{-160+2y})e^{-160+2y}}{(1+e^{-160+2y})^3} \right) t.
\end{aligned}
\tag{65}
$$

The solution in series form is given by

$$
\begin{aligned}
u(x, y, t) &= \frac{-8e^{2x+2y}}{(1+e^{2x+2y})^2} \\
&\quad - 112 \left(\frac{(-1+e^{2x+2y})e^{(2x+2y)}}{(1+e^{2x+2y})^3} + 112 \frac{(-1+e^{-160+2y})e^{-160+2y}}{(1+e^{-160+2y})^3} \right) t,
\end{aligned}
\tag{66}
$$

so that the exact single soliton solution is given by

$$u(x, y, t) = -2\text{sech}^2(x + y - 7t), \tag{67}$$

or equivalently

$$u(x, y, t) = -\frac{8e^{2x+2y-14t}}{(1 + e^{2x+2y-14t})^2}. \tag{68}$$

Exercises 11.2

Use the decomposition method or the substitution $u = f(x - ct)$ to find the solitary wave solutions of the following nonlinear problems:

1. $u_t - 6uu_x + u_{xxx} = 0$, $u(x,0) = -2\operatorname{sech}^2(x)$

2. $u_t + 6uu_x + u_{xxx} = 0$, $u(x,0) = 8\operatorname{sech}^2(2x)$

3. $u_t + 12u^2 u_x + u_{xxx} = 0$, $u(x,0) = \sqrt{2}\operatorname{sech}(2x)$

4. $u_{tt} - u_{xx} + \sin u = 0$, $u(x,0) = 4\arctan(e^{-2x})$

5. $(u_t + 6uu_x + u_{xxx})_x + 3u_{yy} = 0$, $u(x,y,0) = \frac{1}{2}\operatorname{sech}^2(\frac{1}{2}(x+y))$

11.3 Compactons

In 1993, Rosenau and Hyman [100] introduced a class of solitary waves with compact support that are termed *compactons*. Compactons can be defined as solitons with finite wave length or solitons free of exponential tails. In other words, compactons are solitons characterized by the absence of infinite wings and, unlike solitons, the width of the compacton is independent of the amplitude. Rosenau and Hyman [100] discovered that solitary waves may compactify under the influence of nonlinear dispersion which is capable of causing deep qualitative changes in the nature of genuinely nonlinear phenomena. Compactons were proved to collide elastically and reemerge with the same coherent shape. Such solitary wave solutions, which vanish outside a finite core region, are solutions of a two parameter family of genuinely nonlinear dispersive equations $K(n, n)$;

$$u_t + (u^n)_x + (u^n)_{xxx} = 0, n > 1. \tag{69}$$

As stated before, solitons appear as a result of a balance between dispersion and weak nonlinearity. However, when the wave dispersion is purely nonlinear, some novel features may be observed. The most interesting feature of the nonlinear dispersion is the existence of the so-called compactons: solitons with finite wavelength or solitons without exponential tails.

In the following we will discuss the genuinely nonlinear modified forms of the KdV equation and the compactons solutions will be developed.

11.3.1 The K(n,n) Equations in Higher Dimensions

Unlike solitons, compactons are nonanalytic solutions. The points of non analyticity at the compacton edge are related to points of genuine nonlinearity of the equation. In addition, it was shown that in [100] that the inverse scattering tools are inapplicable. The pseudo spectral method was used to obtain the compactons solutions:

$$u(x,t) = \begin{cases} \{\sqrt{\frac{2cn}{n+1}}\cos[\frac{n-1}{2n}(x-ct)]\}^{\frac{2}{n-1}}, & |x-ct| \leq \frac{n\pi}{(n-1)}, n > 1, \\ 0 & \text{otherwise}. \end{cases} \tag{70}$$

However, in [135] an additional general formula for the compactons solutions was derived in the form

$$u(x,t) = \begin{cases} \{\sqrt{\frac{2cn}{n+1}}\sin[\frac{n-1}{2n}(x-ct)]\}^{\frac{2}{n-1}}, & |x-ct| \leq \frac{2n\pi}{(n-1)}, n > 1, \\ 0 & \text{otherwise}. \end{cases} \tag{71}$$

As a result of the work in [100], the focusing branches $(+a)$ of $K(n,n)$ equations

$$u_t + a(u^n)_x + b(u^n)_{xxx} = 0, \tag{72}$$

$$u_t + a(u^n)_x + b(u^n)_{xxx} + k(u^n)_{yyy} = 0, \tag{73}$$

$$u_t + a(u^n)_x + b(u^n)_{xxx} + k(u^n)_{yyy} + r(u^n)_{zzz} = 0, \tag{74}$$

will be investigated here, where $a, b, k, r > 0$ are constants.

The One Dimensional Focusing Branch

Consider the nonlinear dispersive equation

$$u_t + a(u^n)_x + b(u^n)_{xxx} = 0, a, b > 0. \tag{75}$$

Following the discussions in [100] and [135], we assume that the general solution of Eq. (75) is of the form

$$u(x,t) = \rho \sin^{\frac{2}{n-1}}[\sigma(x-ct)], \tag{76}$$

or of the form

$$u(x,t) = \rho \cos^{\frac{2}{n-1}}[\sigma(x-ct)], \tag{77}$$

where ρ and σ are constants that will be determined. Substituting these assumptions into (75) and by solving the resulting equations for ρ and σ we find

$$\sigma = \pm \frac{(n-1)}{2n}\sqrt{\frac{a}{b}},$$

$$\rho = \begin{cases} \left(\frac{2nc}{a(n+1)}\right)^{\frac{1}{n-1}} & n \text{ is even}, \\ \pm\left(\frac{2nc}{a(n+1)}\right)^{\frac{1}{n-1}} & n \text{ is odd}. \end{cases} \tag{78}$$

Consequently, we find the following sets of general compactons solutions:

1. For n even, the general compactons solutions are given by:

$$u(x,t) = \begin{cases} \{\sqrt{\frac{2nc}{a(n+1)}} \sin[\frac{(n-1)}{2n}\sqrt{\frac{a}{b}}(x-ct)]\}^{\frac{2}{n-1}}, & |x-ct| \leq \frac{2n\pi}{\sigma}, \\ \\ 0 & \text{otherwise}. \end{cases} \tag{79}$$

and

$$u(x,t) = \begin{cases} \{\sqrt{\frac{2nc}{a(n+1)}} \cos[\frac{(n-1)}{2n}\sqrt{\frac{a}{b}}(x-ct)]\}^{\frac{2}{n-1}}, & |x-ct| \leq \frac{n\pi}{\sigma}, \\ \\ 0 & \text{otherwise}. \end{cases} \tag{80}$$

2. For n odd, the compactons and anticompactons solutions are defined by

$$u(x,t) = \begin{cases} \pm\{\sqrt{\frac{2nc}{a(n+1)}} \sin[\frac{(n-1)}{2n}\sqrt{\frac{a}{b}}(x-ct)]\}^{\frac{2}{n-1}}, & |x-ct| \leq \frac{2n\pi}{\sigma}, \\ \\ 0 & \text{otherwise}. \end{cases} \tag{81}$$

and

$$u(x,t) = \begin{cases} \pm\{\sqrt{\frac{2nc}{a(n+1)}} \cos[\frac{(n-1)}{2n}\sqrt{\frac{a}{b}}(x-ct)]\}^{\frac{2}{n-1}}, & |x-ct| \leq \frac{n\pi}{\sigma}, \\ \\ 0 & \text{otherwise}. \end{cases} \tag{82}$$

In a similar manner to the analysis introduced above we will investigate the focusing branch of a two dimensional nonlinear dispersive equation.

The Two Dimensional Focusing Branch

We next consider the equation

$$u_t + a(u^n)_x + b(u^n)_{xxx} + k(u^n)_{yyy} = 0, \; n > 1, \tag{83}$$

where $u \equiv u(x,y,t)$, and $a,b,k > 0$ are constants.

It is reasonable to assume that the general compacton solution of the dispersive equation (83) is of the form:

$$u(x,y,t) = \alpha \sin^{\frac{2}{n-1}}[\beta(x+y-ct)]. \tag{84}$$

or of the form

$$u(x,y,t) = \alpha \cos^{\frac{2}{n-1}}[\beta(x+y-ct)], \tag{85}$$

where α and β are constants that will be determined later. Proceeding as before we find

$$\beta = \pm\frac{(n-1)}{2n}\sqrt{\frac{a}{b+k}},$$

$$\alpha = \begin{cases} \left(\frac{2nc}{a(n+1)}\right)^{\frac{1}{n-1}} & n \text{ is even}, \\ \\ \pm\left(\frac{2nc}{a(n+1)}\right)^{\frac{1}{n-1}} & n \text{ is odd}. \end{cases} \tag{86}$$

Consequently, we obtain the following set of the general compactons solutions:
1. For n even: the general compactons solutions are given by:

$$u(x,y,t) = \begin{cases} \{\sqrt{\frac{2nc}{a(n+1)}}\sin[\frac{(n-1)}{2n}\sqrt{\frac{a}{b+k}}(\phi)]\}^{\frac{2}{n-1}} & |\phi| \le \frac{2n\pi}{\beta}, \\ \\ 0 & \text{otherwise}, \end{cases} \tag{87}$$

and

$$u(x,y,t) = \begin{cases} \{\sqrt{\frac{2nc}{a(n+1)}}\cos[\frac{(n-1)}{2n}\sqrt{\frac{a}{b+k}}(\phi)]\}^{\frac{2}{n-1}} & |\phi| \le \frac{n\pi}{\beta}, \\ \\ 0 & \text{otherwise}, \end{cases} \tag{88}$$

where $\phi = x + y - ct$. 2. For n odd: the general compactons and anticompactons solutions are of the form

$$u(x,y,t) = \begin{cases} \pm\{\sqrt{\frac{2nc}{a(n+1)}}\sin[\frac{(n-1)}{2n}\sqrt{\frac{a}{b+k}}(\phi)]\}^{\frac{2}{n-1}} & |\phi| \le \frac{2n\pi}{\beta}, \\ \\ 0 & \text{otherwise}, \end{cases} \tag{89}$$

and

$$u(x,y,t) = \begin{cases} \{\sqrt{\frac{2nc}{a(n+1)}}\cos[\frac{(n-1)}{2n}\sqrt{\frac{a}{b+k}}(\phi)]\}^{\frac{2}{n-1}} & |\phi| \le \frac{n\pi}{\beta}, \\ \\ 0 & \text{otherwise}, \end{cases} \tag{90}$$

where $\phi = x + y - ct$.

In what follows, the focusing branch in a three dimensional space will be examined.

The Three Dimensional Focusing Branch

We next consider the equation

$$u_t + a(u^n)_x + b(u^n)_{xxx} + k(u^n)_{yyy} + r(u^n)_{zzz} = 0, \, n > 1, \tag{91}$$

where $u \equiv u(x, y, z, t)$, and $a, b, k, r > 0$ are constants. Following the analysis presented above leads to

1. For n even, we found

$$u(x,y,z,t) = \begin{cases} \{\sqrt{\frac{2nc}{a(n+1)}}\sin[\frac{(n-1)}{2n}\sqrt{\frac{a}{b+k+r}}(\psi)]\}^{\frac{2}{n-1}} & |\psi| \leq \frac{2n\pi}{\lambda}, \\ \\ 0 & \text{otherwise,} \end{cases} \tag{92}$$

and

$$u(x,y,z,t) = \begin{cases} \{\sqrt{\frac{2nc}{a(n+1)}}\cos[\frac{(n-1)}{2n}\sqrt{\frac{a}{b+k+r}}(\psi)]\}^{\frac{2}{n-1}} & |\psi| \leq \frac{n\pi}{\lambda}, \\ \\ 0 & \text{otherwise,} \end{cases} \tag{93}$$

where $\psi = x + y + z - ct$, and $\lambda = \frac{(n-1)}{2n}\sqrt{\frac{a}{b+k+r}}$.

2. For n odd, we obtained

$$u(x,y,z,t) = \begin{cases} \pm\{\sqrt{\frac{2nc}{a(n+1)}}\sin[\frac{(n-1)}{2n}\sqrt{\frac{a}{b+k+r}}(\psi)]\}^{\frac{2}{n-1}} & |\psi| \leq \frac{2n\pi}{\lambda}, \\ \\ 0 & \text{otherwise,} \end{cases} \tag{94}$$

and

$$u(x,y,z,t) = \begin{cases} \pm\{\sqrt{\frac{2nc}{a(n+1)}}\cos[\frac{(n-1)}{2n}\sqrt{\frac{a}{b+k+r}}(\psi)]\}^{\frac{2}{n-1}} & |\psi| \leq \frac{n\pi}{\lambda}, \\ \\ 0 & \text{otherwise,} \end{cases} \tag{95}$$

where $\psi = x + y + z - ct$.

This completes the analysis for handling the nonlinear dispersive equation $K(n,n)$ for all values of $n > 1$ in higher dimensional spaces. In what follows, we will examine two specific examples of this equation.

Examples

For specific values of n, we apply Adomian decomposition method to the nonlinear dispersive equations. For this purpose, we have chosen to present two test problems, namely $K(2,2)$ and $K(3,3)$.

Example 1. We first consider the initial value problem $K(2,2)$

$$\begin{aligned} u_t + (u^2)_x + (u^2)_{xxx} &= 0, \\ u(x,0) &= \tfrac{4}{3}c\cos^2(\tfrac{1}{4}x). \end{aligned} \tag{96}$$

Solution.

Following Adomian analysis we find

$$u(x,t) = \frac{4}{3}c\cos^2\left(\frac{1}{4}x\right) - L_t^{-1}\left((u^2)_x + (u^2)_{xxx}\right). \tag{97}$$

Substituting the decomposition series for $u(x,t)$ into (97) gives

$$\sum_{n=0}^{\infty} u_n(x,t) = \frac{4}{3}c\cos^2\left(\frac{1}{4}x\right) - L^{-1}\left(\sum_{n=0}^{\infty} A_n + \sum_{n=0}^{\infty} B_n\right) \tag{98}$$

where A_n and B_n are Adomian polynomials that represent the nonlinear operators $(u^2)_x$ and $(u^2)_{xxx}$ respectively. In view of (98), the decomposition technique admits the use of the recursive relation

$$\begin{aligned} u_0(x,t) &= \frac{4}{3}c\cos^2\left(\frac{1}{4}x\right), \\ u_{k+1}(x,t) &= -L^{-1}\left(A_k + B_k\right), \ k \geq 0. \end{aligned} \tag{99}$$

The Adomian polynomials A_n and B_n for $(u^2)_x$ and $(u^2)_{xxx}$ are given by

$$\begin{aligned} A_0 &= F(u_0) = (u_0^2)_x, \\ A_1 &= u_1 F'(u_0) = (2u_1 u_0)_x, \\ A_2 &= u_2 F'(u_0) + \tfrac{1}{2}u_1^2 F''(u_0) = (2u_2 u_0 + u_1^2)_x, \end{aligned} \tag{100}$$

and

$$\begin{aligned} B_0 &= G(u_0) = (u_0^2)_{xxx}, \\ B_1 &= u_1 G'(u_0) = (2u_1 u_0)_{xxx}, \\ B_2 &= u_2 G'(u_0) + \tfrac{1}{2}u_1^2 G''(u_0) = (2u_2 u_0 + u_1^2)_{xxx}. \end{aligned} \tag{101}$$

This in turn gives

$$\begin{aligned} u_0(x,t) &= \tfrac{4}{3}c\cos^2\left(\tfrac{1}{4}x\right), \\ u_1(x,t) &= -L^{-1}(A_0 + B_0) = \tfrac{1}{3}c^2 t \sin\left(\tfrac{1}{2}x\right), \\ u_2(x,t) &= -L^{-1}(A_1 + B_1) = -\tfrac{1}{12}c^3 t^2 \cos\left(\tfrac{1}{2}x\right), \\ u_3(x,t) &= -L^{-1}(A_2 + B_2) = -\tfrac{1}{72}c^4 t^3 \sin\left(\tfrac{1}{2}x\right). \end{aligned} \tag{102}$$

The solution in a series form

$$u(x,t) = \frac{4}{3}c\cos^2\left(\frac{1}{4}x\right) + \frac{1}{3}c^2 t \sin\left(\frac{1}{2}x\right) - \frac{1}{12}c^3 t^2 \cos\left(\frac{1}{2}x\right) - \frac{1}{72}c^4 t^3 \sin\left(\frac{1}{2}x\right) + \cdots, \tag{103}$$

follows immediately, and as a result, the closed form solution

$$u(x,t) = \begin{cases} \frac{4}{3}c\cos^2(\frac{1}{4}(x-ct)), & |x-ct| \le 2\pi, \\ 0 & \text{otherwise}, \end{cases} \tag{104}$$

is readily obtained.

Example 2. We now consider the initial value problem $K(3,3)$

$$\begin{aligned} u_t + (u^3)_x + (u^3)_{xxx} &= 0, \\ u(x,0) &= \sqrt{\frac{3c}{2}}\cos(\frac{1}{3}x). \end{aligned} \tag{105}$$

Solution.

Following the analysis presented above we obtain

$$u(x,t) = \sqrt{\frac{3c}{2}}\cos(\frac{1}{3}x) - L^{-1}\left((u^3)_x + (u^3)_{xxx}\right). \tag{106}$$

Using the decomposition series assumption for $u(x,t)$ gives

$$\sum_{n=0}^{\infty} u_n(x,t) = \sqrt{\frac{3c}{2}}\cos(\frac{1}{3}x) - L^{-1}\left(\sum_{n=0}^{\infty}\tilde{A}_n + \sum_{n=0}^{\infty}\tilde{B}_n\right) \tag{107}$$

where $\tilde{A}_n$ and $\tilde{B}_n$ are Adomian polynomials that represent the nonlinear operators $(u^3)_x$ and $(u^3)_{xxx}$ respectively. In view of (107), we use the recursive relation

$$\begin{aligned} u_0(x,t) &= \sqrt{\frac{3c}{2}}\cos(\frac{1}{3}x), \\ u_{k+1}(x,t) &= -L^{-1}\left(\tilde{A}_k + \tilde{B}_k\right), \ k \ge 0. \end{aligned} \tag{108}$$

Adomian polynomials $\tilde{A}_n$ and $\tilde{B}_n$ can be calculated as before to find

$$\tilde{A}_0 = (u_0^3)_x, \ \ \tilde{A}_1 = (3u_1u_0^2)_x, \ \ \tilde{A}_2 = (3u_2u_0^2 + 3u_0u_1^2)_x, \tag{109}$$

and

$$\tilde{B}_0 = (u_0^3)_{xxx}, \ \ \tilde{B}_1 = (3u_1u_0^2)_{xxx}, \ \ \tilde{B}_2 = (3u_2u_0^2 + 3u_0u_1^2)_{xxx}. \tag{110}$$

This gives

$$\begin{aligned} u_0(x,t) &= \frac{\sqrt{6c}}{2}\cos(\frac{1}{3}x), \\ u_1(x,t) &= -L^{-1}(\tilde{A}_0 + \tilde{B}_0) = \frac{\sqrt{6c^3}}{6}t\sin(\frac{1}{3}x), \\ u_2(x,t) &= -L^{-1}(\tilde{A}_1 + \tilde{B}_1) = -\frac{\sqrt{6c^5}}{36}t^2\cos(\frac{1}{3}x), \\ u_3(x,t) &= -L^{-1}(\tilde{A}_2 + \tilde{B}_2) = -\frac{\sqrt{6c^7}}{324}t^3\sin(\frac{1}{3}x). \end{aligned} \tag{111}$$

The solution in a series form is given by

$$u(x,t) = \frac{\sqrt{6c}}{2}\cos(\frac{1}{3}x) + \frac{\sqrt{6c^3}}{6}t\sin(\frac{1}{3}x) - \frac{\sqrt{6c^5}}{36}t^2\cos(\frac{1}{3}x) - \cdots, \tag{112}$$

and in a closed form is given by

$$u(x,t) = \begin{cases} \frac{\sqrt{3c}}{2}\cos(\frac{1}{3}(x-ct)), & \mid x-ct \mid \le \frac{3\pi}{2}, \\ \\ 0 & \text{otherwise} \end{cases} \tag{113}$$

It is interesting to point out that if we considered the initial condition with a negative sign, then an identical solution to (113) but with an opposite sign is obtainable. The solution in (113) represents compacton, and with the negative solution represents the anticompacton.

Exercises 11.3

Use the decomposition method or any other method to find the compactons solutions for the following nonlinear dispersive equations:

1. $u_t + (u^3)_x + (u^3)_{xxx} = 0$, $u(x,0) = 3\cos(\frac{1}{3}x)$

2. $u_t + (u^4)_x + (u^4)_{xxx} = 0$, $u(x,0) = \left(2\sin(\frac{3}{8}x)\right)^{\frac{2}{3}}$

3. $u_t + (u^3)_x + (u^3)_{xxx} + (u^3)_{yyy} = 0$, $u(x,y,0) = 3\cos(\frac{1}{3\sqrt{2}}(x+y))$

4. $\frac{1}{2}(u^2)_t + (u^2)_x + (u^2)_{xxx} + u^2_{xxxxx} = 0$, $u(x,0) = \sqrt{\cos x}$

5. $\frac{1}{2}(u^2)_t + (u^2)_x + (u^2)_{xxx} + u^2_{xxxxx} + (u^2)_{yyyyy} = 0$, $u(x,y,0) = \sqrt{\cos(\frac{1}{\sqrt{2}}(x+y))}$

11.4 The Defocusing Branch of K(n,n)

As indicated before the defocusing branch

$$u_t - (u^n)_x + (u^n)_{xxx} = 0, \ n > 1, \tag{114}$$

where $a = -1$, was examined in [100] and in [135]. It was revealed in these studies that solutions with solitary patterns having cusps or infinite slopes arise from the nonlinear dispersive equation of the form given in (114).

In [100], it was concluded that while compactons are the essence of the focusing $(+a)$ branch, spikes, peaks and cusps are the hallmark of the defocusing $(-a)$ branch which also supports the motion of kinks. Further, it can be easily observed that the map $(x,t) \rightarrow (ix,-it)$ maps the positive branch into the negative branch and vice versa. Furthermore, the focusing branch and the defocusing branch represent two different sets of models, each leading to a different physical structure.

In this section, exact solutions will be developed for the defocusing branches of the dispersive equations $K(n,n)$ in one, two and three dimensional spaces given by

$$u_t - a(u^n)_x + b(u^n)_{xxx} = 0, \tag{115}$$

$$u_t - a(u^n)_x + b(u^n)_{xxx} + k(u^n)_{yyy} = 0, \tag{116}$$

$$u_t - a(u^n)_x + b(u^n)_{xxx} + k(u^n)_{yyy} + r(u^n)_{zzz} = 0, \tag{117}$$

where $a, b, k, r > 0$ are constants.

The One Dimensional Defocusing Branch

In the analysis that follows, we consider the defocusing $K(n,n)$ equation given by

$$u_t - a(u^n)_x + b(u^n)_{xxx} = 0, a, b > 0. \tag{118}$$

In view of the works in [100] and [135], it is natural to seek a general solution of the dispersive $K(n,n)$ equation (118 in the form

$$u(x,t) = \rho \sinh^{\frac{2}{n-1}}[\sigma(x - ct)], \tag{119}$$

or in the form

$$u(x,t) = \rho \cosh^{\frac{2}{n-1}}[\sigma(x - ct)], \tag{120}$$

where ρ and σ are constants that will be determined. To determine the general formula for the solution, we substitute (119) or (120) into (118), and by solving the resulting equations for ρ and σ, it then follows that
1. For the sinh profile:

$$\sigma = \pm \frac{(n-1)}{2n}\sqrt{\frac{a}{b}},$$

$$\rho = \begin{cases} \left(\frac{2nc}{a(n+1)}\right)^{\frac{1}{n-1}} & n \text{ is even,} \\ \\ \pm\left(\frac{2nc}{a(n+1)}\right)^{\frac{1}{n-1}} & n \text{ is odd.} \end{cases} \tag{121}$$

2. For the cosh profile:

$$\sigma = \pm \frac{(n-1)}{2n}\sqrt{\frac{a}{b}},$$

$$\rho = \begin{cases} -\left(\frac{2nc}{a(n+1)}\right)^{\frac{1}{n-1}} & n \text{ is even.} \\ \\ \pm\left(\frac{-2nc}{a(n+1)}\right)^{\frac{1}{n-1}} & n \text{ is odd.} \end{cases} \tag{122}$$

This gives the following sets of general solutions:
1. For n even, the general solutions are given by:

$$u(x,t) = \{\sqrt{\frac{2nc}{a(n+1)}}\sinh[\frac{(n-1)}{2n}\sqrt{\frac{a}{b}}(x-ct)]\}^{\frac{2}{n-1}}, \tag{123}$$

and

$$u(x,t) = -\{\sqrt{\frac{2nc}{a(n+1)}}\cosh[\frac{(n-1)}{2n}\sqrt{\frac{a}{b}}(x-ct)]\}^{\frac{2}{n-1}}. \tag{124}$$

2. For n odd, the solutions are defined by

$$u(x,t) = \pm\{\sqrt{\frac{2nc}{a(n+1)}}\sinh[\frac{(n-1)}{2n}\sqrt{\frac{a}{b}}(x-ct)]\}^{\frac{2}{n-1}}, \text{ for } c > 0, \tag{125}$$

and

$$u(x,t) = \pm\{\sqrt{\frac{-2nc}{a(n+1)}}\cosh[\frac{(n-1)}{2n}\sqrt{\frac{a}{b}}(x-ct)]\}^{\frac{2}{n-1}}, \text{ for } c < 0. \tag{126}$$

The Two Dimensional Defocusing Branch

In this section we extend our study to the two-dimensional nonlinear dispersive equation defined by

$$u_t - a(u^n)_x + b(u^n)_{xxx} + k(u^n)_{yyy} = 0, \; n > 1, \tag{127}$$

where $u \equiv u(x,y,t)$, and $a, b, k > 0$ are constants.

Following the analysis made above it is natural to seek a solution in the form

$$u(x,y,t) = \alpha \sinh^{\frac{2}{n-1}}[\beta(x+y-ct)]. \tag{128}$$

or in the form

$$u(x,y,t) = \alpha \cosh^{\frac{2}{n-1}}[\beta(x+y-ct)], \tag{129}$$

where α and β are constants. To simplify matters further, we substitute (128) or (129) into (127), and by solving the resulting equations for α and β we find:
1. For the sinh profile:

$$\beta = \pm\frac{(n-1)}{2n}\sqrt{\frac{a}{b+k}},$$

$$\alpha = \begin{cases} \left(\frac{2nc}{a(n+1)}\right)^{\frac{1}{n-1}} & n \text{ is even,} \\ \\ \pm\left(\frac{2nc}{a(n+1)}\right)^{\frac{1}{n-1}} & n \text{ is odd.} \end{cases} \tag{130}$$

2. For the cosh profile:

$$\beta = \pm\frac{(n-1)}{2n}\sqrt{\frac{a}{b+k}},$$

$$\alpha = \begin{cases} -\left(\frac{2nc}{a(n+1)}\right)^{\frac{1}{n-1}} & n \text{ is even,} \\ \\ \pm\left(\frac{-2nc}{a(n+1)}\right)^{\frac{1}{n-1}} & n \text{ is odd.} \end{cases} \tag{131}$$

This gives the following set of the general solutions:

1. For n even: the general solutions are given by:

$$u(x,y,t) = \{\sqrt{\frac{2nc}{a(n+1)}}\sinh[\frac{(n-1)}{2n}\sqrt{\frac{a}{b+k}}(\phi)]\}^{\frac{2}{n-1}}, \tag{132}$$

and

$$u(x,y,t) = -\{\sqrt{\frac{2nc}{a(n+1)}}\cosh[\frac{(n-1)}{2n}\sqrt{\frac{a}{b+k}}(\phi)]\}^{\frac{2}{n-1}}. \tag{133}$$

2. For n odd, we find

$$u(x,y,t) = \pm\{\sqrt{\frac{2nc}{a(n+1)}}\sinh[\frac{(n-1)}{2n}\sqrt{\frac{a}{b+k}}(\phi)]\}^{\frac{2}{n-1}}, \text{ for } c > 0, \tag{134}$$

and

$$u(x,y,t) = \pm\{\sqrt{\frac{-2nc}{a(n+1)}}\cosh[\frac{(n-1)}{2n}\sqrt{\frac{a}{b+k}}(\phi)]\}^{\frac{2}{n-1}}, \text{ for } c < 0 \tag{135}$$

where $\phi = x + y - ct$.

The Three Dimensional Defocusing Branch

We consider the three-dimensional nonlinear dispersive equation defined by

$$u_t - a(u^n)_x + b(u^n)_{xxx} + k(u^n)_{yyy} + r(u^n)_{zzz} = 0, \; n > 1, \tag{136}$$

where $u \equiv u(x,y,z,t)$, and $a, b, k, r > 0$ are constants. We can assume that the general solution of the nonlinear dispersive equation is of the form:

$$u(x,y,z,t) = \gamma\sinh^{\frac{2}{n-1}}[\eta(x + y + z - ct)], \tag{137}$$

or

$$u(x,y,z,t) = \gamma\cosh^{\frac{2}{n-1}}[\eta(x + y + z - ct)], \tag{138}$$

where γ and η are constants that will be determined.

To determine the general formula for the solution, we substitute (137) or (138) into (136) to find
1. For the sinh profile:

$$\eta = \pm\frac{(n-1)}{2n}\sqrt{\frac{a}{b+k+r}},$$

$$\gamma = \begin{cases} \left(\frac{2nc}{a(n+1)}\right)^{\frac{1}{n-1}} & n \text{ is even,} \\[3mm] \pm\left(\frac{2nc}{a(n+1)}\right)^{\frac{1}{n-1}} & n \text{ is odd.} \end{cases} \tag{139}$$

2. For the cosh profile:

$$\eta = \pm\frac{(n-1)}{2n}\sqrt{\frac{a}{b+k+r}},$$

$$\gamma = \begin{cases} -\left(\frac{2nc}{a(n+1)}\right)^{\frac{1}{n-1}} & n \text{ is even,} \\[3mm] \pm\left(\frac{-2nc}{a(n+1)}\right)^{\frac{1}{n-1}} & n \text{ is odd.} \end{cases} \tag{140}$$

This gives the general solutions:
1. For n even, we found

$$u(x,y,z,t) = \{\sqrt{\frac{2nc}{a(n+1)}}\sinh[\frac{(n-1)}{2n}\sqrt{\frac{a}{b+k+r}}(\psi)]\}^{\frac{2}{n-1}}, \tag{141}$$

and

$$u(x,y,z,t) = -\{\sqrt{\frac{2nc}{a(n+1)}}\cosh[\frac{(n-1)}{2n}\sqrt{\frac{a}{b+k+r}}(\psi)]\}^{\frac{2}{n-1}}, \tag{142}$$

where $\psi = x+y+z-ct$.
2. For n odd, we obtained

$$u(x,y,z,t) = \pm\{\sqrt{\frac{2nc}{a(n+1)}}\sinh[\frac{(n-1)}{2n}\sqrt{\frac{a}{b+k+r}}(\psi)]\}^{\frac{2}{n-1}}, \text{ for } c > 0, \tag{143}$$

and

$$u(x,y,z,t) = \pm\{\sqrt{\frac{-2nc}{a(n+1)}}\cosh[\frac{(n-1)}{2n}\sqrt{\frac{a}{b+k+r}}(\psi)]\}^{\frac{2}{n-1}}, \text{ for } c < 0, \tag{144}$$

where $\psi = x+y+z-ct$.

The discovery in [100] that the focusing branch of the nonlinear dispersive equation $K(n,n)$ produces solitary traveling structures, whereas the defocusing branch gives solitary patterns having cusps or infinite slopes is still considered in its early stages. The discovery has attracted a reasonable amount of research work and the subsequent studies will add useful results to the scientific applications.

Exercises 11.4

Use the decomposition method or any other method to find the compactons solutions for the following nonlinear dispersive equations:

1. $u_t - (u^3)_x + (u^3)_{xxx} = 0$, $u(x,0) = 3\sinh(\frac{1}{3}x)$

2. $u_t - (u^4)_x + (u^4)_{xxx} = 0$, $u(x,0) = -\left(2\cosh(\frac{3}{8}x)\right)^{\frac{2}{3}}$

3. $u_t - (u^3)_x + (u^3)_{xxx} + (u^3)_{yyy} = 0$, $u(x,y,0) = 3\sinh(\frac{1}{3\sqrt{2}}(x+y))$

4. $\frac{1}{2}(u^2)_t - (u^2)_x + (u^2)_{xxx} + u^2_{xxxxx} = 0$, $u(x,0) = \sqrt{\sinh x}$

5. $\frac{1}{2}(u^2)_t - (u^2)_x + (u^2)_{xxx} + u^2_{xxxxx} + (u^2)_{yyyyy} = 0$, $u(x,y,0) = \sqrt{\sinh(x+y)}$

Appendix A

Indefinite Integrals

A.1 Fundamental Forms

1. $\displaystyle\int x^n\,dx = \frac{1}{n+1}x^{n+1} + C, n \neq -1.$

2. $\displaystyle\int \frac{1}{x}\,dx = \ln|x| + C.$

3. $\displaystyle\int e^{ax}\,dx = \frac{1}{a}e^{ax} + C.$

4. $\displaystyle\int \frac{1}{a^2 + x^2}\,dx = \frac{1}{a}\tan^{-1}\frac{x}{a} + C.$

5. $\displaystyle\int \frac{1}{\sqrt{a^2 - x^2}}\,dx = \sin^{-1}\frac{x}{a} + C.$

6. $\displaystyle\int \frac{1}{x\sqrt{x^2 - 1}}\,dx = \sec^{-1}x + C.$

7. $\displaystyle\int \cos x\,dx = \sin x + C.$

8. $\displaystyle\int \sin x\,dx = -\cos x + C.$

9. $\displaystyle\int \tan x\,dx = -\ln|\cos x| + C.$

10. $\displaystyle\int \cot x\,dx = \ln\sin x + C.$

11. $\displaystyle\int \tan x \sec x\,dx = \sec x + C.$

12. $\displaystyle\int \sec x\,dx = -\ln(\sec x - \tan x) + C.$

13. $\displaystyle\int \csc x\,dx = -\ln(\csc x + \cot x) + C.$

14. $\displaystyle\int \sec^2 x\,dx = \tan x + C.$

15. $\displaystyle\int \csc^2 x\, dx = -\cot x + C.$

A.2   Trigonometric Forms

1. $\displaystyle\int \sin^2 x\, dx = \frac{1}{2}x - \frac{1}{4}\sin 2x + C.$

2. $\displaystyle\int \cos^2 x\, dx = \frac{1}{2}x + \frac{1}{4}\sin 2x + C.$

3. $\displaystyle\int \sin^3 x\, dx = -\frac{1}{3}\cos x\,(2 + \sin^2 x) + C.$

4. $\displaystyle\int \cos^3 x\, dx = \frac{1}{3}\sin x\,(2 + \cos^2 x) + C.$

5. $\displaystyle\int \tan^2 x\, dx = \tan x - x + C.$

6. $\displaystyle\int \cot^2 x\, dx = -\cot x - x + C.$

7. $\displaystyle\int x \sin x\, dx = \sin x - x\cos x + C.$

8. $\displaystyle\int x \cos x\, dx = \cos x + x\sin x + C.$

9. $\displaystyle\int x^2 \sin x\, dx = 2x\sin x - (x^2 - 2)\cos x + C.$

10. $\displaystyle\int x^2 \cos x\, dx = 2x\cos x + (x^2 - 2)\sin x + C.$

11. $\displaystyle\int \sin x \cos x\, dx = \frac{1}{2}\sin^2 x + C.$

12. $\displaystyle\int \frac{1}{1 + \sin x}\, dx = -\tan(\frac{1}{4}\pi - \frac{1}{2}x) + C.$

13. $\displaystyle\int \frac{1}{1 - \sin x}\, dx = \tan(\frac{1}{4}\pi + \frac{1}{2}x) + C.$

14. $\displaystyle\int \frac{1}{1 + \cos x}\, dx = \tan(\frac{1}{2}x) + C.$

15. $\displaystyle\int \frac{1}{1 - \cos x}\, dx = -\cot(\frac{1}{2}x) + C.$

A.3   Inverse Trigonometric Forms

1. $\displaystyle\int \sin^{-1} x\, dx = x\sin^{-1} x + \sqrt{1 - x^2} + C.$

2. $\displaystyle\int \cos^{-1} x\, dx = x\cos^{-1} x - \sqrt{1 - x^2} + C.$

3. $\int \tan^{-1}x\,dx = x\tan^{-1}x - \dfrac{1}{2}\ln(1+x^2) + C.$

4. $\int x\sin^{-1}x\,dx = \dfrac{1}{4}[(2x^2-1)\sin^{-1}x + x\sqrt{1-x^2}] + C.$

5. $\int x\cos^{-1}x\,dx = \dfrac{1}{4}[(2x^2-1)\cos^{-1}x - x\sqrt{1-x^2}] + C.$

6. $\int x\tan^{-1}x\,dx = \dfrac{1}{2}[(x^2+1)\tan^{-1}x - x] + C.$

7. $\int x\cot^{-1}x\,dx = \dfrac{1}{2}[(x^2+1)\cot^{-1}x + x] + C.$

8. $\int \sec^{-1}x\,dx = x\sec^{-1}x - \ln(x + \sqrt{x^2-1}) + C.$

9. $\int x\sec^{-1}x\,dx = \dfrac{1}{2}[x^2\sec^{-1}x - \sqrt{x^2-1}] + C.$

A.4 Exponential and Logarithmic Forms

1. $\int e^{ax}\,dx = \dfrac{1}{a}e^{ax} + C.$

2. $\int xe^{ax}\,dx = \dfrac{1}{a^2}(ax-1)e^{ax} + C.$

3. $\int x^2 e^{ax}\,dx = \dfrac{1}{a^3}(a^2x^2 - 2ax + 2)e^{ax} + C.$

4. $\int x^3 e^{ax}\,dx = \dfrac{1}{a^4}(a^3x^3 - 3a^2x^2 + 6ax - 6)e^{ax} + C.$

5. $\int e^x \sin x\,dx = \dfrac{1}{2}(\sin x - \cos x)e^x + C.$

6. $\int e^x \cos x\,dx = \dfrac{1}{2}(\sin x + \cos x)e^x + C.$

7. $\int \ln x\,dx = x\ln x - x + C.$

8. $\int x\ln x\,dx = \dfrac{1}{2}x^2(\ln x - \dfrac{1}{2}) + C.$

A.5 Hyperbolic Forms

1. $\int \sinh x\,dx = \cosh x + C.$

2. $\int \cosh x\,dx = \sinh x + C.$

3. $\int x\sinh x\,dx = x\cosh x - \sinh x + C.$

4. $\int x\cosh x\,dx = x\sinh x - \cosh x + C.$

5. $\displaystyle\int \sinh^2 x\, dx = \frac{1}{2}(\sinh x \cosh x - x) + C.$

6. $\displaystyle\int \cosh^2 x\, dx = \frac{1}{2}(\sinh x \cosh x + x) + C.$

7. $\displaystyle\int \tanh x\, dx = \ln \cosh x + C.$

8. $\displaystyle\int \coth x\, dx = \ln \sinh x + C.$

9. $\displaystyle\int \operatorname{sech}^2 x\, dx = \tanh x + C.$

10. $\displaystyle\int \operatorname{csch}^2 x\, dx = -\coth x + C.$

A.6 Other Forms

1. $\displaystyle\int \frac{1}{\sqrt{a^2 - x^2}}\, dx = \arcsin \frac{x}{a} + C.$

2. $\displaystyle\int \frac{1}{a^2 + x^2}\, dx = \frac{1}{a}\arctan \frac{x}{a} + C.$

3. $\displaystyle\int \frac{1}{\sqrt{2ax - x^2}}\, dx = \arccos \frac{a - x}{a} + C.$

4. $\displaystyle\int \frac{1}{a^2 - x^2}\, dx = \frac{1}{2a}\ln \frac{x + a}{x - a} + C.$

Appendix B

Series

B.1 Exponential Functions

1. $e^x = 1 + x + \dfrac{x^2}{2!} + \dfrac{x^3}{3!} + \dfrac{x^4}{4!} + \cdots$.

2. $e^{-x} = 1 - x + \dfrac{x^2}{2!} - \dfrac{x^3}{3!} + \dfrac{x^4}{4!} + \cdots$.

3. $e^{-x^2} = 1 - x^2 + \dfrac{x^4}{2!} - \dfrac{x^6}{3!} + \cdots$.

4. $a^x = 1 + x \ln a + \dfrac{1}{2!}(x \ln a)^2 + \dfrac{1}{3!}(x \ln a)^3 + \cdots, a > 0$.

5. $e^{\sin x} = 1 + x + \dfrac{x^2}{2!} - \dfrac{3x^4}{4!} - \dfrac{8x^5}{5!} - \dfrac{3x^6}{6!} + \cdots$.

6. $e^{\cos x} = e(1 - \dfrac{x^2}{2!} - \dfrac{4x^4}{4!} - \dfrac{31x^6}{6!} + \cdots)$.

7. $e^{\tan x} = 1 + x + \dfrac{x^2}{2!} + \dfrac{3x^3}{3!} + \dfrac{9x^4}{4!} + \dfrac{57x^5}{5!} + \cdots$.

8. $e^{\sin^{-1} x} = 1 + x + \dfrac{x^2}{2!} + \dfrac{2x^3}{3!} + \dfrac{5x^4}{4!} + \cdots$.

B.2 Trigonometric Functions

1. $\sin x = x - \dfrac{x^3}{3!} + \dfrac{x^5}{5!} - \dfrac{x^7}{7!} + \cdots$.

2. $\cos x = 1 - \dfrac{x^2}{2!} + \dfrac{x^4}{4!} - \dfrac{x^6}{6!} + \cdots$.

3. $\tan x = x + \dfrac{x^3}{3} + \dfrac{2x^5}{15} + \dfrac{17x^7}{315} + \cdots$.

B.3   Inverse Trigonometric Functions

1. $\sin^{-1} x = x + \dfrac{1}{2}\dfrac{x^3}{3} + \dfrac{1\cdot 3}{2\cdot 4}\dfrac{x^5}{5} + \dfrac{1\cdot 3\cdot 5}{2\cdot 4\cdot 6}\dfrac{x^7}{7} + \cdots, \; x^2 < 1.$

2. $\tan^{-1} x = x - \dfrac{x^3}{3} + \dfrac{x^5}{5} - \dfrac{x^7}{7} + \cdots.$

B.4   Hyperbolic Functions

1. $\sinh x = x + \dfrac{x^3}{3!} + \dfrac{x^5}{5!} + \dfrac{x^7}{7!} + \cdots.$

2. $\cosh x = 1 + \dfrac{x^2}{2!} + \dfrac{x^4}{4!} + \dfrac{x^6}{6!} + \cdots.$

3. $\tanh x = x - \dfrac{x^3}{3} + \dfrac{2x^5}{15} - \dfrac{17x^7}{315} + \cdots.$

B.5   Inverse Hyperbolic Functions

1. $\sinh^{-1} x = x - \dfrac{1}{2}\dfrac{x^3}{3} + \dfrac{1\cdot 3}{2\cdot 4}\dfrac{x^5}{5} - \dfrac{1\cdot 3\cdot 5}{2\cdot 4\cdot 6}\dfrac{x^7}{7} + \cdots.$

2. $\tanh^{-1} x = x + \dfrac{x^3}{3} + \dfrac{x^5}{5} + \dfrac{x^7}{7} + \cdots.$

Appendix C

Exact Solutions of Burgers' Equation

1. $u(x,t) = 2\tan x$

2. $u(x,t) = -2\cot x$

3. $u(x,t) = -2\tanh x$

4. $u(x,t) = -\frac{2}{x}$

5. $u(x,t) = \frac{x}{t}$

6. $u(x,t) = -2\coth x$

7. $u(x,t) = \frac{x}{t}$

8. $u(x,t) = \frac{x}{t} - \frac{2}{x+t}$

9. $u(x,t) = \frac{x}{t} - \frac{2}{x+nt}$, n is an integer

10. $u(x,t) = \frac{2}{x+n}$, n is an integer

11. $u(x,t) = \frac{x}{t} + \frac{2}{x+t} + \frac{x+t}{2t^2 - t}$

12. $u(x,t) = -\frac{2}{1 \pm e^{-t-x}}$

13. $u(x,t) = \frac{2}{1 \pm e^{-t+x}}$

14. $u(x,t) = 1 + 2k\tan(k(x-t))$

15. $u(x,t) = 1 - 2k\tanh(k(x-t))$

16. $u(x,t) = 1 - \dfrac{1}{x-t}$

17. $u(x,t) = \dfrac{2\sin x}{\cos x \pm e^t}$

18. $u(x,t) = -\dfrac{2\cos x}{\sin x \pm e^t}$

19. $u(x,t) = \dfrac{x}{t} + \dfrac{2}{t}\tan\dfrac{x}{t}$

20. $u(x,t) = \dfrac{x}{t} - \dfrac{2}{t}\cot\dfrac{x}{t}$

21. $u(x,t) = \dfrac{x}{t} - \dfrac{t}{x}$

Appendix D

Padé Approximants for Well Known Functions

D.1 Exponential Functions

1. $f(x) = e^x$

$$[2/2] = \frac{12 + 6x + x^2}{12 - 6x + x^2}$$

$$[3/3] = \frac{120 + 60x + 12x^2 + x^3}{120 - 60x + 12x^2 - x^3}$$

$$[4/4] = \frac{1680 + +840x + 180x^2 + 20x^3 + x^4}{1680 - 840x + 180x^2 - 20x^3 + x^4}$$

2. $f(x) = e^{-x}$

$$[2/2] = \frac{12 - 6x + x^2}{12 + 6x + x^2}$$

$$[3/3] = \frac{120 - 60x + 12x^2 - x^3}{120 + 60x + 12x^2 + x^3}$$

$$[4/4] = \frac{1680 - 840x + 180x^2 - 20x^3 + x^4}{1680 + 840x + 180x^2 + 20x^3 + x^4}$$

D.2 Trigonometric Functions

1. $f(x) = \sin x$

$$[2/2] = \frac{6x}{6 + x^2}$$

$$[3/3] = \frac{60x - 7x^3}{60 + 3x^2}$$

$$[4/4] = \frac{5880x - 620x^3}{5880 + 360x^2 + 11x^4}$$

2. $f(x) = \cos x$

$$[2/2] = \frac{12 - 5x^2}{12 + 5x^2}$$

$$[3/3] = \frac{12 - 5x^2}{12 + 5x^2}$$

$$[4/4] = \frac{15120 - 6900x^2 + 313x^4}{15120 + 660x^2 + 13x^4}$$

3. $f(x) = \tan x$

$$[2/2] = \frac{3x}{3 - x^2}$$

$$[3/3] = \frac{15x - x^3}{15 - 6x^2}$$

$$[4/4] = \frac{105x - 10x^3}{105 - 45x^2 + x^4}$$

4. $f(x) = \sec x$

$$[2/2] = \frac{12 + x^2}{12 - 5x^2}$$

$$[3/3] = \frac{12 + x^2}{12 - 5x^2}$$

$$[4/4] = \frac{15120 + 660x^2 + 13x^4}{15120 - 6900x^2 + 313x^4}$$

5. $f(x) = \tan^{-1} x$

$$[2/2] = \frac{3x}{3 + x^2}$$

$$[3/3] = \frac{15x + 4x^3}{15 + 9x^2}$$

$$[4/4] = \frac{105x + 55x^3}{105 + 90x^2 + 9x^4}$$

D.3 Hyperbolic Functions

1. $f(x) = \sinh x$

$$[2/2] = \frac{6x}{6 - x^2}$$

$$[3/3] = \frac{60x + 7x^3}{60 - 3x^2}$$

$$[4/4] = \frac{5880x + 620x^3}{5880 - 360x^2 + 11x^4}$$

2. $f(x) = \cosh x$

$$[2/2] = \frac{12 + 5x^2}{12 - x^2}$$

$$[3/3] = \frac{12 + 5x^2}{12 - x^2}$$

$$[4/4] = \frac{15120 + 6900x^2 + 313x^4}{15120 - 660x^2 + 13x^4}$$

D.4 Logarithmic Functions

1. $f(x) = \ln(1 + x)$

$$[2/2] = \frac{6x + 3x^2}{6 + 6x + x^2}$$

$$[3/3] = \frac{60x + 60x^2 + 11x^3}{60 + 90x + 36x^2 + 3x^3}$$

$$[4/4] = \frac{420x + 630x^2 + 260x^3 + 25x^4}{420 + 840x + 540x^2 + 120x^3 + 6x^4}$$

2. $f(x) = \ln(1 - x)$

$$[2/2] = \frac{-6x + 3x^2}{6 - 6x + x^2}$$

$$[3/3] = \frac{60x - 60x^2 + 11x^3}{-60 + 90x - 36x^2 + 3x^3}$$

$$[4/4] = \frac{-420x + 630x^2 - 260x^3 + 25x^4}{420 - 840x + 540x^2 - 120x^3 + 6x^4}$$

3. $f(x) = \ln(1+x)/x$

$$[2/2] = \frac{30 + 21x + x^2}{30 + 36x + 9x^2}$$

$$[3/3] = \frac{420 + 510x + 140x^2 + 3x^3}{420 + 720x + 360x^2 + 48x^3}$$

$$[4/4] = \frac{3780 + 6510x + 3360x^2 + 505x^3 + 6x^4}{3780 + 8400x + 6300x^2 + 1800x^3 + 150x^4}$$

4. $f(x) = \ln(1-x)/x$

$$[2/2] = \frac{30 - 21x + x^2}{-30 + 36x - 9x^2}$$

$$[3/3] = \frac{420 - 510x + 140x^2 - 3x^3}{-420 + 720x - 360x^2 + 48x^3}$$

$$[4/4] = \frac{3780 - 6510x + 3360x^2 - 505x^3 + 6x^4}{-3780 + 8400x - 6300x^2 + 1800x^3 - 150x^4}$$

Appendix E

The Error and Gamma Functions

E.1 The Error function

The *error function* erf(x) is defined by :

1. $erf(x) = \dfrac{2}{\sqrt{\pi}} \displaystyle\int_0^x e^{-u^2} du.$

2. $erf(x) = \dfrac{2}{\sqrt{\pi}} \left(x - \dfrac{x^3}{3} + \dfrac{x^5}{5 \cdot 2!} - \dfrac{x^7}{7 \cdot 3!} + \cdots \right).$

The *complementary error function* erfc(x) is defined by :

3. $erfc(x) = \dfrac{2}{\sqrt{\pi}} \displaystyle\int_x^\infty e^{-u^2} du.$

4. $erf(x) + erfc(x) = 1.$

5. $erfc(x) = 1 - \dfrac{2}{\sqrt{\pi}} \left(x - \dfrac{x^3}{3} + \dfrac{x^5}{5 \cdot 2!} - \dfrac{x^7}{7 \cdot 3!} + \cdots \right).$

E.2 The Gamma function $\Gamma(x)$

1. $\Gamma(x) = \displaystyle\int_0^\infty t^{x-1} e^{-t} dt.$

2. $\Gamma(x+1) = x\Gamma(x),\ \Gamma(1) = 1, \Gamma(n+1) = n!,$ n is an integer .

3. $\Gamma(x)\Gamma(1-x) = \dfrac{\pi}{\sin \pi x}.$

4. $\Gamma(1/2) = \sqrt{\pi}.$

Answers

Exercises 1.2

1. (a) 3 $\qquad$ (b) 2 $\quad$ (c) 1 $\quad$ (d) 4

2. (a) linear $\qquad$ (b) nonlinear
   $\quad$ (c) linear $\qquad$ (d) nonlinear

3. (a) inhomogeneous $\quad$ (b) homogeneous
   $\quad$ (c) inhomogeneous $\quad$ (d) homogeneous

Exercises 1.3

1. Hyperbolic $\qquad$ 2. Elliptic $\qquad$ 3. Parabolic

4. Hyperbolic $\qquad$ 5. Parabolic $\qquad$ 6. Heperbolic

7. Elliptic $\qquad$ 8. Elliptic $\qquad$ 9. Hyperbolic

10. Elliptic if $y > 0$, $\quad$ Parabolic if $y = 0$, $\quad$ Hyperbolic if $y < 0$

Exercises 2.2

1. $u(x, y) = x^2 y^2$ $\qquad$ 2. $u(x, y) = x^2 + y^2$

3. $u(x, y) = e^{-xy}$ $\qquad$ 4. $u(x, y) = e^x + e^y$

5. $u(x, y) = e^{x+y}$ $\qquad$ 6. $u(x, y) = x - y$

7. $u(x, y) = \frac{1}{3}x^3 + \frac{1}{3}y^3$ $\qquad$ 8. $u(x, y) = x + e^y$

9. $u(x, y) = y + e^x$ $\qquad$ 10. $u(x, y) = xe^y$

11. $u(x, y) = ye^x$ $\qquad$ 12. $u(x, y) = e^{x-y}$

13. $u(x, y, z) = x + y + z$ $\qquad$ 14. $u(x, y, z) = e^{x+y+z}$

15. $u(x, y, z) = yze^x$ $\qquad$ 16. $u(x, y, z) = e^x + y + z$

17. $u(x, y) = 3x^2 + 2x + 4xy$ $\quad$ 18. $u(x, y) = x^2 + x + 2xy$
   $\qquad\qquad +y + y^2$ $\qquad\qquad\qquad +y + y^2$

19. $u(x, y) = \sin(x + y)$ $\qquad$ 20. $u(x, y) = \cosh(x + y)$

Exercises 2.3

1. $u(x,y) = x^3 + y^3$ 2. $u(x,y) = \cosh x + \cosh y$

3. $u(x,y) = xy$ 4. $u(x,y) = \sin x + \cos y$

5. $u(x,y) = x \sin y + y \sin x$ 6. $u(x,y) = x \cos y - y \cos x$

7. $u(x,y) = xe^y + ye^x$ 8. $u(x,y) = xe^{-y} - ye^{-x}$

9. $u(x,y) = x^2 y^2$ 10. $u(x,y) = xy^2 + yx^2$

11. $u(x,y) = \sin x + \cosh y$ 12. $u(x,y) = xe^y$

13. $u(x,y) = x^2 y^3 + x^3 y^2$ 14. $u(x,y) = x^3 y^4 + x^4 y^3$

15. $u(x,y) = (x+y)^2$ 16. $u(x,y) = (x+y)^2$

17. $u(x,y) = x^2 - xy + y^2$ 18. $u(x,y) = e^x + e^y + x$

Exercises 2.4

1. $u(x,y) = x^3 + y^3$ 2. $u(x,y) = x^2 - y^2$

3. $u(x,y) = (x+y)^2$ 4. $u(x,y) = \cosh x + \cosh y$

5. $u(x,y) = \sin x + \cos y$ 6. $u(x,y) = x^2 y^2$

7. $u(x,y) = \sin x + \sin y$ 8. $u(x,y) = e^x - e^y$

9. $u(x,y) = x^2 + y^3$ 10. $u(x,y) = 1 + x^2 + \sin y$

11. $u(x,y) = 1 + x + \sinh y$ 12. $u(x,y) = \sin x + \sinh y$

Exercises 2.5

1. $u(x,y) = x^2 + y^2$ 2. $u(x,y) = e^x + e^y$

3. $u(x,y) = e^{-xy}$ 4. $u(x,y) = 2x + 3y - \cos x$

5. $u(x,y) = e^{x+y}$ 6. $u(x,y) = xe^y$

7. $u(x,y) = ye^x$ 8. $u(x,y) = \cosh x + \cosh y$

9. $u(x,y) = 2xy$ 10. $u(x,y) = x^2 y^2$

11. $u(x,y) = y - kx$ 12. $u(x,y) = x - y$

13. $u(x,y) = x + \sin y$ 14 $u(x,y) = 4xe^y$

15. $u(x,y) = xy$ 16. $u(x,y) = e^x + e^y$

17. $u(x,y) = e^{xy}$ 18. $u(x,y) = e^{x+y}$

19. $u(x,y) = x \sinh y$ 20. $u(x,y) = y \cosh x$

Exercises 2.6

1. $u(x,t) = \sinh(x+t) \quad v(x,t) = \cosh(x+t)$

2. $u(x,t) = \sin(x+t)$ $v(x,t) = \cos(x+t)$

3. $u(x,t) = \cos(x+t)$ $v(x,t) = \cos(x+t)$

4. $u(x,t) = \sinh(x-t)$ $v(x,t) = \cosh(x-t)$

5. $u(x,y,t) = -w = \sin(x+y-t)$ $v(x,y,t) = \cos(x+y-t)$

6. $u(x,y,t) = \sin(x+y+t)$ $v(x,y,t) = -w = \cos(x+y+t)$

Exercises 3.2

1. $u(x,t) = x + e^{-t}\sin x$ 2. $u(x,t) = 4 + e^{-t}\cos x$

3. $u(x,t) = e^{-t}\sin x$ 4. $u(x,t) = e^{-5t}\sin x$

5. $u(x,t) = e^{-t}\sinh x$ 6. $u(x,t) = e^{-t}\cosh x$

7. $u(x,t) = e^{-t}\sin x + \frac{1}{4}\sin 2x$ 8. $u(x,t) = x^2 + e^{-t}\sin x$

9. $u(x,t) = x^3 + e^{-t}\sin x$ 10. $u(x,t) = 3x^2 + e^{-t}\cos x$

11. $u(x,t) = x^2 + e^{-t}\cos x$ 12. $u(x,t) = x^3 + e^{-t}\cos x$

13. $u(x,t) = 1 + e^{-\pi^2 t}\sin(\pi x)$ 14. $1 + e^{-4\pi^2 t}\sin(\pi x)$

15. $u = e^{-4t}\cos x$ 16. $u = x + e^{-2t}\sin x$

17. $u = e^{-t}\cos x$ 18. $u = 2 + e^{-t}\cos x$

Exercises 3.3.1

1. $u(x,t) = e^{-t}\sin x + 2e^{-9t}\sin(3x)$

2. $u(x,t) = e^{-\pi^2 t}\sin(\pi x) + e^{-4\pi^2 t}\sin(2\pi x)$

3. $u(x,t) = e^{-16t}\sin(2x)$

4. $u(x,t) = e^{-2\pi^2 t}\sin(\pi x)$

5. $u(x,t) = 1 + e^{-t}\cos x$

6. $u(x,t) = 3 + 4e^{-2t}\cos x$

7. $u(x,t) = 1 + e^{-3t}\cos x + e^{-12t}\cos(2x)$

8. $u(x,t) = 2 + 2e^{-16\pi^2 t}\cos(2\pi x)$

9. $u(x,t) = \frac{8}{\pi}\sum_{m=0}^{\infty}\frac{1}{(2m+1)}e^{-4(2m+1)^2 t}\sin(2m+1)x$

10. $u(x,t) = \frac{6}{\pi}\sum_{m=0}^{\infty}\frac{1}{(2m+1)}e^{-2(2m+1)^2 t}\sin(2m+1)x$

11. $u(x,t) = \pi - \frac{8}{\pi}\sum_{m=0}^{\infty}\frac{1}{(2m+1)^2}e^{-(2m+1)^2 t}\cos(2m+1)x$

12. $u(x,t) = (1+\frac{\pi}{2}) - \frac{4}{\pi^2}\sum_{m=0}^{\infty}\frac{1}{(2m+1)}e^{-(2m+1)^2 t}\cos(2m+1)x$

Exercises 3.3.2

1. $u(x,t) = 1 + 2x + 3e^{-\pi^2 t} \sin(\pi x)$

2. $u(x,t) = 1 + 4e^{-t} \sin x$

3. $u(x,t) = x + e^{-4t} \sin 2x$

4. $u(x,t) = 4 - 4x + e^{-9t} \sin(3x)$

5. $u(x,t) = 2 + 3x + e^{-t} \sin x$

6. $u(x,t) = 1 + 2x + 3e^{-4\pi^2 t} \sin(2\pi x)$

Exercises 3.3.3

1. $u(x,t) = e^{-2t} \sin x$

2. $u(x,t) = e^{-5\pi^2 t} \sin(2\pi x)$

3. $u(x,t) = e^{-4t} \sin x$

4. $u(x,t) = e^{-3\pi^2 t} \sin(\pi x)$

5. $u(x,t) = 1 + e^{-2t} \sin x$

6. $u(x,t) = 3 + 3e^{-3\pi^2 t} \sin(\pi x)$

Exercises 4.2.1

1. $u(x,y,t) = e^{-4t} \sin x \sin y$

2. $u(x,y,t) = 2e^{-2t} \sin x \sin y$

3. $u(x,y,t) = e^{-4t} \cos(x+y)$

4. $u(x,y,t) = e^{-6t} \sin(x-y)$

5. $u(x,y,t) = e^{-5t} \sin x \sin y$

6. $u(x,y,t) = e^{-8t} \sin(x+y)$

7. $u(x,y,t) = e^{-4t} \sin x \sin y + \sin x$

8. $u(x,y,t) = e^{-6t} \sin x \sin y + \cos x$

9. $u(x,y,t) = e^{-2t} \sin(x+y) + \cos(x+y)$

10. $u(x,y,t) = e^{-2t} \sin x \sin y + \sin x + \sin y$

11. $u(x,y,t) = e^{-2t} \sin x \sin y + x^2$

12. $u(x,y,t) = e^{-2t} \sin x \sin y + y^2$

Exercises 4.2.2

1. $u(x,y,z,t) = e^{-6t} \sin x \sin y \sin z$

2. $u(x, y, z, t) = 2e^{-3t} \sin x \sin y \sin z$

3. $u(x, y, z, t) = e^{-3t} \sin(x + y + z)$

4. $u(x, y, z, t) = e^{-4t} \sin x \sin y \sin z$

5. $u(x, y, z, t) = 2x^2 + e^{-3t} \sin x \sin y \sin z$

6. $u(x, y, z, t) = y^2 + e^{-3t} \sin x \sin \sin z$

7. $u(x, y, z, t) = \sin x + e^{-2t} \sin(y + z)$

8. $u(x, y, z, t) = x^2 + e^{-3t}(\sin x + \sin y + \sin z)$

Exercises 4.3.1

1. $u(x, y, t) = e^{-13t} \sin 2x \sin 3y$

2. $u(x, y, t) = e^{-6t} \sin x \sin y + e^{-24t} \sin 2x \sin 2y$

3. $u(x, y, t) = e^{-8t} \sin x \sin y + e^{-20t} \sin x \sin 2y + e^{-20t} \sin 2x \sin y$

4. $u(x, y, t) = e^{-2t} \cos x \sin y$

5. $u(x, y, t) = e^{-2t} \sin x \cos y$

6. $u(x, y, t) = e^{-2t} \cos x \sin y + e^{-8t} \cos 2x \sin 2y$

7. $u(x, y, t) = e^{-4t} \sin x \cos y + e^{-16t} \sin 2x \cos 2y$

8. $u(x, y, t) = e^{-13t} \cos 2x \cos 3y$

9. $u(x, y, t) = 1 + e^{-5t} \cos x \cos 2y$

10. $u(x, y, t) = 4 + e^{-32t} \cos 2x \cos 2y$

Exercises 4.3.2

1. $u(x, y, z, t) = e^{-29t} \sin 2x \sin 3y \sin 4z$

2. $u(x, y, z, t) = e^{-3t} \sin x \sin y \sin z + e^{-12t} \sin 2x \sin 2y \sin 2z$

3. $u(x, y, z, t) = e^{-6t} \sin x \sin y \sin 2z + e^{-14t} \sin x \sin 2y \sin 3z$

4. $u(x, y, z, t) = e^{-3t} \cos x \sin y \sin z$

5. $u(x, y, z, t) = e^{-3t} \sin x \cos y \cos z$

6. $u(x, y, z, t) = e^{-3t} \cos x \sin y \cos z$

7. $u(x, y, z, t) = e^{-3t} \sin x \cos y \sin z$

8. $u(x, y, z, t) = 2 + 3e^{-6t} \cos x \cos 2y \cos z$

9. $u(x, y, z, t) = 1 + e^{-3t} \cos x \cos y \cos z + e^{-12t} \cos 2x \cos 2y \cos 2z$

10. $u(x, y, z, t) = 1 + 2e^{-3t} \cos x \cos y \cos z$

$$+ 3e^{-29t} \cos 2x \cos 3y \cos 4z$$

Exercises 5.2.2

1. $u(x,t) = \sin(2x)\cos(4t)$

2. $u(x,t) = \sin x \sin t + \sin x \cos t$

3. $u(x,t) = 2 + \cos x \cos t$

4. $u(x,t) = 1 + x + \sin x \sin t$

5. $u(x,t) = \sin x \cos(3t)$

6. $u(x,t) = \sin x \sin(2t)$

7. $u(x,t) = \cos x \cos t$

8. $u(x,t) = x + \cos x \cos t$

9. $u(x,t) = \cos x + \sin x \sin t$

10. $u(x,t) = \sin x + \sin x \cos t$

11. $u(x,t) = 1 + \sin x \sin(2t)$

12. $u(x,t) = x^4 + \sin x \cos t$

13. $u(x,t) = x^3 + \sin x \sin t$

14. $u(x,t) = 1 + \cos x + \sin x \sin t$

15. $u(x,t) = 2x^2 + \sin x \cos t$

16. $u(x,t) = x^2 + \cos x \sin t$

17. $u(x,t) = \sin x + \cos x \sin t$

18. $u(x,t) = x^2 + \cos x \cos t$

19. $u(x,t) = t^4 + t^3 x + \cos x \sin t$

20. $u(x,t) = t^2 + x^3 + \sin x \sin t$

21. $u(x,t) = x^2 \cosh t$

22. $u(x,t) = x^2 e^t$

23. $u(x,t) = x^4 \sinh t$

24. $u(x,t) = x^3 \cosh t$

Exercises 5.2.3

1. $u(x,t) = 4t + \sin x \sin t$

2. $u(x,t) = \sin(x + t)$

3. $u(x,t) = \cos(x + t)$

4. $u(x,t) = \sin(x - t)$

5. $u(x,t) = 6t + \sin x \cos t$

6. $u(x,t) = 6t + 2xt^2$

7. $u(x,t) = x^2 t + t^3 + e^x \sinh t$

8. $u(x,t) = \cos x \sin t + xe^t$

9. $u(x,t) = x^2 + t^2 + \sin x \sin t$

10. $u(x,t) = t + 2xt + \cos x(\cos t - 1)$

11. $u(x,t) = x^2 + t^2 - \sin x + \sin x \sin t$

12. $u(x,t) = \cos x + \cos x \cos t$

Exercises 5.3.1

1. $u(x,t) = \sin(3x)\sin(3t)$

2. $u(x,t) = \sin x \cos t$

3. $u(x,t) = 2\sin(2x)\sin(2t)$

4. $u(x,t) = \sin x \cos(2t)$

5. $u(x,t) = \sin(2x)\cos(4t)$

6. $u(x,t) = \sin x \sin(3t)$

7. $u(x,t) = 1 + \cos x \sin(3t)$

8. $u(x,t) = 2 + \cos x \cos(2t)$

9. $u(x,t) = \cos x \sin(3t)$

10. $u(x,t) = \cos x \sin t + \cos x \cos t$

11. $u(x,t) = \sum_{n=1}^{\infty} \frac{(-1)^{n+1}}{5n^2} \sin(nx)\sin(nt)$

12. $u(x, t) = \sum_{n=0}^{\infty} \frac{4}{(2n+1)^2} \sin(nx) \sin 2(2n+1)t$

Exercises 5.3.2

1. $u(x, t) = 1 + \sin(\pi x) \sin(\pi t)$

2. $u(x, t) = 2 + x + 2 \sin(\pi x) \cos(\pi t)$

3. $u(x, t) = 3x + 4 \sin(\pi x) \sin(\pi t)$

4. $u(x, t) = 4 - 3x + \sin(\pi x) \sin(\pi t)$

5. $u(x, t) = 3 + 4x + \sin(\pi x) \sin(2\pi t)$

6. $u(x, t) = 1 + x + \sin(2\pi x) \sin(4\pi t)$

7. $u(x, t) = 3x + x^2 + t^2 + \cos(\pi x) \cos(\pi t)$

8. $u(x, t) = 4x + \cos(\pi x) \sin(\pi t)$

9. $u(x, t) = 2x + 2x^2 + 2t^2 + \cos(\pi x) \cos(\pi t)$

10. $u(x, t) = x + \cos(\pi x) \sin(2\pi t)$

11. $u(x, t) = 2x + \cos(\pi x) \cos(2\pi t)$

12. $u(x, t) = x + x^2 + t^2 + \cos(\pi x) \sin(3\pi t)$

Exercises 5.3.3

1. $u(x, t) = 2t + \sin x \sin t$ 2. $u(x, t) = \sin(x + t)$

3. $u(x, t) = \cos(x + t)$ 4. $u(x, t) = \sin(x + 4t)$

5. $u(x, t) = 2t + \sin x \cos t$ 6. $u(x, t) = \cos(x + 2t)$

7. $u(x, t) = \sinh(x + t)$ 8. $u(x, t) = x + e^{-x} \sinh t$

9. $u(x, t) = \cosh x \cosh t$ 10. $u(x, t) = \sinh x \sinh(2t)$

11. $u(x, t) = t + 2xt + \cos x \cos t$ 12. $u(x, t) = 4t + 4xt + \sin x \cos t$

Exercises 6.2.1

1. $u(x, y, t) = \sin x \sin y \sin(2t)$

2. $u(x, y, t) = \sin(2x) \sin(2y) \sin(4t)$

3. $u(x, y, t) = \sin(2x) \sin(2y) \cos(4t)$

4. $u(x, y, t) = 2 + \sin x \sin y \sin t$

5. $u(x, y, t) = 1 + y + \sin x \sin y \sin(2t)$

6. $u(x, y, t) = 1 + x + \sin y \sin t$

7. $u(x, y, t) = \sin x \sin y \sin(2t)$

8. $u(x, y, t) = \sin x \sin y \cos(3t)$

9. $u(x, y, t) = \sin x + \sin x \sin y \sin t$

10. $u(x, y, t) = \cos x + \sin y \sin t$

11. $u(x, y, t) = x^2 + y^2 + \sin x \sin t$

12. $u(x, y, t) = 2x^2 + 2y^2 + 2 \sin x \sin y \cos(2t)$

13. $u(x, y, t) = t^2 + tx + ty + \sin x \sin y \sin t$

14. $u(x, y, t) = t^3 + t^2x + ty + \sin x \sin y \cos(2t)$

15. $u(x, y, t) = y^2 + \sin x \cos t$

16. $u(x, y, t) = x^2 + \sin y \sin t$

17. $u(x, y, t) = \sin x + \sin y \sin t$

18. $u(x, y, t) = \cos x \cos y \sin(2t)$

19. $u(x, y, t) = t^4 + t^2 y + \sin x \sin y \sin t$

20. $u(x, y, t) = t^2 + x^2 + y^2 + \sin x \sin y \sin t$

21. $u(x, y, t) = x^2 y^2 \sinh t$

22. $u(x, y, t) = x^2 y^2 e^t$

23. $u(x, y, t) = x^2 \sinh t + y^2 \cosh t$

24. $u(x, y, t) = x^2 e^{-t} + y^2 e^t$

Exercises 6.2.2

1. $u(x, y, z, t) = \sin 2x \sin 2y \sin 2z \sin 6t$

2. $u(x, y, z, t) = 1 + \sin x \sin y \sin z \sin t$

3. $u(x, y, z, t) = 3 + \sin x \sin y \sin z \cos 3t$

4. $u(x, y, z, t) = \sin x \sin y \sin z \sin 3t$

5. $u(x, y, z, t) = \sin(x + 2y) \sin(z + 2t)$

6. $u(x, y, z, t) = \sin x \sin 2y \sin 3z \cos 4t$

7. $u(x, y, z, t) = \cos(x + y) \sin(z + t)$

8. $u(x, y, z, t) = 1 + z + \sin x \sin y \sin z \sin t$

9. $u(x, y, z, t) = \sin x \sin y + \sin z \sin t$

10. $u(x, y, z, t) = \cos x + \cos y + \sin z \sin t$

11. $u(x, y, z, t) = x^2 + y^2 + z^2 + \sin t$

12. $u(x, y, z, t) = t^2 + tx + ty + tz + \sin t$

13. $u(x, y, z, t) = t^2(x + y + z) + \sin y \sin t$

14. $u(x, y, z, t) = x^2 + y^2 + z^2 + \cos y \cos t$

15. $u(x, y, z, t) = x^2 + \sin y \sin z \cos t$

16. $u(x, y, z, t) = 1 + \cos x \cos y \cos z \sin 2t$

17. $u(x, y, z, t) = 1 + \sin x \sin y \sin z \cos 2t$

18. $u(x, y, z, t) = 1 + \sin x \sin y + \sin z \sin t$

19. $u(x, y, z, t) = t^2 + x^2 + y^2 + z^2 + \cos x \cos y \cos z \cos t$

20. $u(x, y, z, t) = x^4 + y^4 + \cos z \cos t$

21. $u(x, y, z, t) = x^2 y^2 z^2 \cosh t$

22. $u(x, y, z, t) = x^3 \sinh t + \cosh t (y^3 + z^3)$

23. $u(x, y, z, t) = x^2 e^t + y^2 e^{-t} + z^2 e^t$

24. $u(x, y, z, t) = (x^3 + y^3 + z^3) \sinh t$

Exercises 6.3.1

1. $u(x, y, t) = \sin 2x \sin 2y \cos 4t$

2. $u(x, y, t) = \sin x \sin 2y \cos 5t$

3. $u(x, y, t) = \sin x \sin y \sin 2t$

4. $u(x, y, t) = \sin x \sin 2y \sin 5t$

5. $u(x, y, t) = 2 + \cos x \cos y \sin 2t$

6. $u(x, y, t) = 1 + \cos x \cos y \cos 4t$

7. $u(x, y, t) = \sin x \cos y \sin 2t$

8. $u(x, y, t) = \cos x \sin y \cos 2t$

9. $u(x, y, t) = \cos x \sin y \sin 2t$

10. $u(x, y, t) = 3 + \cos x \cos 2y \sin 5t$

11. $u(x, y, t) = \frac{32}{\pi^2} \sum_{n=0}^{\infty} \sum_{m=0}^{\infty} \frac{1}{(2n+1)(2m+1)}$
$\times \sin(2n + 1)x \sin(2m + 1)y \cos(\sqrt{2}\lambda_{nm}t),$

$\lambda_{nm} = \sqrt{(2n + 1)^2 + (2m + 1)^2}$

12. $u(x, y, t) = \frac{48}{\pi^2} \sum_{n=0}^{\infty} \sum_{m=0}^{\infty} \frac{1}{(2n+1)(2m+1)}$
$\times \sin(2n + 1)x \sin(2m + 1)y \sin(\sqrt{2}\lambda_{nm}t),$

$\lambda_{nm} = \sqrt{(2n + 1)^2 + (2m + 1)^2}$

Exercises 6.3.2

1. $u(x, y, z, t) = \sin 2x \sin 2y \sin 2z \sin 12t$

2. $u(x, y, z, t) = \sin x \sin 2y \sin 3z \cos 14t$

3. $u(x, y, z, t) = \sin x \sin y \sin 2z \sin 6t$

4. $u(x, y, z, t) = \cos x \cos 2y \cos 2z \sin 6t$

5. $u(x, y, z, t) = 3 + \cos x \cos y \cos z \sin 6t$

6. $u(x, y, z, t) = 4 + \cos x \cos y \cos z \cos 6t$

7. $u(x, y, z, t) = \sin x \cos y \cos z \sin 3t$

8. $u(x, y, z, t) = \sin x \sin y \cos z \sin 3t$

9. $u(x, y, z, t) = \cos x \sin y \sin z \cos 6t$

10. $u(x, y, z, t) = \sin x \sin y \cos 2z \cos 6t$

11. $u(x, y, z, t) = \sin x \sin y \cos z \sin \sqrt{3}t$

12. $u(x, y, z, t) = \cos x \cos y \cos 2z \sin 6t$

Exercises 7.2.1

1. $u(x, y) = \sinh x \cos y$ 2. $u(x, y) = \cosh x \sin y$

3. $u(x, y) = \cosh x \cos y$ 4. $u(x, y) = \sin 2x \sinh 2y$

5. $u(x, y) = \sin 2x \cosh 2y$ 6. $u(x, y) = \cos 3x \cosh 3y$

7. $u(x, y) = \sinh 2x \cos 2y$ 8. $u(x, y) = \cosh 2x \cos 2y$

9. $u(x, y) = \cos x \cosh y$ 10. $u(x, y) = \sin x \cosh y$

11. $u(x, y) = \sinh x \cos y$ 12. $u(x, y) = \sin x \cosh y$

13. $u(x, y) = x + \sin x \sinh y$ 14. $u(x, y) = y + \sin x \cosh y$

15. $u(x, y) = 1 + \sin x \sinh y$ 16. $u(x, y) = 1 + \cos x \sinh y$

Exercises 7.3.1

1. $u(x, y) = \sin 2x \sinh 2y$ 2. $u(x, y) = \sinh 3x \sin 3y$

3. $u(x, y) = 4 \sinh 2x \sin 2y$ 4. $u(x, y) = \cos x \cosh y$

5. $u(x, y) = \sin x \cosh y$ 6. $u(x, y) = \cos 2x \cosh 2y$

7. $u(x, y) = \sinh x \sin(\pi - y)$ 8. $u(x, y) = \sin 2x \sinh(2\pi - 2y)$

9. $u(x, y) = C_0 + \cos 2x \cosh 2y$ 10. $u(x, y) = C_0 + \cosh 2x \cos 2y$

Exercises 7.3.2

1. $u(x, y, z) = \sin x \sin 2y \sinh \sqrt{5}z$

2. $u(x, y, z) = \sin 6x \sin 8y \sinh 10z$

3. $u(x, y, z) = \sin 2x \sin 2y \sinh \sqrt{8}z$

4. $u(x, y, z) = \sin x \sin 2y \sinh \sqrt{5}(\pi - z)$

5. $u(x, y, z) = \sin 3x \sin 4y \sinh 5(\pi - z)$

6. $u(x, y, z) = \sin 5x \sin 12y \sinh 13(\pi - z)$

7. $u(x, y, z) = \cos x \cos 2y \cosh \sqrt{5}z$

8. $u(x, y, z) = \cos 5x \cos 12y \cosh 13z$

9. $u(x, y, z) = \cos 3x \cos 4y \cosh 5(\pi - z)$

10. $u(x, y, z, t) = \cos 2x \cos 2y \cosh \sqrt{8}(\pi - z)$

11. $u(x, y, z) = \sin 8x \sin 15y \cosh 17z$

12. $u(x, y, z) = \sin 3x \cos 4y \sinh 5z$

Exercises 7.4.1

1. $u(r, \theta) = 2 + 3r \sin \theta + 4r \cos \theta$ 2. $u(r, \theta) = 1 + r^4 \cos 4\theta$

3. $u(r, \theta) = 1 - r^6 \cos 6\theta$ 4. $u(r, \theta) = r^2 \sin 2\theta + r^2 \cos 2\theta$

5. $u(r, \theta) = C_0 + 2r^4 \sin 4\theta$ 6. $u(r, \theta) = C_0 + r(\sin \theta - \cos \theta)$

7. $u(r, \theta) = C_0 + r^2 \sin 2\theta$ 8. $u(r, \theta) = C_0 + r^3 \cos 3\theta$

9. $u(r, \theta) = C_0 + r^2 \sin 2\theta + r^3 \cos 3\theta$ 10. $u(r, \theta) = C_0 + r^2 \cos 2\theta$

Exercises 7.4.2

1. $u(r, \theta) = 1 + \ln r + (r - \frac{1}{r}) \cos \theta + (r - \frac{1}{r}) \sin \theta$

2. $u(r, \theta) = 1 + \ln r + r \cos \theta + r \sin \theta$

3. $u(r, \theta) = 1 + (r - \frac{1}{r}) \sin \theta$

4. $u(r, \theta) = 1 + (r - \frac{2}{r}) \cos \theta + (r - \frac{2}{r}) \sin \theta$

5. $u(r, \theta) = 1 + \ln r + (r - \frac{1}{r}) \cos \theta$

6. $u(r, \theta) = 1 + \ln r + (r - \frac{1}{r}) \sin \theta$

7. $u(r, \theta) = 1 + \ln r + (r + \frac{1}{r}) \sin \theta$

8. $u(r, \theta) = C_0 + (r - \frac{1}{r}) \cos \theta$

9. $u(r, \theta) = C_0 + (3r + \frac{2}{r}) \sin \theta$

10. $u(r, \theta) = C_0 + \ln r + (r + \frac{1}{r}) \cos \theta + (r + \frac{1}{r}) \sin \theta$

11. $u(r,\theta) = C_0 + 2(r - \frac{1}{r})\cos\theta$

12. $u(r,\theta) = C_0 + (3r + \frac{2}{r})\sin\theta$

Exercises 8.2

1. $\begin{aligned}
A_0 &= u_0^4 \\
A_1 &= 4u_0^3 u_1 \\
A_2 &= 4u_0^3 u_2 + 6u_0^2 u_1^2 \\
A_3 &= 4u_0^3 u_3 + 12u_0^2 u_1 u_2 + 4u_0 u_1^3
\end{aligned}$

2. $\begin{aligned}
A_0 &= u_0^2 + u_0^3 \\
A_1 &= u_1(2u_0 + 3u_0^2) \\
A_2 &= u_2(2u_0 + 3u_0^2) + u_1^2(1 + 3u_0) \\
A_3 &= u_3(2u_0 + 3u_0^2) + u_1 u_2(2 + 6u_0) + u_1^3
\end{aligned}$

3. $\begin{aligned}
A_0 &= \cos 2u_0 \\
A_1 &= -2u_1 \sin 2u_0 \\
A_2 &= -2u_2 \sin 2u_0 - 2u_1^2 \cos 2u_0 \\
A_3 &= -2u_3 \sin 2u_0 - 4u_1 u_2 \cos 2u_0 + \frac{4}{3}u_1^3 \sin 2u_0
\end{aligned}$

4. $\begin{aligned}
A_0 &= \sinh 2u_0 \\
A_1 &= 2u_1 \cosh 2u_0 \\
A_2 &= 2u_2 \cosh 2u_0 + 2u_1^2 \sinh 2u_0 \\
A_3 &= 2u_3 \cosh 2u_0 + 4u_1 u_2 \sinh 2u_0 + \frac{4}{3}u_1^3 \cosh 2u_0
\end{aligned}$

5. $\begin{aligned}
A_0 &= e^{2u_0} \\
A_1 &= 2u_1 e^{2u_0} \\
A_2 &= 2(u_2 + u_1^2)e^{2u_0} \\
A_3 &= 2(u_3 + 2u_1 u_2 + \frac{2}{3}u_1^3)e^{2u_0}
\end{aligned}$

6. $\begin{aligned}
A_0 &= u_0^2 u_{0_x} \\
A_1 &= 2u_0 u_1 u_{0_x} + u_0^2 u_{1_x} \\
A_2 &= 2u_0 u_2 u_{0_x} + u_1^2 u_{0_x} + 2u_0 u_1 u_{1_x} + u_0^2 u_{2_x} \\
A_3 &= 2u_0 u_3 u_{0_x} + 2u_1 u_2 u_{0_x} + 2u_0 u_2 u_{1_x} \\
&\quad + u_1^2 u_{1_x} + 2u_0 u_1 u_{2_x} + u_0^2 u_{3_x}
\end{aligned}$

7. $\begin{aligned}
A_0 &= u_0 u_{0_x}^2 \\
A_1 &= 2u_0 u_{0_x} u_{1_x} + u_1 u_{0_x}^2 \\
A_2 &= 2u_0 u_{0_x} u_{2_x} + u_0 u_{1_x}^2 + 2u_1 u_{0_x} u_{1_x} + u_2 u_{0_x}^2 \\
A_3 &= 2u_0 u_{0_x} u_{3_x} + 2u_0 u_{1_x} u_{2_x} + 2u_1 u_{0_x} u_{2_x} \\
&\quad + u_1 u_{1_x}^2 + 2u_2 u_{0_x} u_{1_x} + u_{0_x}^2 u_3
\end{aligned}$

8. $\begin{aligned}
A_0 &= u_0 e^{u_0} \\
A_1 &= (u_0 u_1 + u_1)e^{u_0} \\
A_2 &= (u_0 u_2 + \frac{1}{2}u_0 u_1^2 + u_1^2 + u_2)e^{u_0} \\
A_3 &= (u_0 u_3 + u_0 u_1 u_2 + \frac{1}{6}u_0 u_1^3 + 2u_1 u_2 + \frac{1}{2}u_1^3 + u_3)e^{u_0}
\end{aligned}$

9. A_0 $=$ $u_0 \sin u_0$

A_1 $=$ $u_0 u_1 \cos u_0 + u_1 \sin u_0$

A_2 $=$ $u_0 u_2 \cos u_0 - \frac{1}{2} u_0 u_1^2 \sin u_0 + u_1^2 \cos u_0 + u_2 \sin u_0$

A_3 $=$ $u_0 u_3 \cos u_0 - u_0 u_1 u_2 \sin u_0 - \frac{1}{6} u_0 u_1^3 \cos u_0$
$+ 2u_1 u_2 \cos u_0 - frac12 u_1^3 \sin u_0 + u_3 \sin u_0$

10. A_0 $=$ $u_0 \cosh u_0$

A_1 $=$ $u_0 u_1 \sinh u_0 + u_1 \cosh u_0$

A_2 $=$ $u_0 u_2 \sinh u_0 + \frac{1}{2} u_0 u_1^2 \cosh u_0 + \frac{1}{2} u_1^2 \sinh u_0 + u_2 \cosh u_0$

A_3 $=$ $u_0 u_3 \sinh u_0 + u_0 u_1 u_2 \cosh u_0 + \frac{1}{6} u_0 u_1^3 \sinh u_0$
$2u_1 u_2 \sinh u_0 + \frac{1}{2} u_1^3 \cosh u_0 + u_3 \cosh u_0$

11. A_0 $=$ $u_0^2 + \sin u_0$

A_1 $=$ $2u_0 u_1 + u_1 \cos u_0$

A_2 $=$ $2u_0 u_2 + u_1^2 + u_2 \cos u_0 - \frac{1}{2} u_1^2 \sin u_0$

A_3 $=$ $2u_0 u_3 + 2u_1 u_2 + u_3 \cos u_0 - u_1 u_2 \sin u_0 - \frac{1}{6} u_1^3 \cos u_0$

12. A_0 $=$ $u_0 + \cos u_0$

A_1 $=$ $u_1 - u_1 \sin u_0$

A_2 $=$ $u_2 - u_2 \sin u_0 - \frac{1}{2} u_1^2 \cos u_0$

A_3 $=$ $u_3 - u_3 \sin u_0 - u_1 u_2 \cos u_0 + \frac{1}{2} u_1^3 \sin u_0$

13. A_0 $=$ $u_0 + \ln u_0$

A_1 $=$ $u_1 + \dfrac{u_1}{u_0}$

A_2 $=$ $u_2 + \dfrac{u_2}{u_0} - \dfrac{1}{2} \dfrac{u_1^2}{u_0^2}$

A_3 $=$ $u_3 + \dfrac{u_3}{u_0} - \dfrac{u_1 u_2}{u_0^2} + \dfrac{1}{3} \dfrac{u_1^3}{u_0^3}$

14. A_0 $=$ $u_0 \ln u_0$

A_1 $=$ $u_1(1 + \ln u_0)$

A_2 $=$ $u_2(1 + \ln u_0) + \dfrac{1}{2} \dfrac{u_1^2}{u_0}$

A_3 $=$ $u_3(1 + \ln u_0) + \dfrac{u_1 u_2}{u_0} - \dfrac{1}{6} \dfrac{u_1^3}{u_0^2}$

15. A_0 $=$ $u_0^{\frac{1}{2}}$

A_1 $=$ $\frac{1}{2} u_1 u_0^{-\frac{1}{2}}$

A_2 $=$ $\frac{1}{2} u_2 u_0^{-\frac{1}{2}} - \frac{1}{8} u_1^2 u_0^{-\frac{3}{2}}$

A_3 $=$ $\frac{1}{2} u_3 u_0^{-\frac{1}{2}} - \frac{1}{4} u_1 u_2 u_0^{-\frac{3}{2}} + \frac{1}{16} u_1^3 u_0^{-\frac{5}{2}}$

16. $A_0 \;=\; u_0^{-1}$
   $\quad A_1 \;=\; -u_1 u_0^{-2}$
   $\quad A_2 \;=\; -u_2 u_0^{-2} + \frac{1}{2} u_1^2 u_0^{-3}$
   $\quad A_3 \;=\; -u_3 u_0^{-2} + u_1 u_2 u_0^{-3} - \frac{1}{2} u_1^3 u_0^{-4}$

Exercises 8.3

1. $y = \tan 3x$

2. $y = \tanh 4x$

3. $y = 1 + x$

4. $y = 1 - \ln(1 + ex),\; -1 < ex \le 1$

5. $y = x - 2$

6. $y = 1 + \frac{1}{1-x}$

7. $y = x$

8. $y = -x$

9. $y = \frac{\pi}{2} + (1 - \frac{\pi}{2})x - \frac{1}{2!}(1 - \frac{\pi}{2})x^2 + \frac{1}{3!}\frac{\pi}{2}(1 - \frac{\pi}{2})x^3 + \cdots$

10. $y = 1 + x + x^2 + \frac{4}{3}x^3 + \frac{7}{6}x^4 + \cdots$

11. $y = 2 + 2x + 3x^2 + \frac{12}{3}x^3 + \cdots$

12. $y = 1 - x + \frac{3}{2}x^2 - \frac{8}{3}x^3 + \cdots,\; y = e^{-xy}$

13. $y = \cos x$

14. $y = \tan x$

15. $y = \tanh x$

16. $y = 1 - e^{-x}$

17. $y = 1 - \frac{1}{2}x^2 + \frac{1}{12}x^4 - \frac{1}{72}x^6 + \cdots$

18. $y + 1 + \frac{1}{2}x^2 + \frac{1}{8}x^4 + \frac{3}{80}x^6 + \cdots$

19. $y = \frac{\pi}{2} + \frac{1}{2}x^2 - \frac{1}{240}x^6 + \frac{1}{480 \times 90}x^{10} + \cdots$

20. $y = 1 + \frac{e}{2}x^2 + \frac{e^2}{12}x^4 + \frac{13e^3}{720}x^6 + \cdots$

21. $y = x^3$

22. $y = x + \sin x$

23. $y = 1 - \cos x$

24. $y = \sinh x$

Exercises 8.4

1. $u = x^2 + xy$

2. $u = y^2 + xy$

3. $u = x + t$

4. $u = 1 + xt$

5. $u = \frac{x}{1-t}$

6. $u = t + \sin x$

7. $u = 2x^2 \tanh t$

8. $u = x^3 \tanh t$

9. $u = 3x$, for $t = 0$, $\frac{1}{6t}(\sqrt{1 + 36xt} - 1)$, for $t > 0$

10. $u = 2y + \arctan x$

11. $u = t + \tanh x$

12. $u = t + \tan x$

13. $u = x^2 y^2$

14. $u = y + e^x$

15. $u = x + \ln y$

16. $u = y + \ln x$

17. $u = y e^{-x}$

18. $u = y \cos x$

19. $u = x + y + \ln(1 + y)$

20. $u = x + \frac{1}{y}$

21. $u = y + \arctan x$

22. $u = t + \sinh x$

23. $u = t + \sin x$

24. $u = t + \cos x$

25. $u = \sinh x - t \sinh x \cosh x + \frac{1}{2}(2 \sinh x \cosh^2 x + \sinh^3 x)t^2 + \cdots$

26. $u = \cos x + t \sin x \cos x + (\sin^2 x \cos x)t^2 + \cdots$

27. $u = x^2 t - \frac{2}{3}x^2 t^3 + \frac{8}{15}x^2 t^5 + \cdots$

28. $u = x - t + \frac{1}{2}t^2 + \cdots$

29. $u = x - x^2 t + 2x^3 t^2 + \cdots$

30. $u = x - xt + \frac{3}{2}xt^2 + \cdots$

Exercises 8.5

1. $u(x,t) = x + t$ $\qquad$ $v(x,t) = x - t$

2. $u(x,t) = e^{-x+t}$ $\qquad$ $v(x,t) = e^{x-t}$

3. $u(x,t) = e^{x+t}$ $\qquad$ $v(x,t) = e^{-x-t}$

4. $u(x,t) = e^{2x+3t}$ $\qquad$ $v(x,t) = e^{-2x-3t}$

5. $u(x,y,t) = x + y + t$ $\quad$ $v(x,y,t) = x - y + t,\ w = -x + y + t$

6. $u(x,y,t) = e^{x+y-t}$ $\quad$ $v(x,y,t) = e^{x-y+t},\ w = e^{-x+y+t}$

7. $u(x,y,t) = e^{x+y-t}$ $\quad$ $v(x,y,t) = e^{x-y+t},\ w = e^{-x+y+t}$

8. $u(x,y,t) = x + y + e^t$ $\quad$ $v(x,y,t) = x - y + e^{-t},\ w = -x + y + e^{-t}$

Exercises 9.2

1. $u(x,t) = t + e^{-x}$ $\qquad\qquad$ 2. $u(x,t) = t^2 + xt$

3. $u(x,t) = 1 + x^2t^2$ $\qquad\qquad$ 4. $u(x,t) = t + \sin x$

5. $u(x,t) = \frac{x}{t-1}$ $\qquad\qquad$ 6. $u(x,t) = x\tanh t - \operatorname{sech} t$

7. $u(x,t) = (1+x)\tanh t$ $\qquad\qquad$ 8. $u(x,t) = x + e^t$

9. $u(x,t) = 4x$ for $t = 0$,

$\quad = \frac{1}{8t}(\sqrt{1 + 64xt} - 1)$ for $t > 0$

10. $u = x^2 - 2x^3t + \frac{5}{2}x^4t^2 + \cdots$

Exercises 9.3

1. $u(x,y) = y + e^{x+y}$ $\qquad\qquad$ 2. $u(x,y) = xy + e^{x+y}$

3. $u(x,y) = -(x+y) + e^{x+y}$ $\quad$ 4. $u(x,y) = x^2 + e^{x+y}$

5. $u(x,y) = e^{x+y}$ $\qquad\qquad$ 6. $u(x,y) = (x+y)e^{x+y}$

7. $u(x,y) = \frac{x+y}{2} - \ln(e^x + e^y)$ $\quad$ 8. $u(x,y) = \frac{x-y}{2} - \ln(e^x + e^{-y})$

9. $u(x,y) = \frac{x}{2} - \ln(e^x + e^y)$ $\qquad$ 10. $u(x,y) = \frac{y}{2} - \ln(e^x + e^y)$

11. $u(x,y) = \frac{x}{5} - \frac{2}{5}\ln(e^x + e^y)$ $\quad$ 12. $u(x,y) = -\ln(e^x + e^y)$

Exercises 9.4

1. $u(x,t) = \cos x \cos t$ $\quad$ 2. $u(x,t) = \cos x \sin t$

3. $u(x,t) = \sin x \sin t$ $\quad$ 4. $u(x,t) = t\sin x$

5. $u(x,t) = t\cosh x$ $\qquad$ 6. $u(x,t) = 1 + xt$

7. $u(x,t) = 1 + xt$ $\qquad$ 8. $u(x,t) = x^3t^3$

9. $u(x,t) = t^2 + x^2$ $\qquad$ 10. $u(x,t) = x\cos t$

11. $\phi_3 = \frac{\pi}{6} + \frac{1}{4}t^2 + \frac{\sqrt{3}}{96}t^4$

12. $\phi_3 = \frac{\pi}{4} + \frac{1}{2\sqrt{2}}t^2 + \frac{1}{48}t^4 - \frac{1}{720\sqrt{2}}t^6$

13. $\phi_3 = t + \frac{1}{3}t^3$

14. $\phi_3 = \pi + t - \frac{1}{3}t^3$

15. $\phi_3 = \frac{3\pi}{2} + t - \frac{1}{4}t^2$

Exercises 9.5

1. $u(x,t) = \frac{x}{1+t}$

2. $u(x,t) = \frac{x}{t-1}$

3. $u(x,t) = \frac{2x}{1+2t}$

4. $u(x,t) = \frac{2x}{2t-1}$

5. $u(x,t) = \frac{1}{1+x}\left(1 - \frac{1}{(1+x)^2}t + \cdots\right)$

6. $u(x,t) = \frac{x}{t-1}$

7. $u(x,t) = \frac{2x}{1+2t}$

8. $u(x,t) = 4\tan 2x$

9. $u(x,t) = \frac{x}{t} + \frac{x+t}{2t^2-t}$

10. $u(x,t) = \frac{x}{t} - \frac{2}{x+3t}$

Exercises 9.6

1. $u(x,t) = e^{x-t}$

2. $u(x,t) = e^{x+t}$

3. $u(x,t) = e^x + e^{-t}$

4. $u(x,t) = e^{2x-t}$

5. $u(x,t) = e^x - e^{-t}$

6. $u(x,t) = \sinh x + e^{-t}$

7. $u(x,t) = \cosh x - e^{-t}$

8. $u(x,t) = e^x + e^{-3t}$

9. $u(x,t) = e^{2x} + e^{-2t}$

10. $u(x,t) = \sinh 2x + e^{-2t}$

Exercises 9.7

1. $u(x,t) = e^{i(2x-4t)}$

2. $u(x,t) = \sin x e^{-it}$

3. $u(x,t) = \cosh x e^{it}$

4. $u(x,t) = 1 + \cos 3x e^{-9it}$

5. $u(x,t) = \sin 2x e^{-4it}$

6. $u(x,t) = e^{i(2x-3t)}$

7. $u(x,t) = e^{i(t-x)}$

8. $u(x,t) = e^{i(2x+3t)}$

9. $u(x,t) = e^{i(2x-6t)}$

10. $u(x,t) = e^{i(3x+8t)}$

Exercises 9.8

6. $u(x,t) = \frac{x}{1+6t}$

7. $u(x,t) = \frac{2}{x^2}$

8. $u(x,t) = \frac{1}{6}\left(\frac{x-2}{2-t}\right)$

9. $u(x,t) = \frac{2}{(x-3)^2}$

10. $u(x,t) = \frac{1}{6}\left(\frac{x-4}{3-t}\right)$

Exercises 9.9

1. $u(x,t) = \sin(x+t)$ 2. $u(x,t) = \sin x \cos t$

3. $u(x,t) = \cos x \cos t$ 4. $u(x,t) = 1 + \cos(x+t)$

5. $u(x,t) = 2 + \sin x \sin t$ 6. $u(x,t) = \sin 2x \cos t$

7. $u(x,t) = e^{x+t}$ 8. $u(x,t) = \sin 2x \sin 2t$

9. $u(x,t) = e^{x-t}$ 10. $u(x,t) = \sin(x+2t)$

11. $u(x,t) = (x - \sin x)e^{-t}$ 12. $u(x,t) = \frac{x^6}{6!}\sin t$

13. $u(x,t) = (x - \cos x)e^{-t}$ 14. $u(x,t) = \frac{5}{6!}x^6\sin t$

15. $u(x,t) = \frac{6}{7!}x^7(\sin t + \cos t)$

Exercises 10.2

1. $y(x) = (1 - \frac{1}{2!}x^2 + \frac{1}{4!}x^4) + \epsilon(\frac{1}{3!}x^3 - \frac{1}{30}x^5) + \epsilon^2(\frac{1}{96}x^6)$

2. $y(x) = (1 + \frac{1}{2!}x^2) + \epsilon(\frac{1}{2!}x^2 + \frac{1}{2}x^3) + \epsilon^2(\frac{1}{8}x^4)$

3. $u(t) = (1 - \frac{1}{2!}t^2 + \frac{1}{4!}t^4 - \frac{1}{6!}t^6) + \epsilon(\frac{2}{3!}t^3 - \frac{4}{5!}t^5) + \epsilon^2(-\frac{4}{4!}t^6)$

4. $u(t) = (1 - \frac{1}{2!}t^2 + \frac{1}{4!}t^4 - \frac{1}{6!}t^6) + \epsilon(-\frac{6}{5!}t^5$

5. $\phi_3 = 1 + \frac{1}{\epsilon}\sin(\epsilon x) - \frac{1}{2\epsilon^2}(1 - \cos(2\epsilon x)$

6. $y(x) = x - \frac{x^3}{3} + \frac{x^5}{5\cdot 2!} - \frac{x^7}{7\cdot 3!}$, $y = \frac{\sqrt{\pi}}{2}\operatorname{erf}(x)$

7. $y(x) = \frac{4}{\sqrt{\pi}}\left(x - \frac{x^3}{3} + \frac{x^5}{5\cdot 2!} - \frac{x^7}{7\cdot 3!}\right)$, $y = 2\operatorname{erf}(x)$

8. $y(x) = x + \frac{2x^3}{3} + \frac{4x^5}{15} + \frac{8x^7}{105}$

9. $y(x) = x^4\frac{1}{90}x^{10} + \frac{1}{190\cdot 120}x^{16}$

10. $y(x) = x^2 + \frac{1}{3}x^3 + \frac{1}{12}x^4 + \frac{1}{30}x^6 + \frac{13}{630}x^7 + \frac{1}{1350}x^{10}$

Exercises 10.3

1. $\phi_4 = \cos x(1 - t + \frac{t^2}{2!} - \frac{t^3}{3!})$

2. $\phi_4 = \cos x(1 - \frac{t^2}{2!} + \frac{t^4}{4!} - \frac{t^6}{6!})$

3. $\phi_4 = x^3(t - \frac{1}{3}t^3 + \frac{2}{15}t^5 - \frac{17}{315}t^7)$

4. $\phi_4 = 4x(1 - 16xt + 512x^2t^2 - 20480x^3t^3)$

5. $\phi_4 = 2(1 - \frac{1}{2}t^2 + \frac{5}{24}t^4 - \frac{61}{120}t^6) + x(t - \frac{1}{3}t^3 + \frac{2}{15}t^5 - \frac{17}{315}t^7)$

6. $\phi_3 = x - xt + \frac{3}{2}xt^2$

7. $\phi_3 = x^2(t - \frac{2}{3}t^3 + \frac{8}{15}t^5)$

8. $\phi_3 = x - (1 + x^2)t + (2x - x^3)t^2$

9. $\phi_3 = \cosh x(1 + (it) + \frac{1}{2!}(it)^2)$

10. $\phi_3 = 1 + \sinh x(\frac{t^2}{2!} - \frac{t^3}{3!} + \frac{t^4}{4!})$

Exercises10.4

1. (a) $[2/2] = \dfrac{1 + 7x + 11x^2}{1 + 5x + 5x^2}$

$[3/3] = \dfrac{1 + 10x + 31x^2 + 29x^3}{1 + 8x + 19x^2 + 13x^3}$

(b) $\sqrt{5} \approx 2.2$, $\sqrt{5} \approx 2.230769$

2. (a) $[2/2] = \dfrac{1 + 17x + 61x^2}{1 + 11x + 19x^2}$

$[3/3] = \dfrac{1 + 24x + 171x^2 + 337x^3}{1 + 18x + 87x^2 + 97x^3}$

(b) $\sqrt{13} \approx 3.210526$, $\sqrt{13} \approx 3.474227$

3. (a) $[2/2] = \dfrac{6x}{6 + x^2}$

$[3/3] = \dfrac{60x - 7x^3}{60 + 3x^2}$

4. (a) $[2/2] = \dfrac{12 + 6x + x^2}{12 - 6x + x^2}$

$[3/3] = \dfrac{120 + 60x + 12x^2 + x^3}{120 - 60x + 12x^2 - x^3}$

5. (a) $[2/2] = \dfrac{-30 + 21x - x^2}{30 - 36x + 9x^2}$

$[3/3] = \dfrac{420 - 510x + 140x^2 - 3x^3}{-420 + 720x - 360x^2 + 48x^3}$

6. (a) $[2/2] = \dfrac{-15 + x^2}{-15 + 6x^2}$

$[3/3] = \dfrac{-15 + x^2}{-15 + 6x^2}$

7. (a) $[3/3] = \dfrac{-15x + 4x^3}{-15 + 9x^2}$

$[4/4] = \dfrac{105x - 55x^2}{105 - 90x^2 + 9x^4}$

8. (a) $[3/3] = \dfrac{60x + 7x^3}{60 - 3x^2}$

$[4/4] = \dfrac{5880x + 620x^3}{5880 - 360x^2 + 11x^4}$

9. (a) $[3/3] = \dfrac{1}{1+x}$

$[4/4] = \dfrac{1}{1+x}$

10. (a) $[3/3] = \dfrac{15 + 4x^2}{15 + 9x^2}$

$[4/4] = \dfrac{945 + 735x^2 + 64x^4}{945 + 1050x^2 + 225x^4}$

11. (a) $[3/3] = \dfrac{120 + 60x + 28x^2 - x^3}{120 - 60x + 28x^2 + x^3}$

$[4/4] = \dfrac{240 + 120x - 140x^2 - 100x^3 - 49x^4}{240 - 120x - 140x^2 + 100x^3 - 49x^4}$

12. (a) $[3/3] = \dfrac{120 + 60x - 28x^2 + x^3}{120 - 60x - 28x^2 - x^3}$

$[4/4] = \dfrac{240 + 120x - 500x^2 - 220x^3 + 111x^4}{240 - 120x - 500x^2 + 220x^3 + 111x^4}$

Exercises 10.5

1. $u(x) = .2 + 1.6x + 4.6x^2 - .933333x^3 - 44.466667x^4$
$- 101.506667x^5 + 157.857778x^6 + 1291.071746x^7 + O(x^8)$

$$u_{[4/4]} = \frac{.2 + .922403x + 1.802939x^2 + 1.820630x^3 + .841658x^4}{1 - 3.387987x + 13.118589x^2 - 13.255201x^3 + 15.045083x^4}$$

2. $u(x) = .1 + .45x + .875x^2 + .716667x^3 - .540104x^4$
$- 2.260417x^5 - 2.507231x^6 + .377294x^7 + O(x^8)$

$$u_{[4/4]} = \frac{.1 + .221916x + .201328x^2 + .087574x^3 + .014724x^4}{1 - 2.280845x + 3.527084x^2 - 2.205410x^3 + .956694x^4}$$

3. $u(x) = \alpha + \frac{1}{4}(\alpha + \alpha^2)r^2 + \frac{1}{4}(\alpha + 3\alpha^2 + 2\alpha^2)r^4$
$+ \frac{1}{2034}(\alpha + 9\alpha^2 + 16\alpha^3 + 8\alpha^4)r^6 + O(r^8)$

$\alpha = -2.392$

Exercises 11.2

1. $u(x, t) = -2\operatorname{sech}^2(x - 4t)$

2. $u(x, t) = 8\operatorname{sech}^2 2(x - 16t)$

3. $u(x, t) = \sqrt{2}\operatorname{sech} 2(x - 4t)$

4. $u(x, t) = 4\arctan\left[\exp\left(-2(x - \tfrac{\sqrt{3}}{2}t)\right)\right]$

5. $u(x, y, t) = \frac{1}{2}\operatorname{sech}^2\left[\frac{1}{2}(x + y - 4t)\right]$

Exercises 11.3

1. $u(x,t) = 3\cos\left[\frac{1}{3}(x - 6t)\right]$

2. $u(x,t) = \left\{2\sin\left[\frac{3}{8}\left(x - \frac{5}{2}t\right)\right]\right\}^{\frac{2}{3}}$

3. $u(x,y,t) = 3\cos\left[\frac{1}{3\sqrt{2}}(x + y - 6t)\right]$

4. $u(x,t) = \cos^{\frac{1}{2}}(x - 2t)$

5. $u(x,y,t) = \cos^{\frac{1}{2}}\left[\frac{1}{\sqrt{2}}(x + y - 2t)\right]$

Exercises 11.4

1. $u(x,t) = 3\sinh\left[\frac{1}{3}(x - 6t)\right]$

2. $u(x,t) = -\left[2\cosh\{\frac{3}{8}(x - \frac{5}{2}t)\}\right]^{\frac{2}{3}}$

3. $u(x,y,t) = 3\sinh\left[\frac{1}{3\sqrt{2}}(x + y - 6t)\right]$

4. $u(x,t) = \sinh^{\frac{1}{2}}(x - 2t)$

5. $u(x,y,t) = \sinh^{\frac{1}{2}}(x + y - 4t)$

Bibliography

[1] K. Abbaoui and Y. Cherruault, Convergence of Adomian's method applied to differential equations, *Computers Math. Applic.*, 28(5), 103 – 109, (1994).

[2] M.J.Ablowitz and P.A.Clarkson, *Solitons, Nonlinear Evolution Equations and Inverse Scattering*, Cambridge University Press, Cambridge, (1991).

[3] M. Abramowitz and I. A. Stegun, *Handbook of Mathematical functions*, NBS, Appl. Math. Series 55, Washington, DC, (1964).

[4] G. Adomian, A new approach to nonlinear partial differential equations, *J. Mathematical Analysis and Applications*, 102, 420–434,(1984).

[5] G. Adomian, *Nonlinear Stochastic Operator Equations*, Academic Press, San Diego, (1986).

[6] G. Adomian and R. Rach, Equality of partial solutions in the decomposition method for linear or nonlinear partial differential equations, *Computers Math. Appl.*, 19(12), 9 – 12, (1990).

[7] G. Adomian, A review of the decomposition method and some recent results for nonlinear equation, *Math. Comput. Modelling* 13(7), 17 – 43, (1992).

[8] G. Adomian and R. Rach, Noise terms in decomposition series solution, *Computers Math. Appl.*, 24(11), 61 – 64, (1992).

[9] G. Adomian and R. Rach, A further consideration of partial solutions in the decomposition method, *Computers Math. Appl.*, 23(1), 51–64, (1992).

[10] G. Adomian and R. Rach, Analytic solution of nonlinear boundary value problems in several dimensions by decomposition, *Journal of Mathl. Anal. and Applic.*, 174, 118 – 137, (1993).

[11] G. Adomian, *Solving Frontier Problems of Physics, The Decomposition Method*, Kluwer, Boston, (1994).

[12] G. Adomian, Solution of coupled nonlinear partial differential equations by decomposition, *Computers Math. Applic*, 31(6), 117–120, (1996).

[13] R.P. Agarwal and D. O'Regan, Singular boundary value problems for superlinear second order ordinary delay differential equations, J. Differential Equations, 130, 333–335, (1996).

[14] W. F. Ames, *Nonlinear Partial Differential Equations in Engineering*, Vol I, Academic Press, New York, (1965).

[15] V.K.Andreev, O.V.Kaptsov, V.V.Pukhnachov and A.A.Radinov, *Applications of Group-Theoretical methods in Hydrodynamics*, Kluwer, Boston, (1998).

[16] T. Badredine, K. Abbaoui and Y. Cherruault, Convergence of Adomian's method applied to integral equations, *Kybernotes*, 28(5), 557–564, (1999).

[17] G.A.Baker, *Essentials of Pade' Approximants*, Academic Press, London, (1975).

[18] G.A.Baker and P. Graves-Morris, *Essentials of Pade' Approximants*, Cambridge University Press, Cambridge, (1996).

[19] N. Bellomo and R. Monaco, A comparison between Adomian's decomposition methods and perturbation techniques for nonlinear random differential equations, *Journal of Mathl. Anal. and Applic.*, 110, 495 – 502, (1985).

[20] C.M. Bender and S.A. Orszag, *Advanced Mathematical Methods for Scientists and Engineering*, McGraw-Hill, New York, (1978).

[21] E.R.Benton and G.W.Platzman, A table of solutions of one dimensional Burgers equation, *Quart. Appl. Math.* 30, 195 – 212, (1972).

[22] P. Berg and J. McGregor, *Elementary Partial Differential Equations*, Holden-Day, New York, (1966).

[23] P.Bhatnnagar, *Nonlinear Waves in One-Dimensional Dispersive Systems*, Clarendon Press, Oxford, (1979).

[24] D. Bleecker and G. Csordas, *Basic Partial Differential Equations*, Chapman and Hall, New York, (1995).

[25] G. W. Bluman and S. Kumei, *Symmetries and Differential Equations*, Springer, New York, (1989).

[26] H. D. Boer and F. V. Keulen, Padé approximants applied to a nonlinear finite element solution strategy, *Communications in Numerical Methods in Engineering*, 13, 593 – 602, (1997).

[27] W. E. Boyce and R. C. DiPrima, *Elementary Differential Equations and Boundary Value Problems*, Wiley, New York, (1992).

[28] J. Boyd, Pade' approximant algorithm for solving nonlinear ordinary differential equation boundary value problems on an unbounded domain, *Computers in physics*, 11(3), 299 – 303, (1997).

[29] J. Boyd, *Weakly Nonlocal Solitary Waves and Beyond-All-Orders Asymptotics*, Kluwer, Boston, (1998).

[30] J.M. Burgers, A mathematical model illustrating the theory of turbulence, *Adv. Appl. Mech.* 1, 171 – 199, (1948).

[31] H. Cabannes, *Lecture Notes in Physics, Pade' Approximants and its Applications to Mechanics*, Springer, New York, (1976).

[32] F. Calogero and A. Degasperis, *Spectral Transform and Solitons I*, North-Holland, New York, (1982).

[33] L. Casasus and W. Al-Hayani, The decomposition method for ordinary differential equations with discontinuities, *Appl. Math. Comput.*, (2001) to appear.

[34] Y. Cherruault, Convergence of Adomian's method, *Mathl. Comput. Modelling*, 14, 83 – 86, (1989).

[35] Y. Cherruault, Convergence of Adomian's method, *Kybernotes*, 18(20), 31–38, (1990).

[36] Y. Cherruault. G. Saccomandi and B. Some, New results for convergence of Adomian's method applied to integral equations, *Mathl. Comput. Modelling* 16(2), 85 – 93, (1992).

[37] Y. Cherruault and G. Adomian, Decomposition methods: a new proof of convergence, *Mathl. Comput. Modelling*, 18, 103 – 106, (1993).

[38] J.M.Cooper, *Introduction to Partial Differential Equations with MATLAB*, Birkhauser, Boston, (1998).

[39] B. K. Datta, A new approach to the wave equation-an application of the decomposition method, *Journal of Mathl. Anal. and Applic.*, 142, 6 – 12, (1989).

[40] B. K. Datta, *Introduction to Partial Differential Equations*, New Central Book Agency, Calcutta, (1993).

[41] H.T. Davis, *Introduction to Nonlinear Differential and Integral Equations*, Dover Publications, New York, (1962).

[42] L. Debnath, *Nonlinear Waves*, Cambridge University Press, Cambridge, (1983).

[43] L. Debnath, *Nonlinear Partial Differential Equations for Scientists and Engineers*, Birkhauser, Boston, (1997).

[44] E. Deeba and S. Khuri, A decomposition method for solving the nonlinear Klein-Gordon equation, *J. Computational Physics*, 124, 442–448, (1996).

[45] E. Deeba and S. Khuri, The decomposition method applied to Chandrasekhar H-equation, *Appl. Math. Comput.*, 77, 67–78, (1996).

[46] R. Dennemeyer, *Partial Differential Equations and Boundary Value Problems*, McGraw Hill, New York, (1968).

[47] P.T. Dinda and M. Remoissenet, Breather compactons in nonlinear Klein-Gordon systems, Physical Review E, 60(3) (1999) 6218–6221.

[48] P.G.Drazin and R.S.Johnson, *Solitons: an introduction*, Cambridge University Press, Cambridge, (1996).

[49] L. Dresner, *Similarity Solutions of Nonlinear Partial Differential Equations*, Pitman, New York, (1983).

[50] D. G. Duffy, *Advanced Engineering Mathematics*, CRC, New York, (1998).

[51] S. Dusuel, P. Michaux and M. Remoissenet, From kinks to compactonlike kinks, *Physical Review E*, 57(2) (1998) 2320–2326.

[52] B. Epstein, *Partial Differential Equations*, McGraw Hill, New York, (1962).

[53] D. J. Evans and B.B.Sanugi, Numerical solution of the Goursat problem by a nonlinear trapezoidal formula, *Appl. Math. Lett.*, 1(3), 221 – 223, (1988).

[54] D. J. Evans and W.S. Yousif, A note on solving the fourth order parabolic equation by the AGE method, *Intern. J. Computer Math.*, 40, 93 – 97, (1991).

[55] J.D.Faires and R.L.Burden, *Numerical Methods*, PWS-KENT, Boston, (1993).

[56] S. J. Farlow, *Partial Differential Equations for Scientists and Engineers*, Dover, New York, (1993).

[57] R. A. Fisher, The wave of advance of advantageous genes, *Ann. Eugenics*, 7, 335 – 369, (1936).

[58] L. Gabet, The theoretical foundation of the Adomian method, *Computers Math. Applic.*, 27(12), 41–52, (1994)

[59] P.R.Garabedian, *Partial Differential Equations*, Chelsea, New York, (1986).

[60] W.B Gong and H.Yang, Rational approximants for some performance analysis problems, *IEEE Transaction on Computers*, 44(12), (1995).

[61] C.K. Hayes, A new traveling-wave solution of Fisher's equation with density-dependent diffusivity,*J. Math. Biol.*, 29,531 – 537, (1991).

[62] R. Hirota, Direct methods in soliton theory, In *Solitons* (Bullogh, R. K. and Caudrey, P.J., eds). Springer, Berlin (1980).

[63] S. Hood, New exact solutions of Burgers's equation-an extension to the direct method of Clarkson and Kruskal, *J. Math. Phys.*, 36(4), (1995).

[64] F. John, *Partial Differential Equations*, Springer-Verlag, New York, (1982).

[65] O.V.Kaptsov, Construction of exact solutions of the Boussinesq equation, *J. Applied Mechanics and Theoretical Physics*, 39(3), 389–392, (1998).

[66] O.V. Kaptsov, Determining equations in diffusion problems, *Russ. J. Numer. Anal. Math. Modelling*, 15(2), 163–166 (2000).

[67] D. Kaya, A new approach to solve a nonlinear wave equation, *Bull. Malaysian Math. Soc.*, 21, 95–100 (1998).

[68] D. Kaya, On the solution of a Kortweg-de Vries like equation by the decomposition method, *Intern. J. Computer Math.*, 72, 531–539 (1999).

[69] D. Kaya and H. Bulut, The decomposition method for approximate solution of a Burgers equation, *Bull. Inst. Math. Academia Sinica*, 28(1) 34–42, (2000).

[70] D. Kaya, An application of the decomposition method on second order wave equations, *Intern. J. Computer Math.*, 75, 51–57 (2000).

[71] J. Kevorkian, *Partial Differential Equations*, Brooks/Cole, Pacific Grove, (1990).

[72] Kivshar, Yuri, Compactons in discrete lattices, Nonlinear Coherent Structures in Physics and Biology, 329 (1994) 255–258.

[73] R.Knobel, *An Introduction to the Theory of Waves*, AMS, (2000).

[74] G. L. Lamb, *Elements of Soliton Theory*, John Wiley, New York, (1980).

[75] B.J.Laurenzi, An analytic solution to the Thomas-Fermi equation,*J. Math. Physics*, 31(10), 2535 – 2537, (1990).

[76] J. Lighthill, *Waves in Fluids*, Cambridge University Press, Cambridge, (1978).

[77] J. D. Logan, *An Introduction to Nonlinear Partial Differential Equations*, Wiley-Interscience, New York, (1994).

[78] A. Ludu and J.P. Draayer, Patterns on liquid surfaces:cnoidal waves, compactons and scaling, Physica D, 123 (1998) 82–91.

[79] A. Ludu, G. Stoitcheva, and J.P. Draayer, Similarity analysis of nonlinear equations and bases of finite wavelength solitons, *International J. of Modern Physics E*, 9(3), 263–278, (2000).

[80] K. Maleknejad and M. Hadizadeh, A new computational method for Volterra-Fredholm integral equations, *Computers Math. Applic.*, 37, 1–8, (1999).

[81] J.H.Mathews, *Numerical Methods for Mathematics, Science, and Engineering*, Prentice-Hall, New Jersey, (1992).

[82] T. Mavoungou and Y. Cherruault, Convergence of Adomian's method and applications to non-linear partial differential equations, *Kybernotes*, 21(6), 13–25, (1992).

[83] T. Mavoungou and Y. Cherruault, Numerical study of Fisher's equation by Adomian's method, *Mathl. Comput. Modeling*, 19(1), 89 – 95, (1994).

[84] M. McAsey and L. A. Rubel, Some closed-form solutions of Burgers' equation, *Studies in Applied Mathematics*, 88, 173 – 190, (1993).

[85] R. C. McOwen, *Partial Differential Equations*, Prentice Hall, New Jersey, (1996).

[86] K.S. Miller, *Partial Differential Equations in Engineering Problems*, Prentice-Hall, New Jersey, (1953).

[87] D.A.Morales, An n-dimensional interpretation of the δ-expansion for the Thomas-Fermi equation, *J. Math. Physics*, 35(8), 3916 – 3921, (1994).

[88] J.D.Murray, *Nonlinear Differential Equation Models in Biology*, Oxford University Press, Oxford, (1977).

[89] J.D.Murray, *Mathematical Biology*, Springer, New York, (1993).

[90] J.J.C.Nimmo and N.C.Freeman, The use of Backlund transformations in obtaining the N-soliton solutions in terms of a Wronskian, *J. Phys. A:Math. General*, 17, 1415–1424, (1983).

[91] S. Olek, An accurate solution to the multispecies Lotka-Volterra equations, *SIAM Review*, 36(3), 480–488, (1994).

[92] P.J. Olver and P. Rosenau, Tri-Hamiltonian duality between solitons and solitary-wave solutions having compact support, *Physical Review E*, 53(2),1900–1906, (1996).

[93] I. Petrovsky, *Lectures on Partial Differential Equations*, Interscience, New York, (1954).

[94] M. A. Pinsky, *Partial Differential Equations and Boundary-Value Problems with Applications*, McGraw Hill, New York, (1998).

[95] R. Rach, On the Adomian decomposition method and comparisons with Picards method, J. Math. Anal. and Applic. 128, 480–483, (1987).

[96] R. Rach, G.Adomian and R.E.Meyers, A modified decomposition, *Computers Math. Applic.*, 23(1) 17–23, (1992).

[97] R. Rach, A. Baghdasarian, and G. Adomian, Differential Equations with Singualar Coefficients, *Appl. Math. Lett.*, 47(2/3), 179-184, (1992).

[98] A. Ralston and H.S.Wilf, *Mathematical Methods for Digital Computers*,Wiley, New York, (1965).

[99] A. Re'paci, Nonlinear dynamical systems: on the accuracy of Adomian's decomposition method, *Appl. Math. Lett.*, 3, 35–39, (1990).

[100] P. Rosenau and J.M. Hyman, Compactons: Solitons with finite wavelengths,*Phys. Rev. Lett.*, 70(5),564–567, (1993).

[101] P. Rosenau, Nonlinear dispersion and compact structures, *Phys. Rev. Lett.*, 73(13), 1737–1741, (1994).

[102] P. Rosenau, On nonanalytic solitary waves formed by a nonlinear dispersion, *Phys. Lett. A*, 230(5/6),305–318, (1997).

[103] P.Rosenau, Compact and noncompact dispersive structures, *Physics Letters A*, 275(3),193–203, (2000).

[104] N. T. Shawagfeh, Nonperturbative approximate solution for Lane-Emden equation, *J. Math. Phys.*, 34(9), 4364–4369, (1993).

[105] N.T. Shawagfeh and G. Adomian, Non-perturbative analytical solution of the general Lotka-Volterra three-species system, *Appl. Math. Comput.*, 76, 251–266, (1996).

[106] N. T. Shawagfeh, Analytic approximate solution for a nonlinear oscillator equation, *Computers Math. Applic.*, 31(6), 135–141, (1996)

[107] V. I. Smirnov, *Integral Equations and Partial Differential Equations*, Addison Wesley, Reading, (1964).

[108] G. D. Smith, *Numerical Solution of Partial Differential Equations*, Clarendon Press, New York, (1978).

[109] I.N. Sneddon, *Elements of Partial Differential Equations*, McGraw Hill, New York, (1957).

[110] A. Sommerfeld, *Partial Differential Equations in Physics*, Academic Press, New York, (1964).

[111] I. Stakgold, *Boundary Value Problems of Mathematical Physics*, SIAM, (2000).

[112] G. Stephenson, *Partial Differential Equations for Scientists and Engineers*, Imperial College Press, London, (1996).

[113] S. Tang and R. O. Weber, Numerical study of Fisher's equation by a Petrov-Galerkin finite element method, *J. Austral. Math. Soc. Ser.*, B 33 ,27 – 38, (1991).

[114] K.G Tebeest, Numerical and analytical solutions of Volterra's population model, *SIAM Rev.*, 39(3),484 – 493, (1997).

[115] S.V.Tonningen, Adomian's decomposition method: a powerful technique for solving engineering equations by computer, *Computers in Education Journal*, 5(4), 30–34, (1995).

[116] S. N. Venkatarangan and K. Rajalashmi, A modification of Adomian's solution for nonlinear oscillatory systems, *Computers Math. Applic.*, 29(6),67 – 73, (1995).

[117] S.N. Venkatarangan and K. Rajalashmi, Modification of Adomian's decomposition method to solve equations containing radicals, *Computers Math. Applic.*, 29(6),75 – 80, (1995).

[118] A.M.Wazwaz, Two turning points of second order, *SIAM J. Appl. Math.* , 50, 883 – 892, (1988).

[119] A.M.Wazwaz, Resonance and asymptotic solution of a singular perturbation problem, *IMA J. of Appl. Math.* , 49, 231 – 244, (1992).

[120] A.M.Wazwaz, Exact special solutions with solitary patterns for the nonlinear dispersive K(m,n) equations,*Chaos, Solitons and Fractals*, 13(1), 161–170, (2001).

[121] A.M.Wazwaz, The decomposition method for approximate solution of the Goursat problem, *Appl. Math. Comput.*, 69, 299 – 311, (1995).

[122] A.M.Wazwaz, A new approach to the nonlinear advection problem, an application of the decomposition method,*Appl. Math. Comput.*, 72, 175 – 181, (1995).

[123] A.M.Wazwaz,*A First Course in Integral Equations*,WSPC, New Jersey, (1997).

[124] A.M.Wazwaz, Necessary conditions for the appearance of noise terms in decomposition solution series,*Appl. Math. Comput.*, 81, 199 – 204, (1997).

[125] A.M.Wazwaz, A study on a boundary-layer equation arising in an incompressible fluid,*Appl. Math. Comput.*, 87, 265 – 274, (1997).

[126] A.M.Wazwaz, Equality of partial solutions in the decomposition method for partial differential equations, *Appl. Math. Comput.*, 65, 293 – 308, (1997).

[127] A.M.Wazwaz, A new algorithm for solving differential equations of the Lane-Emden type, *Appl. Math. Comput.*, 118(2/3), 287–310, (2001).

[128] A.M.Wazwaz, A computational approach to soliton solutions of the Kadomtsev-Petviashvili equation, *Appl. Math. and Comput.*, 123(2) 205–217, (2001).

[129] A.M.Wazwaz, A reliable modification of Adomian's decomposition method,*Appl. Math. and Comput.*, 92, 1–7, (1998).

[130] A .M.Wazwaz, A comparison between Adomian decomposition method and Taylor series method in the series solutions,*Appl. Math. Comput.*, 79,37 – 44, (1998).

[131] A.M.Wazwaz, A reliable technique for solving the wave equation in an infinite one-dimensional medium,*Appl. Math. Comput.*, 79, 37 – 44, (1998).

[132] A.M.Wazwaz, Analytical approximations and Pade' approximants for Volterra's population model,*Appl. Math. and Comput.*, 100, 31–25, (1999).

[133] A.M.Wazwaz, Construction of solitary wave solutions and rational solutions for the KdV equation by Adomian decomposition method, *Chaos, Solitons and Fractals*,12(12), 2283–2293, (2001).

[134] A.M.Wazwaz, The modified decomposition method and Pade' approximants for solving Thomas-Fermi equation,*Appl. Math. and Comput.*, 105, 11–19, (1999).

[135] A.M. Wazwaz, A study of nonlinear dispersive equations with solitary-wave solutions having compact support, *Mathematics and Computers in Simulation*, 56, 269–276, (2001).

[136] A.M. Wazwaz, Construction of soliton solutions and periodic solutions of the Boussinesq equation by the modified decomposition method, *Chaos, Solitons and Fractals*, 12(8), 1549–1556, (2001).

[137] A.M.Wazwaz, On the solution of the fourth order parabolic equation by the decomposition method, *Intern, J, Computer Math.*, 57, 213–217, (1995).

[138] A.G. Webster, *Partial Differential Equations of Mathematical Physics*, Dover, New York, (1955).

[139] H. Weinberger, *A First Course in Partial Differential Equations*, Blaisdell, Waltham, (1965).

[140] G.B.Whitham, *Linear and Nonlinear Waves*, John Wiley, New York, (1976).

[141] T.P.Witelski, An asymptotic solution for traveling waves of a nonlinear-diffusion Fisher's equation,*J. Math. Biol.*, 33,1 – 16, (1994).

[142] E. Yee, Application of the decomposition method to the solution of the reaction-convection-diffusion equation, Appl. Math. Comput. 56(1993) 1–27.

[143] N.J.Zabusky and M.D.Kruskal, Interaction of solitons incollisionless plasma and the recurrence of initial states, *Phys. Rev. Lett.*,15, 240–243, (1965).

[144] D. Zwillinger, *Handbook of Differential Equations*, Academic Press, New York, (1992).

Index

Printed and bound by CPI Group (UK) Ltd, Croydon, CR0 4YY

23/10/2024

01777679-0006